Energy Conservation in Existing Buildings Deskbook

ENERGY CONSERVATION IN EXISTING BUILDINGS DESKBOOK

Edited by:
Albert Thumann, P.E., C.E.M.

Published by
THE FAIRMONT PRESS, INC.
700 Indian Trail
Lilburn, GA 30247

Library of Congress Cataloging-in-Publication Data

Thumann, Albert
Energy conservation in existing buildings deskbook / edited by Albert Thumann.
p. cm.
Includes index.
ISBN 0-88173-138-2
1. Buildings--Energy conservation--Handbooks, manuals, etc.
I. Thumann, Albert.
TJ163.5.B84E5275 1991 697--dc20 91-3412
CIP

Energy Conservation In Existing Buildings Deskbook / Edited By. Albert Thumann.

Published by The Fairmont Press, Inc.
700 Indian Trail
Lilburn, GA 30247

Printed in the United States of America

10 9 8 7 6 5 4 3 2

ISBN 0-88173-138-2 FP

ISBN 0-13-278250-2 PH

Distributed by Prentice-Hall, Inc.
A Simon & Schuster Company
Englewood Cliffs, NJ 07632

Prentice-Hall International (UK) Limited, London
Prentice-Hall of Australia Pty. Limited, Sydney
Prentice-Hall Canada Inc., Toronto
Prentice-Hall Hispanoamericana, S.A., Mexico
Prentice-Hall of India Private Limited, New Delhi
Prentice-Hall of Japan, Inc., Tokyo
Simon & Schuster Asia Pte. Ltd., Singapore
Editora Prentice-Hall do Brasil, Ltda., Rio de Janeiro

TABLE OF CONTENTS

TABLE OF CONTENTS (CONT'D)

LIST OF FIGURES

Figure Numbers		Page No.

WHERE ENERGY IS USED IN BUILDINGS

TABULATING PRESENT ENERGY USE

PRESENT AND FUTURE ENERGY CONSUMPTION (WINDOWS AND DOORS)

LIST OF FIGURES (CONT'D)

Figure Numbers | Page No.

LIST OF FIGURES (CONT'D)

LIST OF FIGURES (CONT'D)

Figure Numbers Page No.

LIST OF FIGURES (CONT'D)

LIST OF FIGURES (CONT'D)

Figure Numbers Page No.

LIST OF FIGURES (CONT'D)

PREFACE

The results of energy conservation have truly been outstanding. Recently the Association of Energy Engineers (AEE) conducted a survey of its members. Eleven hundred energy professionals responded and the results indicate that efficiency products have an excellent performance rating. With the crisis is the Gulf, attention has once again focused on energy management as the key element in a policy strategy to reduce dependence on imported oil. In addition, environmental concerns such as global warming have given new meaning to energy conservation. Reducing energy consumption results in minimizing the requirements for new power-generating stations.

This manual contains the tools to reduce energy consumption in buildings. It is based on the classic reference "Architects and Engineers Guide to Energy Conservation In Existing Buildings." Companies who have used these procedures have slashed utility bills by 40% or more. Hopefully, the information contained herein will help companies review their programs and find opportunities to reduce energy consumption further.

Albert Thumann

ACKNOWLEDGEMENTS AND DISCLAIMER

This manual was prepared in part as a joint venture under contract with the Office of Conservation and Solar Energy, U.S. Department of Energy. The condensed final version of the manual was prepared by Ham-Mer Consulting Engineers of Austin, Texas, and Benbassat and Sporidi's Company, Consulting Engineers of Silver Spring, Maryland. An earlier version of the manual was compiled by Dubin-Mindell-Bloome Associates, Consulting Engineers of New York, also under contract. The manual has been extensively tested under actual field conditions. It incorporates data, information, and suggestions from numerous other technical firms and Federal agencies, manufacturers, professional engineers, architects, professional societies, and trade organizations. Their contributions are gratefully acknowledged.

The contracting firms assume responsibility for the accuracy of the manual which was developed as part of the Federal Energy Management Program's effort to provide technical assistance to the Federal agencies; however, the contents of the manual do not necessarily reflect the views or policies of the Federal Government. The manual is not an official body of standards, specifications, or regulations, nor do references to trade names of products constitute an endorsement or recommendation for use.

SECTION I
INTRODUCTION

Section I
INTRODUCTION

This manual has been carefully designed to enhance and simplify the work of those architects and engineers whose practice includes the analysis and modification of existing buildings to reduce both fuel consumption and operating costs. In effect, the manual offers a field-tested approach to identifying, analyzing, and recommending action on the wide range of opportunities and options available to reduce energy use in most existing buildings. Besides a review of the principles of energy use and conservation, the manual provides a step-by-step methodology for assessing and improving the year-around energy performance of buildings, as well as a series of forms, charts, and nomographs designed to serve as day-to-day tools in the energy professional's toolbox.

A major assumption underlying the manual is that energy conservation is not only necessary but profitable—for the owners and tenants of buildings, and for the engineers and architects who supervise the conservation programs. A second assumption is that energy conservation is a very serious matter which requires the input of trained professionals, whether on staff in buildings, institutions, and government agencies, or from outside engineering and architectural firms.

This manual further assumes that the crisis in the Gulf—with the subsequent need to improve the thermal efficiency of America's buildings—has blurred the traditional distinction between what "architects" do and what "engineers" do in certain areas of their respective practices. Energy conservation demands the integrated talents of both professions. It confronts them with the challenge and the opportunity to enlarge their activities or practices through an aggressive pursuit of energy conservation programs on behalf of their organizations or clients. In doing so, each profession will borrow a little of the other's expertise, if only to better perform their own tasks. Architects, for example, must now be able to knowledgeably discuss the energy conservation potential of mechanical, electrical, and plumbing systems, as well as the technical details of applying "alternative energy" devices. Conversely, engineers must be knowledgeable of the energy conservation potential of building envelope and interior modifications and sensitive to the human and aesthetic implications of operating and retrofit modifications.

What this implies is a whole new range of professional options for the architect and the engineer alike. Indeed, many professionals have already made it their business to learn complimentary professional techniques and to expand their activities and practices accordingly. The manual should prove useful to both professions in the marketing of their talents and services to clients or employers. Meanwhile, to simplify problems of terminology, such terms as "energy consultant" and "energy professional" are used throughout the manual to refer to whomever is performing the energy conservation work, whether that be an architect, an engineer, or a skilled maintenance professional.

Through a careful application of the forms, nomographs and procedures described in this manual, the energy consultant should be able to make an accurate assessment of the energy consumption of most existing buildings and to identify specific "opportunities" for reducing that consumption. A cost-benefit analysis may then be prepared for each conservation opportunity, enabling the owner of the building to make implementation decisions based on projected energy savings, investment costs, operational changes, and "payback" periods. Each of these considerations is accorded a detailed treatment in the pages that follow.

SECTION II
ENERGY MANAGEMENT

A. The New Discipline: Benefits and Impact

Economic growth and shifts in the composition of economic activity over the past 2 decades have resulted in major increases in commercial activity in the United States. These increases have led to faster growth in energy consumption in the commercial sector than in any other end-use sector during that period.

The commercial sector currently accounts for more than 15 percent of the Nation's primary energy use, with almost 70 percent of its primary energy base delivered in the form of electricity. The commercial sector spends about $70 billion each year to consume about 13 quads of primary energy. Slightly more than 4 million commercial buildings account for more than 92 percent of the energy consumed in the commercial sector; the remainder is used for street lighting and miscellaneous public services.

The commercial sector experienced strong growth in energy consumption throughout the 1960's and into the 1970's (more than 5 percent per year), with growth slowing to 2 percent per year after 1973. More than 40 percent of the total growth in U.S. electricity consumption since 1972 has been attributable to the commercial sector.

The diversity of building types and functions within this sector results in a broad range of energy use (Figure 2-1). Although the overall intensity of energy use in the sector (measured as total energy use per unit of floor area) has declined slightly in recent years, the intensity of electricity use has increased substantially.

Figure 2-1. Commercial Building Types

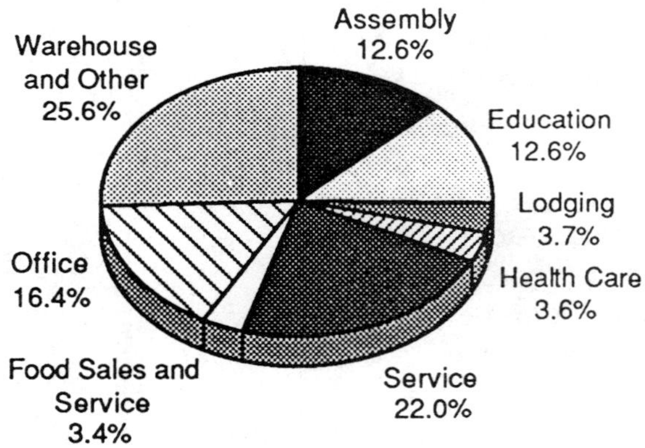

Source: *RCM-4*

The fastest growing use for electricity in this sector is air-conditioning, and this creates "peaks" in electricity demand on hot summer days, forcing utilities to maintain generating capacity throughout the year to satisfy these occasional short-term requirements. The cost of such generating plants must be covered by electricity rates.

Growth in demand for electricity by the commercial sector is a major factor in the supply constraints currently or expected by utilities in several regions of the United States.

The average commercial building constructed today uses 15 percent less energy per square foot than the commercial stock as a whole. Incorporation of innovative automatic control systems, efficient lighting, a shift to cogeneration of heat and electricity, and other advances (such as switchable glazings and evacuated insulation) are all ways to improve the energy performance of these buildings.

During the past decade, a wide variety of energy-efficient products and system modifications have been developed and commercialized, but not widely adopted. For example, natural-gas-fired furnaces and boilers are available with such features as electric ignition and condensing heat exchangers. These units are 10 to 30 percent more efficient than standard furnaces and boilers. New heat pump and chiller compressor designs, using variable speed and output capability, require 25 percent less electricity than many commercial-building heating, ventilating, and air-conditioning systems now in place. Similar efficiency improvements are available in equipment that provides lighting, refrigeration, and other energy services. The replacement of existing inefficient equipment and retrofits to improve the thermal integrity of existing commercial buildings offers an opportunity for energy savings in cost-effective applications.

Many energy-saving technologies and products are available to the building industry. New products and architectural and engineering design tools are being developed that will add to the current options available for use in reducing the energy requirements of both new and existing buildings.

Thousands of existing public and private buildings—from schools and banks to department stores and churches—have been retrofitted or otherwise modified to produce energy savings as high as 50 percent a year. The average reduction is 30 percent, and the front-end costs of such reductions are often so little that they are recovered in 3 to 6 months of reduced utility bills. Indeed, there are dozens of energy conservation opportunities which cost an owner nothing but the labor of a hand removing a fluorescent tube or adjusting a pump control.

This experience has proven, moreover, that dramatic reductions in fuel consumption do not imply reductions in the comfort of tenants, customers, employees, and other occupants of modified buildings. On the contrary, a thoughtful program of energy management frequently enhances the comfort level of an office or a hospital while trimming monthly utility bills.

Energy conservation has become, over the past few years, an established fact of life, buttressed by experience and continuing research by government, corporate, and academic institutions. In the process, it has given birth to a whole new discipline among professional engineers and architects—a new specialty called "energy management."

Numerous firms have added or retrained personnel exclusively to handle the growing demand for energy management services, which can range from a one-day "energy-audit" to a months-long contract to study and implement a wide variety of energy conservation opportunities. Architects, for example, as suggested in the Introduction, have found it profitable to broaden their understanding of electrical, mechanical, and plumbing technology to supplement their skills with respect to a building's shell and interior spaces. This has enabled them to recommend to their clients a complete package of "energy conservation opportunities," along with the appropriate cost-benefit analyses.

The growth of the industry can be reflected by the membership spiral the Association of Energy Engineers has experienced. Since its inception in 1977 as a non-profit international society, the Association of Energy Engineers has grown to 8000 members in 52 countries. The reason the industry has grown despite the ups and downs of oil prices is that energy conservation makes good economic sense. A 1990 survey conducted by the Association based on 1100 responses indicates that 68% have a lighting retrofit program, 71% have installed an energy management system and 61% have installed efficient motors.

The performance ratings of several energy conservation systems are illustrated in Figures 2-2 and 2-3.

Figure 2-2. Rating of the Overall Performance and Cost Effectiveness of Energy Management Systems

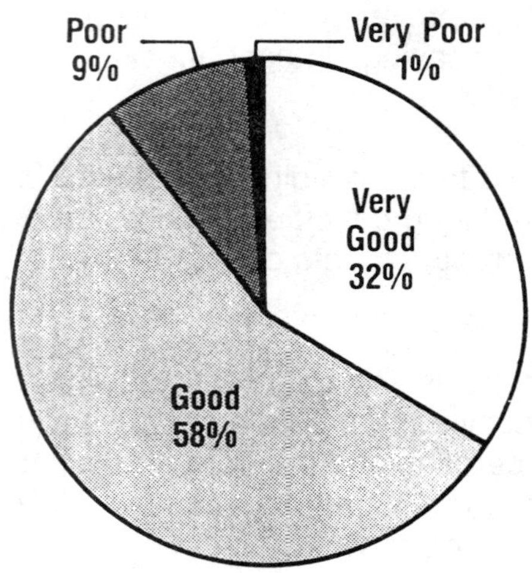

Figure 2-3. Rating of the Overall Performance and Cost Effectiveness for Cogeneration

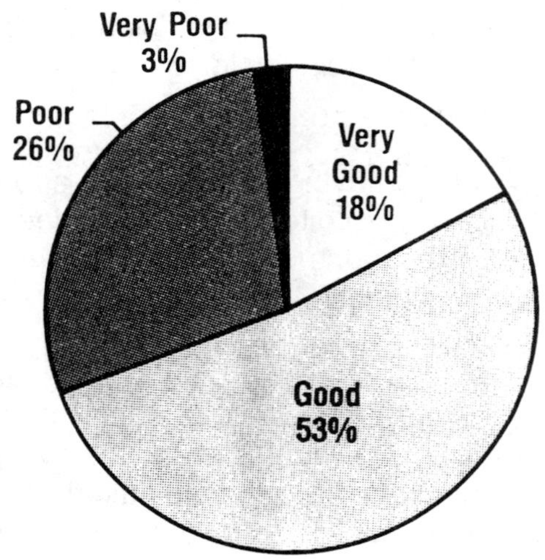

B. Why a Professional Program?

What is an "energy management" program? What is involved in such a program? How much does it cost? How long does it take? Why can't a building owner and/or custodial staff accomplish such a program independently? Why and under what circumstances should a building owner hire a professional energy consultant? What are the criteria by which a consultant establishes fees and other charges? What can the building owner expect in return?

An energy management program, as the term implies, is a systematic, on-going strategy for controlling a building's fuel-consumption patterns in such a manner as to reduce the waste of energy and dollars to the absolute minimum permitted by the climate in which the building in located, as well as by the condition of the building, its functions, occupancy schedules, and other factors. In short, an effective energy management program establishes and maintains an efficient balance between a building's annual functional energy requirements and its annual actual energy consumption—no more, no less.

The complexity of such a program, and the degree of professional expertise required for its design and implementation, will vary of course in relation to the complexity of the building itself. Certain smaller buildings in relatively good condition can be kept energy-efficient by the owner and/or custodian alone. This would apply, for example, to one-story retail outlets of 2,000 to 5,000 square feet, adequately insulated and equipped with lighting and comfort-control systems which are properly sized and responsive to simple, periodic adjustment. In that case, the role of a professional energy consultant might well be limited.

In general, however, those buildings whose size and complexity offer the greatest potential for significant energy savings—such as banks, office buildings, supermarkets, schools, and apartment complexes—will require a more substantial input by a qualified energy professional, at least in the early stages of the program. A possible rule of thumb here might be adapted from the laws of various states requiring that buildings which exceed a certain structural complexity and gross square footage be designed by registered architects: any building so defined might also be considered to require the services of a professional energy consultant. At any rate, it is in regard to these larger and more complicated buildings that the concept of an energy management program is most suitably applied, for such a program generally exacts a degree of on-going professionalism with respect to energy concerns which is not available to the owners and managers of more modest facilities.

The importance of an orderly and continuing energy management program, with clearly defined goals and spheres of responsibility, cannot be over-emphasized. It is through such a programmatic approach to energy conservation, as opposed to once-over-lightly analysis of a building's energy-use patterns, that the full potential of modern conservation techniques can be fully realized. Such a program is based on a team concept (discussed in Chapter IV), which typically involves not only the owner and/or manager of the building but other key persons as well, including office managers, division heads, building accountants, and maintenance supervisors, in addition to the professional energy consultant. Among this team's responsibilities are to oversee the various stages in the conservation program, including the survey and the setting of specific energy conservation goals (also discussed in Chapter IV), and to enlist the cooperation of other tenants and employees in the building or complex.

It is useful at this point to make a clear distinction between an energy management program and an energy study of a particular building, because the differences between the two have a bearing on such questions as the role and fees of the professional energy consultant. An energy study is a crucial early step in the fashioning of a full-scale energy management program. Indeed, it is the study which produces the data on which such a program is based. Until this study has been completed, usually under the personal direction of the energy consultant, neither the owner, the manager, nor other members of the building team will know what their options are for reducing energy waste, nor will they know the costs and benefits of those Energy Conservation Opportunities (ECO's).

An energy study of the building in question is thus the first and generally the most important function of the professional energy consultant. It is a feasibility study. It begins with a detailed, step-by-step analysis of the building's energy-use factors and costs, such as insulation values, occupancy schedules, boiler efficiencies, lighting levels, and records of utility and fuel expenditures. It includes the identification of specific energy conservation opportunities, along with a cost-benefit analysis of each one. When the study has been completed by the energy consultant, it should provide the client with a thorough and detailed basis for deciding which ECO's to implement, the magnitude of savings to be expected, and the energy conservation goals to be established and achieved by the energy management team.

Because the study is thus a preparation for the larger management program, it is executed by the energy consultant as quickly and efficiently as possible. How long it will take is naturally dependent on the size and complexity of the building, as well as the condition of its shell and energy systems. Most thorough energy studies of multi-story buildings require at least 30 days, and larger buildings can take from 120 to 180 days.

It is also due to the preliminary nature of the energy study that most experienced energy consultants have departed from the traditional "percentage-fee" arrangements of architects and engineers—at least in connection with the study phase of the energy management program. Ordinarily, design professionals set their fees as a percentage of the total cost of a construction project. But that system doesn't work for most energy studies, because the total cost of an energy management program cannot be determined until after the initial study has been submitted and decisions have been made as to which ECO's, if any, to implement. The "percentage fee" arrangement for preliminary studies has thus been replaced in most instances by a flat fee based on an estimate of the time and effort which a particular study is likely to require. (In certain other cases, particularly in studies of privately owned buildings as opposed to public, an hourly fee is charged for the energy study, following the example of professional consultants in such other fields as education and finance management.) Of course, once the building owner or manager has committed to a full-scale energy management program, including the implementation of recommended ECO's, the energy consultant might revert to a "percentage-fee" based on the cost of the projected alternations to the building. This, however, should not be committed until the extent of the implementations has been determined.

As suggested above, the energy study can be and often is the most important step in the entire energy management program. It identifies all of the energy conservation opportunities in a particular building, and provides the basis for an on-going conservation effort. Much of the remainder of this manual is therefore devoted to the steps and procedures involved in conducting a successful energy study.

C. The Positive Investment Opportunity

The need to conserve energy is often viewed by owners and managers of existing buildings as a kind of burden imposed on them by rising utility costs and the obligation to join in a serious national effort to preserve scarce resources. Many owners who agree that energy conservation is necessary are nonetheless reluctant to embark on programs in their own facilities. They fear an interruption of "business-as-usual," which may bring changes in customary schedules and procedures, as well as a decline in the comfort of their buildings. They would prefer to avoid such intrusions, especially if "energy conservation" requires an expenditure of capital which could be invested more profitably in some other manner.

This, as suggested earlier, is an erroneous perception of the true nature and potential of energy management and conservation programs. Such programs, first of all, generally do not result in a noticeable change in the "climate" and comfort levels of affected buildings. Second, in the great majority of American buildings which presently consume more fuel than necessary to operate those buildings efficiently, the possibility of a systematic energy conservation program is more properly viewed as a positive investment opportunity which will generate return at a higher rate of profit than virtually any other investment which the owner could make.

Consider, for example, an investment by a bank in a new computer system designed to reduce the time required for processing customer deposits and withdrawals. Say the system costs $520,000 to install and permits a reduction in the bookkeeping staff of four employees, thus reducing the annual payroll by $40,000. This represents a "payback" period of 13 years, which is not unreasonable for a permanent improvement in the efficiency of a traditional business operation. Such investments—aimed at increasing profits by increasing the efficiency of the workforce—are a basic and long-accepted feature of the American business and management community.

A comprehensive energy management program is no less an investment in improved efficiency, and the returns are often higher, meaning greater annual profits. Suppose, for example, that the bank mentioned above invested $600,000 in a variety of energy conservation measures which reduced its annual fuel and utility expenditures from $1.2 million to $1 million. This represents a "payback" period of only 3 years, which is not at all unusual. Suppose further that the bank prorated its investment over a 20-year period, so that its "monthly payments" on the energy investment, including interest, came to $2700. Assuming that the bank's utility bills were reduced from $100,000 to $83,000 per month, due to the building's improved energy efficiency, the bank would realize an immediate return on its investment of $14,300 per month, or 600 percent.

It is because of such potential "earnings" that energy conservation programs in most existing building should be viewed—and marketed—as positive investment opportunities, not as inconvenient interruptions of "business-as-usual."

SECTION III
AN OVERVIEW OF ENERGY USE AND CONSERVATION

A. Where Energy is Used In Buildings

The business of a professional energy manager is of course to reduce the volume of fuel consumed or wasted per annum by an owner-client's building or complex of buildings, thus saving the owner's money (or increasing profits) while conserving the nation's fuel resources. In approaching this task, it is useful to know the types of buildings and the systems within and features of those buildings which tend on the average to consume more fuel than necessary. While each individual building must be analyzed separately, studies have revealed certain recurring patterns of energy consumption based on type of structure and climate. A summary of these patterns is provided in Figure No. 3-1 below.

FIGURE 3-1 COMPARATIVE ENERGY USE BY SYSTEM

		Heating & Vent.	Cooling & Vent.	Lighting	Power & Process	Domestic Hot Water
Schools	A	4	3	1	5	2
	B	1	4	2	5	3
	C	1	4	2	5	3
Colleges	A	5	2	1	4	3
	B	1	3	2	5	4
	C	1	5	2	4	3
Office	A	3	1	2	4	5
	B	1	3	2	4	5
	C	1	3	2	4	5
Comm.-ercial Stores	A	3	1	2	4	5
	B	2	3	1	4	5
	C	1	3	2	4	5
Reli-gious Bldgs.	A	3	2	1	4	5
	B	1	3	2	4	5
	C	1	3	2	4	5
Hos-pitals	A	4	1	2	5	3
	B	1	3	4	5	2
	C	1	5	3	4	2

Climatic Zone A: Fewer than 2500 degree days
Climatic Zone B: 2500 - 5500 degree days
Climatic Zone C: 5500 - 9500 degree days

Heating consumes the most energy in office buildings and schools in cold climates. In mild zones, the electrical load for lighting, which is relatively independent of climate, may exceed either heating or cooling in schools, hospitals and religious buildings. The amount of energy required

FIGURE 3-2 ENERGY USED BY IDENTICAL OFFICE BUILDINGS

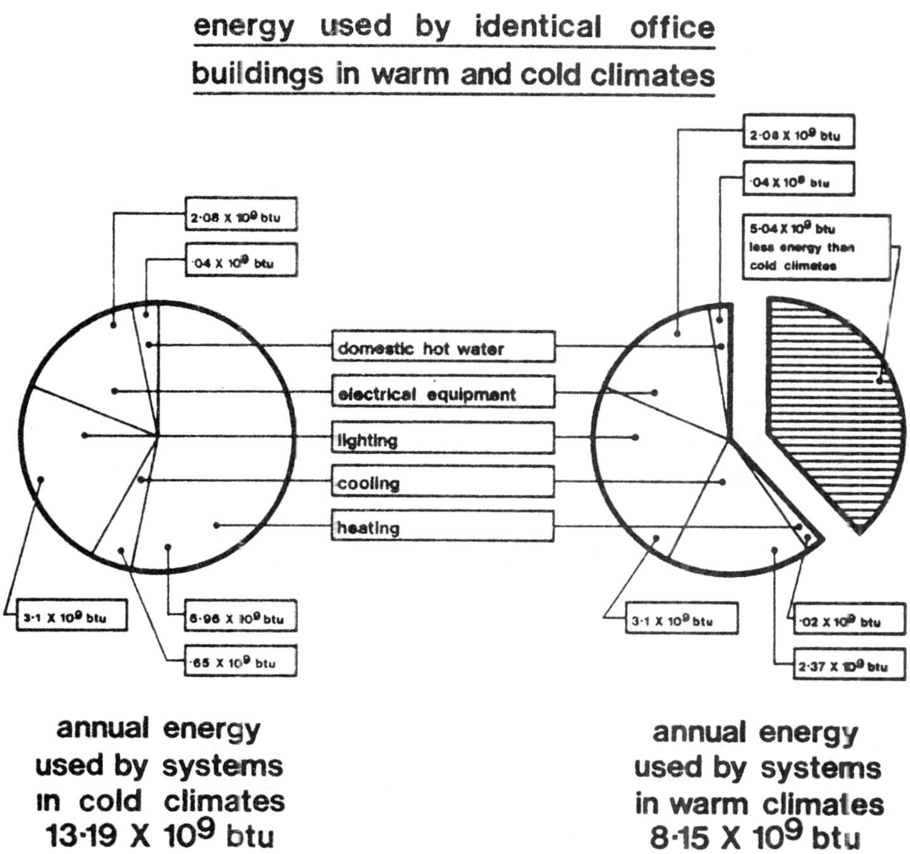

Based on the NBSLD program from the National Bureau of Standards which calculated energy consumption for identical buildings in Manchester, New Hampshire and Orlando, Florida.

Gross floor area - 126,00Ft.^{2}
Occupants -600 people
Operation -8 a.m. to 6 p.m. weekdays only
Ventilation -15,000cfm occupied times only
Infiltration -4,500 cfm continuous (1/4 air change/hour)
Lighting and power -3.4 Watts/Ft^2 occupied times only
Indoor conditions --Heating 68°F occupied
58°F Unoccupied
Cooling 78°F occupied only
Outdoor conditions--Variable - NOAA weathertape

for domestic hot water is especially significant for all zones in hospitals, housing, and athletic and cooling facilities in schools and colleges.

Religious buildings and public halls, which frequently include meeting rooms, offices and school facilities, are likely to consume energy in the same pattern as office buildings in the same geographic location. They will, however, consume less energy per square foot of floor area than office buildings.

Figure 3-2 Illulstrates the relative magnitude of energy consumption for an office building located in a cold climate (New Hampshire) versus consumption for an identical building in a warm climate (Florida).

Since no one building in existence is quite like any other, there is no one rule for determining which part of a building's comfort system accounts for the bulk of that system's energy demand. The seasonal combustion efficiency of typical burner-boiler units, for example, ranges from a high of 78 percent to a low of 30 percent, with any number of factors bearing on that efficiency. The design of a building and the climate in which it is located, its orientation, the type and condition of walls and glazing, the load-demand on conversion equipment, the condition of plumbing and mechanical systems-- all will have a greater or lesser impact on fuel consumption in any given building, with subsequent variations in terms of energy conservation possibilities.

The role of the shell in a building's pattern of energy use will be considered in a later section. As for energy-using systems within the building, those which consume the highest percentage of energy nationwide are ranked below according to the relative magnitude of their demand:

(1) Heating and ventilating

(2) Lighting

(3) Cooling and ventilating

(4) Equipment and machinery

(5) Domestic hot water

The relative magnitude of energy use by the HVAC and lighting systems varies according to climate, building design and construction, mode of operation, and efficiency of equipment. In climatic zones with mild winters, the seasonal cooling load may be heavier than the seasonal heating load and may consume more energy. In cold climates, heating usually consumes the most energy.

The energy required for industrial buildings, exclusive of process loads, is generally similar in all zones to that required for commercial buildings. In mid-climates of 2500-6000 degree days, the magnitude of energy use by the various systems depends upon the type of mechanical and electrical system and the particular characteristics of the building.

B. Principles of Energy Use in Buildings

The number one principle of energy use in existing buildings-- and perhaps the most important to remember in conducting a professional energy study --is that the various features and systems contributing to a building's total annual energy demand are interrelated in very significant ways. A reduction in the building's lighting level, for example, may affect not only the building's cooling load but demands on heating and humidity-control equipment, in turn affecting the load demand on pumps, fans, boilers, and chillers. A reduction in the heat produced by equipment on a lower floor may have a "ripple effect" on the floor above as well as on adjacent zones-- possibly an undesirable effect. It is due to such "thermal interrelationships," which become more complex in larger and more complicated buildings, that a study of the building's patterns of energy use must be approached methodically and systematically (the details of such an approach are discussed in Chapter IV). This principle should be kept in mind while reviewing other major principles of energy use (many of which will be familiar to energy professionals with experience in the design and construction of new buildings). Those other principles are as follows:

1. ENERGY LOADS

The energy requirements of a building are expressed as three loads-- the building load, the distribution (or parasitic) load, and the primary equipment load.

a. Building Load: The building load is the amount of energy in Btu or kilowatts, required to maintain the indoor environment at desired levels, if the building's electrical and mechanical systems were 100 percent efficient. The size of the building load is governed by several factors:

(1) Location and orientation: The geographic location of a building and detailed site conditions determines the climatic conditions to which it is exposed. The climate and site conditions in turn affect the annual heating and cooling loads through the amount of isolation, range of air temperatures, humidity levels, and air movement that occur. For example, the angle and quantity of sunlight which strikes the building exterior determines the solar heat gain at any given moment. Solar heat gain may be detrimental in the summer by increasing the cooling load or beneficial in the winter by reducing the heating load. Sunshine can also reduce the lighting load in certain buildings.

(2) The environment maintained inside the building: The temperature, lighting, and ventilation levels in a building have a major effect on the building load. However, the humidity level and the quantity and temperature of hot water can also be significant factors. Research indicates that until recently, American buildings have been heated, cooled, lighted, and

ventilated to a greater extent than health and comfort have required-- often to the point of discomfort. Comfort standards have accordingly been revised, and this manual takes these changes into account. (Professionals should be warned, however, that building owners are often reluctant to adopt conservation measure which affect the indoor environment. These measures affect tenants directly, and the acceptability of the new conditions can easily be obscured by the complaint that "they're taking something away from us." Insensitivity to this psychological aspect of energy conservation can jeopardize an otherwise acceptable program.)

(3) Heat loss and gain through the building shell: This is affected by practically every feature of a building's exterior and interior skins, from the type, thickness, and color of surface materials to insulation, glazing and efficiency of doors and solar thermal control devices such as awnings and drapes. The major problem generally derives from infiltration of outside air (whether hot or cold) through walls, windows, and doors.

(4) Function and period of occupancy: The function of a building largely determines the number of hours per week it will be used. Long periods of occupancy, for example, will multiply the energy load required by a given environment. On the other hand, machines and human bodies produce heat which may allow reductions in cold-weather heating loads.

b. The distribution loads are often called parasitic loads since they do not contribute directly to the comfort and other requirements of the building occupants. Distribution loads arise mainly in pumps and fans and their associated piping and duct work. A surplus load is caused by poorly-insulated pipes, the resistance of a duct or piping system to the flow of air or water, and excessive circulation of air or water. The size of the distribution load affects the energy consumed by the primary heating and cooling equipment.

c. The primary equipment load is caused by primary energy-conversion equipment such as boilers and chillers. Surplus load is caused by their inefficiency. This inefficiency results from the design of the equipment, from the way it is maintained, and from the number of times the equipment is turned on and off. While a boiler, for example, is designed to convert a latent chemical fuel energy (in gas, oil, coal, etc.) into heat, the full potential heat-energy can rarely be extracted because the optimal fuel-to-air ratio cannot be maintained at all times in all parts of the firebox. Additional heat is lost up the chimney and through radiation from the boiler surfaces.

Poor maintenance can further reduce efficiency. Soot and scale reduce the efficiency of heat transfer in a boiler as dust on lamps reduces their light output. Frequent variations in demand will also reduce equipment efficiency.

2. ANALYSIS OF LOADS ON A SEASONAL BASIS

In addition to the factors cited above, the primary equipment load is strongly affected by daily and seasonal variations in the occupancy schedules and functions of the building itself. Once a furnace is fired and burning, for example, it may achieve a 90 percent efficiency (assuming it is fueled by oil or natural gas-- coal-fired units are less efficient). But that efficiency will decline by 10 to 30 percent when the furnace is monitored on a seasonal basis, due primarily to stack losses between firing periods. There will be a similar decline in the seasonal efficiency of most other units of primary equipment, and it is for that reason that any analysis of a building's energy consumption must be conducted on a twelve-month basis, thus accounting for seasonal variations.

3. THE CUMULATIVE NATURE OF LOAD REDUCTIONS

Due to the interrelationships between the many factors determining a building's energy demand-- as mentioned above --an alteration of one load factor, such as insulation, will impact on other factors, such as air requirements for cooling, and a series of alterations will have a cumulative effect on total building load which must be understood and accounted for in the conservation program. To maximize the potential of this "cumulative" principle, the energy professional will generally want to pursue a standard sequence of operations in implementing fuel conservation measures:

(1) effect operational ECO's;
(2) reduce the building loads;
(3) reduce distribution losses from the central plant to the terminal units;
(4) improve the efficiency of primary energy conversion equipment.

Consider, for example, the "cumulative" effect of a reduction in lighting load through removal of lamps. This automatically reduces cooling load and will allow distribution flow rates to be reduced, which in turn will reduce fan and pump energy consumption. Thus, a change in the building load will improve the efficiency of the distribution system even before specific improvements in that system have been considered.

The cumulative effect of loads becomes particularly significant where the peak building load can be reduced. Distribution systems are designed to supply the worst possible case, the peak building load, even though these peak loads occur for relatively short periods of time (usually less than 5 percent of the year). Ducts are sized and fans are selected to have the necessary capacity. In many cases, however, the fans are not only selected for a certain capacity, they are operated at that maximum capacity all the time. If the peak building load can be reduced, the output of the fans can be reduced. The same rule applies to water distribution and pump sizes, cooling requirements and minimum chilled water temperature. In each case, reduction of a peak building load will often lead to a disproportionate reduction in the distribution load.

(Note: Domestic hot water and commercial refrigeration loads are related only to the demands of the particular systems, and little if any interaction occurs with other systems. An exception of this may occur with the implementation of thermal reclamation techniques, to be discussed later, where waste heat from one system is used to reduce the heating load of another system.)

SECTION IV
A PHASED METHODOLOGY

A. Introduction

It is now appropriate to discuss the actual phases involved in developing and implementing a successful energy management program for a particular client and that client's building or complex of buildings. In approaching this discussion, it is useful to recall the above distinction between an energy management program and an energy study. As stated in Chapter II, this manual is concerned primarily with the proper techniques and procedures for conducting a thorough study-- the heart of the energy management program --but the importance of that study and the way it fits into the large on-going program is best understood through a brief review of the phases comprising the entire program. As with the study itself (to be further clarified in Chapter IX where a model building is actually studied) the energy management program consists of a series of very definite phases which must be executed in sequence to assure maximum effectiveness. Following this summary of the phased methodology for the energy management program, the four phases making up the study itself (Phases Two through Five below) will be examined in greater detail. In fact, each of the four phases in the study will be accorded its own separate chapter, with still another chapter devoted to the Model Building Energy Study.

B. Phase One: Select the Energy Management Team

Once the client has agreed to consider an energy management program for his or her facility, it is advisable to assemble and activate an energy management team, even though the nature and extent of the energy management program will not be known until the subsequent study has been completed. A typical energy management team consists of: (1) the building owner and/or manager or representative; (2) the building accountant; (3) the building engineer or maintenance supervisor; (4) a representative from each major tenant-firm or division in the building, particularly when such divisions correspond with different "functional areas" (e.g., data processing, tying pool, executive suite, parking garage, restaurant, floors or departments in a department store, mailing room, etc.); (5) the professional energy consultant. To these may be added, when possible and appropriate, the original building architect and engineer and a representative from each of the major fuel and utility companies serving the building.

Part of the idea in assembling this team is to achieve as complete a cross-section as possible of persons having direct first-hand experience with the thermal performance and characteristics of the building. Among the early functions of the team is to provide a body of specific suggestions and ideas for reducing the building's energy consumption. Later, when the energy management program is being implemented, the team will be responsible for helping to win the cooperation of other employees and tenants in the building. (This effort may be enhanced by setting specific "energy conservation goals" for different floors, departments, divisons, etc., perhaps on a friendly competitive basis.)

At any rate, the energy consultant should solicit and record ideas from the team, along with such data as utility bills and construction records (to be discussed in Chapter V). Finally, in preparation for conducting the energy study, the energy consultant might want to break out a smaller component of the team to assist directly with the study itself. This "energy study sub-group" will typically include the building accountant and engineer, plus what other persons may be appropriate.

C. Phase Two: Survey the Building

This phase of the program actually consists of several major sub-phases, including an audit and graphic display of the building's percent energy consumption, gleaned from fuel and utility bills, as well as a physical survey of the shell, interior spaces, power and distribution systems of the building itself. It is during this survey phase that the energy consultant begins to note potential energy conservation opportunities, which may or may not be supported by additional evidence and cost-benefit analysis. The details of this important phase of the energy study will be presented in Chapter V.

D. Phase Three: Tabulate Present Energy Use

A thorough survey of the client's building will result in two sets of data pertaining to the building's energy-use. One of these will be a detailed accounting of fuel and electricity consumption on an annual basis, as determined from utility statements. The other will be a detailed breakdown of the energy consumed by individual zones ("functional areas") and features of the shell and mechanical systems of the building itself, as determined from the physical inspection. The latter items are called "energy-users." A total of the energy consumption represented by each "user," projected on an annual basis, should correspond with the total consumption of the building as inferred from the utility bills. (Substantial discrepancies between the two totals will necessitate a review and adjustment of preceding calculations.)

These totals, along with the calculations and subtotals from which they are drawn, will account for all the energy consumed by the building and pinpoint where and how that energy is used. As with Phase Two above, this phase of the energy study, including suggested forms and formulas, will be discussed at greater length in Chapter VI.

E. Phase Four: Identify Energy Conservation Opportunities

At this point, the energy consultant will have established not only a comprehensive picture of the building's present energy use but a tentative list of energy conservation opportunities (ECO's), compiled during the execution of Phases One-Three above. Now the consultant may systematically analyze the feasibility of these tentative ECO's, along with others which may be applicable, using the methods, nomographs, and calculations presented in Chapter VII (which offers a specific treatment of 52 different ECO's commonly available in existing American buildings). A major requirement here is that potential ECO's must be analyzed in such a manner as to account for <u>thermal and functional interrelationships</u> between the various loads and features of the building's shell and energy systems (e.g., the effect of a reduction in lighting levels on heating and cooling loads).

F. Phase Five: Summarize Costs and Benefits

Having identified, grouped, and determined the feasibility of all potential ECO's in the client's building, the energy consultant must now assess the impact of those ECO's, both singly and collectively, on the building's future energy use. In other words, how much energy will be saved by the implementation of each ECO, each group of ECO's, and all the ECO's combined, as measured against the building's present energy use?

Once these potential energy savings have been established (using procedures described in Chapter VII), they can be converted to dollar savings, thus providing the basis for a cost-benefit analysis of the potential ECO's, including a comparison of implementation costs (initial dollar investment required for each potential ECO) with projected savings per year and estimated "payback" periods. This information is then submitted to the owner and/or the energy management team in the form of a list of conclusions and recommendations by the energy consultant. It is on the basis of this last step in the energy study (detailed in Chapter VII) that decisions are made on the scope,cost, and extent of the energy management program to be implemented.

G. Phase Six: Set Goals

Most energy management programs are best designed and accomplished on the basis of specific energy conservation goals which the team will establish after having decided on the scope of the program itself, including the implementation of ECO's. These goals can and should be expressed in a variety of ways, depending on the nature of the building in question, the types of ECO's to be implemented, the types of activity and personnel housed by the building, seasonal fluctuation in load demand, etc. As suggested above, for example, the desired goals for the entire building can be broken down according to zones, divisions, floors, departments, and/or "functional areas," with appropriate employees or staff persons assigned the responsibility of helping to achieve those goals.

Not only annual goals should be set, but seasonal and monthly goals as well, especially in climates with distinct seasonal weather variations or in buildings with marked seasonal fluxation in activities and occupancy such as schools which often close for summer and other holidays. The setting of these goals may well be accompanied by the installation of energy-metering and monitoring equipment in each location where specific goals have been assigned (as discussed in Phase Seven below), so that the persons involved may keep a "running account" of their progress toward the desired goals. This can then be adjusted for heating and cooling degree days and changes in operating hours for comparison to the base goals established. Care should be taken, finally, that the goals imposed are not so ambitious that they cant' be met by the persons responsible for them. In this regard, it is often prudent to establish more modest goals for the first month or two, see those goals accomplished, and then set progressively higher goals until the maximum energy reductions are being accomplished regularly. The goal thereafter, of course, is to maintain those reductions.

(NOTE: Another approach to "energy conservation goals" is to establish a specific energy budget for the building in question, with proportionate budgets assigned to the various zones or "functional areas" to be monitored during the energy management program. Consequently, the terms "energy budget" and "energy goals" are used more or less interchangeably at various points in the remainder of this manual.)

H. Phase Seven: Implement and Monitor the Program

The requirements of this phase of the program will obviously depend on the scope of the program itself, as well as on whether the energy consultant is retained by the client following completion of the preliminary energy study. Assuming that the consultant is retained, it will typically be the consultant's responsibility to prepare working drawings and specifications for all proposed construction work and to help the owner in taking competitive bids. Once the appropriate contracts are signed, usually under the supervision of the owner's attorney, the energy consultant generally will oversee all construction work and certify payments to contractors.

Upon completion of the contracted alterations, the energy consultant will manage or initiate some or all of the following post-construction services (in addition to helping establish an energy budget or set of goals as discussed above):

1. Develop specific operating programs and schedules to monitor and control the operating systems for maintaining the environmental conditions conserving the maximum amount of energy, reducing operating costs, and utilizing manpower most effectively.

2. Train operating personnel and supervise the educational program to acquaint the building occupants with new operating conditions and goals.

3. Supervise the preparation of log books, which record all changes, energy use on a monthly basis (by sub-systems), and costs, in order to document the program and assure continuity of effort. Monthly consumption, adjusted for degree days, cooling hours, and operational changes, will indicate the effectiveness of the current program.

4. Install suitable monitoring procedures to record the performance of modified equipment or procedures. These, where appropriate, may include:

   a. Institute maintenance of equipment operation logs to identify hours of operation and responsibility for start-stop.

   b. Install elapsed time meters to record hours of operation of prime movers.

   c. Install electric watthour meters to record usage of electric energy for specific operations or departments.

d. Install fuel meters to record consumption of fuel for specific operational purposes.

e. Install Btu meters, steam flow meters or condensate meters to identify heat input or output of processes of building environmental apparatus.

f. Recording thermometers may be installed to verify performance of critical automatic temperature controls.

g. Recording ammeters may be used to monitor relative load changes in electric prime movers and power distribution feeders.

5. Develop reporting procedures and summaries to monitor the operating and maintenance personnel execution of energy conservation program elements. Daily, weekly, and monthly logs of equipment use, meter readings and performance evaluations should be summarized and reported to operating personnel. Continued reference to performance goals and actual achievement is necessary to sustain interest and reward successful effort and skill at the level at which operating decisions are actually made.

   a. Start-stop equipment logs and elapsed time meters should be totalized monthly and performance compared to the energy budget allocation.

   b. Submeters on electricity, fuel, steam, condensate, chilled water, hot water, etc., should be totalized monthly and performance compared to the energy budget allocations.

6. Prepare comparative statements of energy use semi-annually to show the actual purchases of electricity and fuel compared to the energy budget and the adjusted record of energy purchased in the past. Graphic presentations may provide a continuing synopsis of the effects of energy conservation programs.

I. Summary

To summarize the phases in the typical energy management program discussed above: (1) Select the Energy Management Team; (2) Survey the Building; (3) Tabulate Present Energy Use; (4) Identify Energy Conservation Opportunities; (5) Summarize Costs and Benefits: (6) Establish an Energy Budget; (7) Implement and Monitor the Program.

Again, it is important to bear in mind that the energy study (Phases Two-Five above) is only a preparatory component within the larger energy management program. It is not the program itself and should not be approached or negotiated with the client as such. Furthermore, while it is strongly recommended that the seven phases of the program be executed in sequence (and especially the four phases of the preliminary study), there will naturally be overlaps and crossovers between some of the phases. The "identification of ECO's" for example, though treated above as a separate operation, will actually commence as early as the first meeting with the building owner and/or energy management team and continue through the entire building survey. In fact, the energy consultant will _always_ be looking for potential ECO's, particularly during the physical inspection of the building, through it won't be until Phase Five that _potential_ ECO's are determined to be _actual_ ECO's.

Likewise, after the first few weeks or months of the implementation-monitoring phase of the program, it may be necessary to revise projections regarding the impact of certain ECO's and even to discontinue some of them, "fine tune" others, and supplement them with additional ECO's. Still another exception to the above sequence of phases may obtain in connection with a building which is large enough to justify an energy management program but has so few employees or permanent occupants that there is no basis for an energy management team. Finally, the above procedures apply equally to single buildings and to complexes or clusters of buildings, though of course it is possible that separate energy budgets and set of ECO's will be demanded by an energy management program for a multi-unit facility.

SECTION V
SURVEYING THE BUILDING

Introduction

The object of this first phase of the energy study is to calculate the total annual energy consumption of the building in question and then to identify the particular zones and features of the building which account for that energy use-- including, of course, the wasted energy which the energy management program is designed to eliminate. The building survey thus consists of two major subphases: (1) an audit of the building owner's annual expenditures for fuel, utility services, and related items (maintenance of power and distribution systems, replacement of lights and filters, etc.); (2) a detailed physical examination and measurement of the building itself, including the shell, distribution systems, electrical and mechanical equipment. These subphases are further divided into a series of definite steps which, for best results, ought to be executed strictly in sequence. They are summarized below and explained at greater length in the pages that follow:

(1) organize a conservation journal;
(2) determine plans for future renovations and/or expansions of the building;
(3) collect and analyze building energy-use data;
(4) prepare a graphic summary of annual energy consumption patterns;
(5) conduct the physical survey.

Step One: Organize a Conservation Journal

This will become the "bible" of the energy management program. Into this journal will go the results of each of the steps in the building survey, including the forms discussed below, as well as suggestions from the energy management team, potential ECO's encountered during the survey, conclusions and recommendations presented to the owner-client, and such other information as may be necessary or desirable. (It is often useful, for example, to record the names and phone numbers of outside persons with whom the energy consultant may wish to confer, such as utility representatives and the building's original architect and engineer.)

Step Two: Determine Plans for Future Renovations and/or Expansions of the Building

This information, along with suggestions for potential ECO's, should constitute an early entry in the conservation journal. Plans for renovation and/or expansion will obviously have a bearing not only on the types of ECO's to be considered but on the scope of the larger energy management program. Some ECO's which would not be economically feasible for the building as it stands may become more attractive in view of projected building alterations. Indeed, this may occasionally work in reverse: an owner may find it profitable to hasten plans for expansion or renovation in order to coordinate that construction with the implementation of a series of ECO's, thus saving front-end money on both investments.

Step Three: Collect and Analyze Building Energy-Use Data

This is the "paper" phase of the building survey, aimed at providing a detailed assessment of the building's total annual energy consumption. Among the documents which the energy consultant will need or find useful at this point are: (1) original construction documents, blueprints, equipment specifications, etc. (some of which will figure more significantly in the physical survey of the building); (2) subsequent construction documents, records of alterations, maintenance expenditures, etc.; (3) as complete and detailed a record as possible of monthly fuel and utility expenditures for at least the past two years; (4) an explanation of the rate systems and billing procedures of each of the utility and fuel companies serving the client's building (here the consultant may wish to confer with the utilites themselves to determine the availability of alternative billing systems which might be profitably applied in conjunction with or following the implementation of such ECO's as load-shedding, off-peak electricity consumption, etc.).

Having become familiar with the building's energy systems through an examination of construction documents and equipment specifications, the energy consultant should now study and analyze the fuel and utility bills. (It is at this point, in particular, that the consultant may begin working closely with the building accountant and/or other persons familiar with the building's energy costs.)

The form to be used in summarizing the recording the appropriate data from fuel and utility statements is the Utility Cost/Consumption Survey Form, shown below in Figure No. 5-1 (with instructions for its use in Chapter VII, "Documents").

FIGURE 5-1 UTILITY CONSUMPTION/COST SURVEY FORM

UTILITY CONSUMPTION/COST SURVEY FORM

JOB: HB&S BUILDING   DATE: 9-9-78

Month Year	Elec 1 Kwh	Elec Demand	Elec 1 $	Fuel Gas mcf	Fuel Gas 1 $	Fuel	Fuel $	Degree Days-Htg
Jan 77	218671	623.2	9935	2061	3607			1006
Feb	225262	646.8	10130	2030	3553			894
Mar	307901	680.7	12749	1883	3295			679
Apr	339913	696.1	13613	1740	3045			132
May	385220	730.0	14837	1207	2112			0
June	403900	840.1	15341	1061	1857			0
July	456283	887.2	16756	990	1733			0
Aug	461200	887.2	16888	1363	2385			0
Sept	445450	865.0	16463	1526	2670			181
Oct	385500	810.5	14845	1892	3311			761
Nov	315630	721.3	12958	1691	2959			662
Dec 77	216212	636.8	10274	2252	3941			1119
TOTAL	4161142	887.2	164789	19696	34468			5434
Jan 78	212318	626.8	10168	1926	3370			974
Feb	208900	640.2	10076	2116	3703			912
Mar	316502	681.8	12982	2042	3574			603
Apr	341206	698.1	13649	1520	2660			261
May	402916	730.0	15315	1260	2205			0
June	400480	835.6	15249	1026	1796			0
July	482916	887.2	17474	1015	1776			0
Aug								
Sept								
Oct								
Nov								
Dec								

The form provides space for a two-year history of the building's total energy consumption, and it is suggested that data be entered for the two years immediately passed (unless those years included periods of atypical or unusual energy-consumption patterns, in which case a more typical period should be substituted). These years will then be averaged in Step Four below. (Keep in mind that the information on utility bills may in some instances be checked against records of electricity and fuel consumption as logged by the building staff from on-site electric and flow meters.)

Step Four: Prepare a Graphic Summary of Annual Energy Consumption Patterns

Using the information on the form above, plus whatever other data may be appropriate (such as records of equipment maintenance and replacement), the energy consultant should now prepare a visual graph or chart of the building's total annual energy consumption and costs. This graph will be used as a quick point of reference throughout the remainder of the energy study, and particularly as the basis for considering the viability of potential ECO's (whose likely effect on energy-use can also be "graphed" for a visual comparison with the original "base-year" graph as shown in Figure 5-2).

FIGURE 5-2 HB&S BUILDING UTILITY CONSUMPTION PROFILE

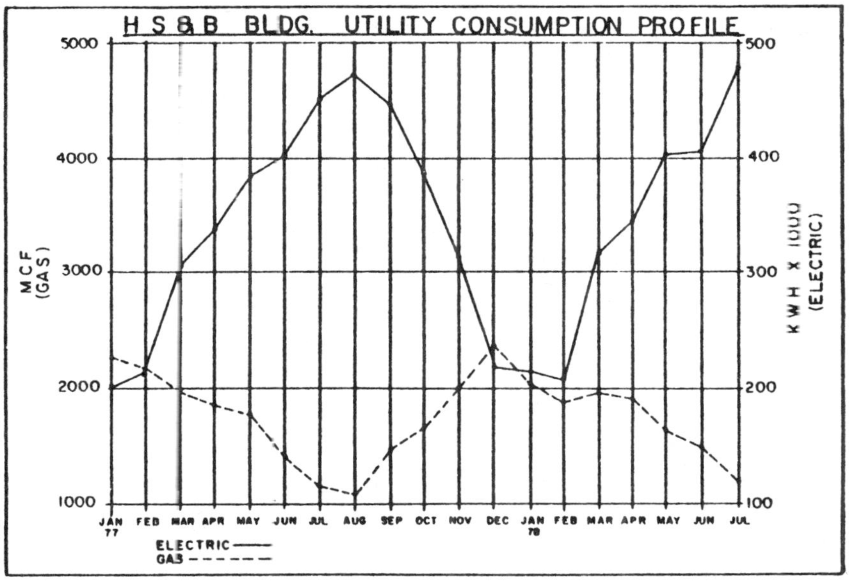

Since the graph should represent a "base year" of energy consumption for the building under survey, it is suggested that an average be taken from the two years of energy consumption entered on the Cost/Consumption Form. The idea, of course, is to obtain an accurate, month-to-month profile of a typical year in the energy performance of the building. The graph should represent not only the building's total energy consumption but also the yearly consumption patterns for each individual energy source, including electricity, gas, fuel oil, coal, etc.

To these may be added a graphic summary of such other factors as the consultant finds useful to have in graphic form. It will eventually be necessary, for ex- example, to determine the building's energy consumption patterns on a daily, weekly, and seasonal basis (as discussed below), and the energy consultant may find it helpful to render these patterns on separate graphs plotted against heating and cooling degree days, to supplement the "base year" profile. Finally, as suggested above, the consultant may wish to add to the "base year" graph a visual summary of equipment and fixture maintenance and replacement patterns.

Step Five: Conduct the Physical Survey

The energy consultant now has a clear idea of the total volume of energy consumed on a monthly and yearly schedule by the building under survey, as well as the dollar-costs of that energy. In terms of locating sources of energy waste, however, the annual consumption patterns are of little value until the consultant has determined which zones of the builidng, which features of the shell, which functions of the occupants and units of the building's energy system are responsible for which percentages of the building's total energy use. It is thus necessary to conduct a thorough survey of the building itself, armed in advance with a study of construction documents and equipment specifications. The purpose of this survey, in effect, is to take an inventory of the energy demand of each separate "energy user" in the building. (Here again the work will be simplified by the assistance of someone familiar with the building's operation, in this case the building engineer or maintenance supervisor.)

The physical survey of the building is essentially an analysis of the efficiency at which the various components of the building are functioning, both individually and in their relationships to one another. Given the climate and condition of the building, which of its features are performing efficiently and which are not? Are the windows tight? Do they lose heat excessively in winter and gain heat in summer? Might their efficiency be improved by shading, double-glazing, interior drapes? What is the rate of heat transmission through the roof? Can it be reduced economically? Are fans, compressors, and pumps adequately sized and operated for the loads demanded of them? Are pipes and duct-work sufficiently insulated? Is the boiler firing at maximum efficiency? Is it properly sized for the average daily building load? Are chillers and refrigeration equipment properly sized? Are lighting levels in task-areas unnecessarily and even uncomfortably high? What is the flow-rate of hot water in bathrooms? Is it necessary for comfort and cleanliness? What is the degree of "human error" responsible for wasted energy? Might certain tasks be rearranged or rescheduled to provide a more efficient utilization of the building's energy systems?

These are a few of the quetions which the energy consultant may fruitfully bear in mind during the physical survey of the building, for attached to each question is a potential energy conservation opportunity. (It is strongly recommended, in fact, that the energy consultant have studied and become familiar with the possible ECO's discussed in Chapter IX before undertaking the physical survey of the building.)

Another question which must be answered is how best to approach the survey itself. Since most large buildings are too complex to inspect as a single unit, how is the survey to be organized? Should the building be "divided" and examined on a floor-by-floor basis? Room by room? Should it be approached in terms of major power systems and units of mechanical equipment such as HVAC and lighting systems? Is the shell to be treated as a separate category?

In practice, most successful building surveys will borrow a little from each of these approaches. But perhaps the soundest overall approach is to organize the survey in terms of the building's "functional areas." What these are and where they are located will differ to a greater or lesser extent in every building surveyed, as for example in a courthouse or a bank versus a department store. A "functional area" is generally defined as any zone or area in the building which is clearly distinct from other areas with regard to the activities performed there, routine operating schedules, number and density of occupants, types of equipment and machinery in use, and other factors bearing on relative energy load demands. A computer room in an office building, for example, would be considered a discrete "functional area" to be surveyed separately from an office suite or a snack bar, a bank lobby or a retail area. There will of course be situations where some or even most of the "functional" areas in a building share highly similar types of activity and operating schedules-- as in schools and certain office buildings --in which case most of the "functional areas" will be defined in terms of floors, rooms, or wings of the building.

At any rate, having divided the building into its "functional areas," the energy consultant may now conduct the actual physical survey. This will require a systematic method for collecting and recording the data being sought ("U" values of walls, lamp lumen output, hot-water flow rates, fan horsepower, etc.), and it is suggested that standardized "survey forms" be employed for this purpose (corresponding with the Utility Consumption/Cost Survey Form discussed above). Since the structure must be surveyed in terms both of building load and distribution load it is further suggested that a different form be used for assessing each type of load. The forms presented below (Figures 5-3 and 5-4 have been field-tested and found useful in most typical building surveys. The Building Load Survey Form (Figure 5-3) is designed for use particularly in conjunction with the "functional area" approach described above, while the Distribution Load Survey Form (Figure 5-4) may be applied either to "functional areas" or to the energy distribution system serving the building as a whole, whichever is most appropriate. (Detailed instructions for using these forms will be found in Chapter VIII, "Documents," blank forms will be found in Appendix B.

FIGURE 5-3 BUILDING LOAD SURVEY FORM

BUILDING LOAD SURVEY FORM

JOB: HB&S Building　　AREA: Office　　WINTER: 75 - 30%　　SUMMER: 75 - 50%　　PECO NO: 0　　DATE: 9-9-78

GLASS					WALLS				ROOF OR FLOORS	
Facing	SqFt	Shading Ext	Shading Interior	Type U= 1.13	Facing	SqFt	Shade	Type U= 0.20	SqFt	Type U= 0.20
N	1152	0	Blinds	Single	N	8368	0	Light	10300	6" Concrete
S	3608	0	S.F.-.64	Metal	S	4587	0	4" Brick		Built-Up
E	4576	0		Casom.	E	9464	0	4" Block		Roofing
W	5456	0		Avg. Fit	W	10609	0	Plas. Finish		2x2 Lay-in Ceiling
TOTAL	14792				TOTAL	33028			10300	

LIGHTING Area	Type	Occupied Kw	Occupied Fc	Occupied Hrs/Wk	Occupied Kwh/Yr	Custodial Kw	Custodial Fc	Custodial Hrs/Wk	Custodial Kwh/Yr
Office - Exterior	2x4-200W Flourescent	142.4	150	40	296192	142.4	150	20	148096
Office - Interior	2x4-200W Flourescent	110.5	150	40	229840	110.5	150	20	114920
Halls	150W. Incandescent Spots	21.6	40	40	44928	21.6	40	20	22464
Restrooms	2x4-200W Flourescent	7.8	75	40	16224	7.8	75	20	8112
	TOTAL	282.3		TOTAL	587184	282.3		TOTAL	293592

PEOPLE Area	No	Hrs/Wk	Infiltration Item	Crack Usage	Figure	Factor	Cfm
Office - Exterior	241	40	Doors	250 Pas/Hr	8 - 15	400	400
Office - Interior	181	40	Windows	14144 LF	8 - 14	.55	7780
						TOTAL	8180

DOMESTIC WATER Use	Temp Hot	Temp Cold	Gal/Meals or Person	People/ Meals	Gallons Per Day	Btu/Yr E6	%Eff. & Conversion	MCF /Yr
Handwash	110	60	3	422	1266	136.4	60 x 1	227

DISTRIBUTION LOADS Item	Cfm	Gpm	Oa	Hp	Voltage	Amperage Rated	Amperage Opr	Kw	TOTAL Hrs/Wk	Kwh Per Year
9-Exterior A-H Units	83500	456	8350	87.5	460/3	123	106	66.1	168	577450
8-Interior A-H Units	36000	216	3600	40.0	460/3	61	48	27.6	168	241100
2-Toilet Exhaust Fans	3200			1.0	230/1	8.5	7.5	0.8	168	6730

FIGURE 5-4 DISTRIBUTION LOAD SURVEY FORM

DISTRIBUTION LOAD SURVEY FORM

JOB: HB&S Building　　AREA: Building　　PECO NO: 0　　DATE: 9-9-78

Area Served	Item	Cfm	Gpm	O A	Hp	Voltage	Amperage Rated	Amperage Opr.	Kw	Hrs/ Wk	Kwh/ Yr
	CHW Pump 1		366		10	460/3	14	14	8.8	168	78877
	CHW Pump 2		366		10	460/3	14	14	8.8	168	78877
	Cond. Pump 1		450		15	460/3	21	21	13.2	168	115315
	Cond. Pump 2		450		15	460/3	21	21	13.2	168	115315
	Cool Twr. Fan				20	460/3	27	27	17.6	Auto	96096
	Boiler Fan 1				2	460/3	2.8	2.8	1.8	Auto	9828
	Boiler Fan 2				2	460/3	2.8	2.8	1.8	Auto	9828
	Elevator 1				40	460/3	56	Varies	Varies	40	16000
	Elevator 2				40	460/3	56	Varies	Varies	40	16000
	Elevator 3				40	460/3	56	Varies	Varies	40	16000
	Elevator 4				40	460/3	56	Varies	Varies	168	25000
	Dom Wtr Pump				2	460/3	2.8	2.8	1.8	168	15725
	Chiller - 1		366		150	460/3	210	-	132	-	-
	Chiller - 2		366		150	460/3	210	-	132	-	-

SECTION VI
TABULATING PRESENT ENERGY USE

A. Introduction

The survey of the owner-client's building concludes the data-gathering phase of the energy study, leaving the energy consultant with two separate versions of the building's energy-use paterns: one derived from utility statements and one derived from the physical study of the building itself. The consultant will also have at least a rough idea of potential ECO's to be applied to the building, and in order to refine those ideas into specific recommendations it will now be necesary to convert the data from the utility bills. (The data on the Building Load Survey and Distribution Load Survey Forms are expressed in watts for lighting, horsepower for motors and pumps, Btu for boilers, etc., while the data on the Utility Consumption/Cost Survey Form are expressed in dollars, kilowatt hours, gallons, tons, etc. These differences in "language" must be meshed into common "language" of energy-units, such as Btu ,through the application of a series of energy conversion tables to be discussed below, along with an Energy Tabulation Form to be used for the final tabulation of the build. g's total present energy-use.)

This tabulation will reveal not only the building's total annual energy consumption-- expressed both in dollars and in common units of electric and fuel energy --it will also reveal the relative percentages of that energy-use consumed by each individual "energy-user" studied in the physical survey. (It is during this tabulation, incidentally, that the energy consumed by the building's reheat control system, if it has one, is specifically quantified and accounted for.) This in turn will provide the basis for a cost-benefit analysis of potential ECO's and a set of specific recommendations for the next phase of the energy management program.

Still another reason for conducting this uniform tabulation of present energy-use is to furnish the energy management team-- and the energy consultant in particular --with a reliable "bookkeeping system." If the total consumption of all the building's "energy-users," for example, does not equal the building's total energy-use as revealed on the utility bills, the consultant has probably made an error in calculating the energy-use of a particular "user" or functional area. In that case, the consultant must review the building data and accompanying measurements and calculations, perhaps re-surveying part of the building itself, in order to "make the books balance." It is very important to have an accurate accounting of energy consumption rates for each and every "energy-user," because this is the basis for all succeeding stages in the energy management program, including the monitoring phase which follows the implementation of whatever ECO's are selected for the building. Once the "books" have been "balanced," however, the energy study is on a firm foundation, and recommendations for energy conservation opportunities may be made with confidence. The following sections will further explain the process of conducting the energy tabulation: (1) Energy Conversions; (2) Heating and Cooling Tabulation; (3) Re-Heat Energy Tabulation.

B. Energy Conversions

The only common denominator for all types and sources of energy is the "British thermal unit," better known as the Btu. As mentioned above, most utility companies measure natural gas in thousands of cubic feet (Mcf), oil in gallons, electricity in kilowatt hours, coal in tons, etc. In order to be of use in an energy conservation study, these various energy sources must be converted to the common "language" of Btu. Furthermore, once the consultant has established the savings--in Btu--of a given body of energy conservation opportunities, those figures must then be converted back into the energy units which appear on the utility bills. Only then can the savings in dollars and cents be calculated. It is necessary, therefore, that the energy consultant become acquainted with and have access to the appropriate energy conversion units and factors, which are listed in Figure 6-1.

FIGURE 6-1 ENERGY CONVERSION UNITS

Motor

$$kw = \frac{hp \times .746}{\text{Motor Efficiency}}$$

Domestic Water

Btu = Gallons x 8.3 x temperature difference
1 gallon = 8.3 lbs.

Crude Oil

1 Barrel of oil = 5,800,000 Btu

Fuel Oil

Number 1	137,400 Btu/gal
Number 2	139,600 Btu/gal
Number 3	141,800 Btu/gal
Number 4	145,100 Btu/gal
Number 5	148,800 Btu/gal
Number 6	152,400 Btu/gal

Propane	91,500 Btu/gal
Natural Gas	950 to 1150 Btu/cu ft
Steam	1012 Btu/lb (based on 10 psig supply condensate returned at 180°F
Electricity	1 kw = 3413 Btu

Coal

Bituminous	11,500 to 14,000 Btu/lb
Sub-Bituminous	9,500 to 11,000 Btu/lb
Lignite, Brown Coal	6,300 to 8,300 Btu/lb

C. Heating and Cooling Tabulation

Heating and cooling loads for most buildings will of course vary on a daily, weekly, monthly, and yearly basis, due to weather, internal loads of people, lights, etc., and because of varying equipment efficiencies. The charts and nomographs in Chapter VIII make it possible to calculate the yearround heating and cooling requirements of the building envelope, as well as the outside air loads.

The climatic conditions to which a building is exposed over extended periods of time must be identified with the aid of climatic maps discussed in Chapter VIII. The values derived from the climatic maps will be entered on the "Climatic Zone Form" Figure No. 6-2, (to be used in conjunction with the appropriated charts and nomographs, likewise discussed in Chapter VIII. The geographic location of the building should be plotted on each of the maps and relavant conditions read from the contour lines. If the building is located in an area between contour lines, the values should be interpolated. (Detailed information pertaining to the local microclimate of a particular building is generally available from the local weather bureau.)

FIGURE 6-2 CLIMATIC ZONE FORMS

CLIMATIC ZONE FORM

Job: HB & S BUILDING   Date: 9-9-78

Design Temperature: Summer 98 Dbt 79 Wbt Winter -2 Dbt

Wind Velocity 7 Miles/Hr.

MAP	SOURCE	VALUE
Latitude	Fig 8-1	38°
Heating Degree Days	Fig 8-1	4,900
Solar Radiation in Langleys	Fig 8-2	375
Degree Hours Less Than 54F Wbt/68F Dbt	Fig 8-3	24,000
Degree Hours Greater Than 78F Dbt	Fig 8-4	6,500
Degree Hours Greater Than 66F Wbt	Fig 8-5	5,100
Degree Hours Greater Than 85F Dbt	Fig 8-6	1,800

REMARKS

The "Annual Heating and Cooling Tabulation Form" , Figure 6-3, provides a basis for calculating the heating and cooling load of each functional area in the building under survey. The energy-use associated with glass, wall and roof areas, (tabulated on the Building Load Survey Form), along with the outside air (from the Distribution Load Survey Form), is obtained by reference to the pertinent charts and nomographs in Chapter VIII. (The figure number for the appropriate chart or nomograph is indicated on the sample survey form in Chapter VIII, and an explanation of the use of each chart and nomograph is included with the charts and nomographs themselves.) Through the application of conversion units factors, discussed above, the annual energy consumption of lighting and fans has been calculated and listed on the Building Load Survey and the Distribution Load Survey Forms. These items, however, in addition to using energy, also produce heat which must be removed by the air conditioning system. The Heating and Cooling Tabulation Form provides a place for listing this additonal cooling load.

Because people themselves produce heat which must be removed by the cooling system, there is also a place on the form to insert the number of people (obtained from the Building Survey), as well as a method for calculating the volume of heat per year which must be absorbed. Once the total Btu for heating and cooling have been established, the form provides a place for converting this energy into the units which appear on utility bills.

FIGURE 6-3 ANNUAL HEATING AND COOLING TABULATION FORM

ANNUAL HEATING AND COOLING TABULATION FORM

JOB: HB&S Building AREA Office PECO NO: 0 DATE: 9-9-78

ITEM	UNITS SqFt-Cfm	COOLING Fig No 8-	COOLING Factor	COOLING Shade Factor	COOLING Total Btu/Yr-E6	HEATING Fig No 8-	HEATING Factor	HEATING Total Btu/Yr-E6
North Glass	1152	16 or 17	55E3	.64	40.6	19 or 20	85E3	97.9
South Glass	3608	16 or 17	135E3	.64	311.7	19 or 20	60E3	216.5
East Glass	4576	16 or 17	135E3	.64	395.4	19 or 20	67E3	306.6
West Glass	5456	16 or 17	135E3	.64	471.4	19 or 20	67E3	365.5
Glass Conduction	14792	18	8E3		118.3	--	--	--
North Wall	8368	22 or 23	1.5E3		12.6	25 or 26	17E3	142.3
South Wall	4587	22 or 23	2.5E3		11.5	25 or 26	12E3	55.0
East Wall	9464	22 or 23	2.5E3		23.7	25 or 26	15E3	142.0
West Wall	10609	22 or 23	2.5E3		26.5	25 or 26	15E3	159.1
Wall Conduction	33028	24	3.0E3		99.1	--	--	--
Roof Solar	10300	27	5.5E3		56.7	--	17E3	175.1
Roof Conduction	10300	24	1.5E3		15.5	28	--	--
Outside Air*	11380	30	75.3E6		856.9	31	150E6	1707.0
Humidification*	11380	--	--			32	75E6	853.5
*Factor is Btu/1000Cfm		Sub-Total Cooling			2439.9	Total Heating		4220.5

Item	Kwh/Yr	Conversion Factor	%Cooling Season/Yr	
Lighting - Exterior	444288	3413 x 1.0	50	758.2
- Interior	436488	3413 x 1.0	100	1489.7
Fans - Exterior	577450	3413 x 1.0	50	985.4
Chill Water Pumps	241100	3413 x 1.0	100	822.9

People	No	Hrs/Yr	Btu/ Person	%Cooling Season/Yr	
Office - Exterior	241	2080	450	50	112.8
- Interior	181	2080	450	100	169.4
			TOTAL	COOLING	6778.3

Summary	
Heating	
Conv. Factor	1 x E6
Efficiency	75%
Mcf/Yr	5625
Cooling	
Conv. Factor	3413
COP	4.0
Kwh/Yr	496506

D. Total Energy Use Tabulation

Each of the building's "energy-users" has now been identified, and the volume of energy which each consumes in the various areas of the building has been tabulated. The Building Load Survey Form tabulates the lighting and domestic water energy-use; the Distribution Load Survey Form tabulates the fuel consumed by fans, pumps, and other items of distribution equipment; the Heating and Cooling Load Calculation Form tabulates the energy consumed by the building's heating and cooling plant. All that remains now is to consolidate these tabulations into a statement of total building energy-use. This is the purpose of the Energy Tabulation Form. A total of the energy consumed in each functional area will equal the total consumed by all the energy-use items. This total, plus "miscellaneous use" (discussed below) and re-heat energy (if applicable), represents and should equal the total energy-use as determined from utility bills.

A "miscellaneous use" of two to five percent of the building's total energy consumption will occur in most buildings. There are always "users" such as typewriters, adding machines, copiers, etc., which represent relatively small amounts of energy not justifying separate tabulation. These small sources of load-demand should be recognized, however, with an estimate as to their percentage of total building energy-use being left to the judgment of the energy consultant.

FIGURE 6-4 ELECTRICAL ENERGY TABULATION FORM

ENERGY TABULATION FORM

JOB: HB&S Building FUEL Electrical UNITS Kwh/Yr PECO NO 0 DATE: 9-9-78

AREA / ITEM	Drug Store	Office	Bldg. & Park.					TOTALS
Lighting	84304	880776	76440					1041520
A-H Units	52416	818550						870966
Cooling	30933	496506	33548					560987
CHW Pumps			157754					157754
Condenser Water Pumps			230630					230630
Cooling Tower			96096					96096
Elevators			73000					73000
Exhaust Fans		6730						6730
Boiler Fans			19656					19656
Domestic Water Pump			15725					15725
SUB-TOTAL	167653	2202562	702849					3073064
MISC.	8380	108670	32950					150000
REHEAT	0	938078	0					938078
TOTAL	176033	3249310	735799					4161142
Utility Bill Energy Use − (Base Year − Avg. Year / Avg. Year Degree Days) x Yearly Building Heating Use = Average Yearly Energy Use								4161142

If the building under survey does not employ a re-heat temperature control system, the above tabulation will complete the summary of the building's present energy-use as determined from the physical survey. The total energy consumption on the Energy Tabulation Form should equal the corresponding total as derived from utility and fuel statements in Step Three of the building survey. (If the totals do not correspond, an error has been made and the requisite data and calculations in the building survey will have to be reviewed and adjusted.) If, on the other hand, the building is equipped with a re-heat control system, a further calculation must be made in order to arrive at a final tabulation of the building's present energy-use. This procedure is explained below.

E. Re-Heat Use Tabulation

If the building under survey employs a re-heat temperature control system, the energy consumed by that system must now be accounted for in the tabulation of present energy-use. There are no special forms or nomographs for making this calculation, nor has the energy thus consumed been specifically represented on any of the forms employed in the building survey and energy-use tabulations up to this point. The energy consumed by re-heat _is_ included, however, in the _totals_ on the Utility Consumption/Cost Survey Form. In most instances, therefore, the specific volume of energy used for re-heat can be ascertained by subtracting the total energy-use as determined by the physical survey of the building from the total energy-use portrayed on the Utility Consumption/Cost Survey Form. (All other energy-use has presumably been accounted for by other "users" and functional areas.)

Furthermore, because a re-heat system uses both heating and cooling energy, the accuracy to the above calculation can be checked by comparing the re-heat cooling energy with the re-heat heating energy (both expressed in Btu). The heating energy thus required should be approximately 75 percent of the cooling energy required, because the heating energy is used only to offset the sensible heat removed by the chiller and not the latent heat. With this in view, the following procedure (as indicated on the Energy Tabulation Form) is recommended:

(1) Sum up the total consumption of all the individual "energy-users" in the building.
(2) Determine the volume of re-heat by subtracting this total from the total energy consumption as represented on the Utility Consumption/Cost Survey Form.
(3) Convert the energy in both cases to Btu per year.

If the heating Btu are approximately 75 percent of the cooling Btu, then it is safe to assume that the entire tabulation is accurate within practical limits. If these totals do not approximately balance, however, it will be necessary to review all of the relevant figures, building data, and tabulations, making adjustment until the "books" can be "balanced."

SECTION VII
ANALYZING ENERGY CONSERVATION OPPORTUNITIES

A. Introduction

Once the energy consultant has "balanced the books" between the physical survey of the building and the annual energy consumption cost analysis based on utility bills, the survey phase of the energy study has been completed. Not only has the building's total "base-year" energy performance been tabulated, graphed, and recorded in the Conservation Journal, but each of the building's individual "energy-users" has likewise been identified, measured for relative efficiency, and judged in terms of its potential as an energy conservation opportunity. In fact, as suggested above, the energy consultant will probably have accumulated by now a fairly complete list of potential ECO's which have been recorded at various points during the building survey. (It is well to remember in this connection, incidentally, that anything which uses or causes energy to be used in the building represents a potential for energy conservation -- whether that be a poorly insulated cooling duct, lights too bright for the functions they are lighting, or southerly windows unshaded in summer.)

What remains now is to determine which of those potential ECO's (or "Project ECO's," as defined below) might be translated into actual ECO's for use in the implementation phase of the energy management program (as discussed in Chapter IV). This generally means conducting a cost-benefit analysis of each potential ECO and/or group of ECO's aimed at answering three specific questions for the building owner: (1) How much energy and how much money per year will be saved by implementing a particular ECO? (2) How much money will it cost the owner to implement the ECO? (3) How long in months or years will it take the owner to recover the cost of implementing the ECO through savings on utility bills (the "payback period")?

It is the purpose of this chapter in the manual to enable the energy consultant to make such cost-benefit analyses in an orderly and reliable manner. The 52 ECO's discussed in the following pages should correspond with most of the ECO's available in typical American buildings. By applying the procedures, charts, and nomographs in Chapter VIII to each of the corresponding potential ECO's in the owner's building, the energy consultant should be able to establish with certainty which of the potential ECO's ought to be recommended for implementation-- within the general confines of the owner's budget --and which should be dropped or considered for implementation at a later stage in the energy management program.

B. Tabulating Future Energy Use

In approaching the cost-benefit analysis of potential ECO's, the energy consultant is advised to recall what was said previously with regard to the cumulative nature of implementing ECO's. It is virtually impossible to change the consumption pattern of one "energy user" without affecting other "users". Furthermore, the potential savings from a particular ECO will vary in relation to other ECO's which might logically be implemented first. In a building where incandescent lighting is operated 24 hours a day, for example, it may well be cost-effective to replace the incandescent with a fluorescent system. But suppose the lighting period itself could be reduced to 12 hours per day without adversely affecting the functions in the building? In that case, the energy savings to be derived from fluorescent lighting would have to be calculated on a 12 hour rather than a 24 hour basis, thus reducing potential savings by 50 percent and possibly eliminating the ECO. What this means is that in analyzing ECO's for cost-benefit, the ECO represented by turning off lights should be analyzed before the ECO represented by a fluorescent system.

Such interrelationships between potential ECO's will occur throughout the entire process of analyzing them for implementation. It is very important, therefore, to analyze the cost-benefit ratio of each of those ECO's as nearly as possible in the order in which they might logically be acted upon. As a general rule of thumb, first calculate the operational ECO's, then those ECO's that reduce the building load, then those that reduce distribution or parasitic loads, and finally those that deal with primary equipment loads. In most energy management programs, the order of implementation will be determined both by cost and by the "cululative principle" discussed above, providing the energy consultant with a built-in method for "grouping" the ECO's in a report of conclusions and recommendations to the building owner and/or the energy management team.

By these criteria, the first "group" of ECO's to be considered will be those costing nothing or relatively little to implement. Generally these are "operational" ECO's which can be effected by "turning off and turning down" the various energy-users in the building which are wasting energy. Turning off lights in empty rooms and turning off fans, pumps, chillers, etc., when they are not needed will reduce the energy consumed and the owner's expenses without a capital investment.

The second "group" of options to be considered are those ECO's that reduce building loads while imposing only a minimal ("low-cost") capital investment. These might include; a reduction in lighting levels by replacing or rearranging fixtures; a reduction in ventilation flowrates by replacing or modifying dampers; a reduction in heating and cooling levels; etc.

Once the "operational" and generally "lowcost" ECO's have been analyzed those potential ECO's which will require greater capital expense are addressed. These generally will deal with the building shell and other elements contributing to the building load and requiring more extensive modification and high first cost. These might include such options as double-glazing windows, applying solar film to glass areas, insulating walls and roof, installing more efficient lobby doors, etc. Non-shell items which increase the building load should also be considered, including further adjustments in lighting, the insulation or shielding of high-temperature office machinery, etc. Most of these potential ECO's will require a moderate to substantial captial investment and so should be studied in careful relation to the owner's budget restrictions and expectations regarding "payback" periods. This process is equally applicable to the third and fourth "groups" of potential ECO's, having to do with the building's <u>distribution load</u> and <u>primary equipment load</u> respectively (discussed in Chapter III above).

There is seldom only one order in which implementation of ECO's can follow.

Often there are two or more ECO's available but only one can be implemented (an either-or situation). Which one is selected will effect following ECO's-- maybe even eliminating some. In such cases the implications of each can be presented to the client and a decision requested.

C. "ECO's Versus "PECO's"

It is important at this point to make a distinction between "project energy conservation opportunities," as discussed in the remainder of this chapter, and the 52 "Energy Conservaton Opportunities" presented in Chapter IX. A "project energy conservation opportunity," or "PECO," is simply one of those 52 "Energy Conservation Opportunities," or "ECO's" which has been selected by the energy consultant as showing enough promise for the specific building under survey to be analyzed for costs and benefits. Thus, "PECO #1" for a specific building may be drawn from "ECO #27" or "ECO #43" or any of the other "ECO's" in Chapter IX.

D. Accounting for the "Cumulative Principle"

Due to the "cumulative principle" affecting the choice and implementation of PECO's, it will be necessary to analyze the cost-benefit of each succeeding PECO in such a manner as to account for whatever changes in the building's energy-use patterns are likely to result from the implementation of previous PECO's. There is a simple and systematic method for keeping such a "running account" of energy-use reductions from PECO to PECO. First, identify all the tabulation forms displaying the building's present energy use (before the consideration of any PECO's) as "Number Zero." Make a complete set of copies of all these "Number Zero" forms as part of the permanent record of the energy management program. Then, in analyzing the cost-benefit, payback period, etc., for the first ECO under consideration ("PECO" #1, meaning "Project Energy Conservation Opportunity," as discussed above,) calculate and indicate the effect of that PECO on all the relevant "Number Zero" forms pertaining to present energy use.

If PECO #1 is turning off air handling units during unoccupied periods, for example, that will affect many of the other functional areas and "energy-users" in the building, as well as total energy-use. The first change to make on the original tabulation forms will pertain to "operating hours." The original figure should be erased and the new figure written in. This will cause a reduction in the energy-use of the functional area under consideration, which means erasing the original figure pertaining to that energy-use and inscribing the new figure as determined from the requisite calculations (discussed in the following pages). This in turn will necessitate changes for all other energy-use factors affected by the PECO, such as outside air volume, fan horsepower required for cooling, etc.

When all of the interrelated changes have been made on the tabulation forms, including a new grand total of annual energy consumption, the energy consultant will in effect have produced an entirely new tabulation of the building's energy-use (this one relating to the building's future energy-use if PECO #1 were implemented). These revised tabulation forms should then be labeled "PECO #1." They should be copied and placed in the permanent record with the "Number Zero" forms. Then the consultant may commence the analysis of "PECO #2," using as a basis the revised forms for "PECO #1," and working through all of the necessary calculations, erasures, and inscriptions of new energy-use figures resulting from "PECO #2." At the end of this process, the energy consultant will have yet another entirely new tabulation of the building's total energy use-- this one reflecting the combined effect of PECO's #1 and #2. These forms should be labeled "PECO #2," copied, and placed in the permanent record, whereupon the analysis of PECO #3 may be started, now using the forms for PECO #2 as the point of departure. This procedure should be followed for all of the PECO's remaining to be analyzed.

The following PECO Energy Tabulation Form Figure 7-1 is designed to help consolidate and calculate the energy savings for each PECO. (Detailed instructions for using this form will be found in Chapter VIII.)

FIGURE 7-1 PECO ENERGY TABULATION FORM

PECO ENERGY TABULATION FORM

JOB: HD&S Building

PECO NO: 1
DATE: 9-9-78

FUEL	AREA / ITEM	Drug Store	Office	Bldg. & Park					TOTALS
ELECTRICAL	Before ECO Energy Use	176033	3249310	735799					4161142
	After ECO Energy Use	150593	1756295	525034					2431922
	Kwh Savings	25540	1493015	210765					1729220
	Before ECO Demand Use	24.0	376.8	486.4					887.2
	After ECO Demand Use	24.0	376.8	486.4					887.2
	Kw Savings	0	0	0					0
GAS	Before ECO Energy Use	371.5	18673						19044.5
	After ECO Energy Use	275.2	8086						8361.2
	Mcf Savings	96.3	10587						10683.3

E. Conclusions and Recommendations: The Final Survey Report

Having worked through the cost-benefit analyses of all feasible PECO's, the energy consultant will now have a final version of the energy-use tabulation forms which clearly state the future energy use of the owner's building, assuming the implementation of the PECO's under consideration. (If, for example, 19 different PECO's have been analyzed for implementation, then it will be the forms labeled "PECO #19" which reveal the total future energy use of the building in question, as opposed to the total present energy use indicated on the "Number Zero" forms and associated charts and graphic displays for the building's "base year.")

This provides the material for a final Survey Report to the building owner and/or energy management team which typically includes the following components: (1) an item-by-item, sidebyside comparison of the building's present energy use and cost-factors (broken down according to "functional areas," energy-users, types of fuel, etc.) with the building's future energy use (broken down into the same categories); (2) a comparative statement of the totals regarding present versus future energy use and associated costs; (3) the location (in the building), description, illustration, and explanation of each selected PECO in sufficient depth to establish its technical and economic feasibility. (4) an accompanying statement of the estimated capital investment of each PECO being recommended, along with a projection of the "payback period" for each one.

It is generally advisable to "group" the recommended PECO's according to the categories discussed above, i.e. "No Capital Expenditure," "Low Capital Expenditure," "Moderate Capital Expenditure," and "Large Capital Expenditure." These categories, when combined in the report with "payback periods" and associated graphs and charts, will provide the client with a logical sequence of implementation steps. Such a format will also emphasize the "self-financing" character of the energy management program, whereby the money saved by implementing "No Cost" and "Low Cost" PECO's may be "reinvested" in those PECO's requiring more substantial expenditures.

F. Survey "Tools" and Examples of Above Procedures

What follows in the remainder of this manual is a set of "tools" which the energy consultant will need for conducting a successful energy study and for putting together a subsequent energy management program for a particular building or complex of buildings. Chapter VIII contains the various documents required for a comprehensive building analysis-- climatic zone maps, survey and tabulation forms, nomographs, charts, and tables --each accompanied by detailed instructions for its use. Chapter IX contains a thorough explanation of 52 energy conservation opportunities commonly found in existing U.S. buildings. These are followed in Chapter X by a "Model Building Energy Study," designed to provide the reader with detailed examples of the procedures described above and in Chapter X as applied to a hypothetical office building in the American midwest.

SECTION VIII
DOCUMENTS

A. Climatic Zone Maps

For the most part, these maps will be used in conjunction with the Climatic Zone Form which immediately follows them. The maps and form together will invariably be applied quite early in a typical building survey, and certainly not later than the "Heating and Cooling Tabulation" discussed in Chapter VI. Indeed, without establishing the microclimatic behavior of a buildings immediate environment, it would be virtually impossible to determine either the building's present energy efficiency or the most promising Energy Conservation Opportunities to be applied to that building-- particularly the building envelope.

In making judgments concerning the effect of solar radiation through a buildings windows, for example, the energy consultant will need to estimate the annual mean daily solar radiation from Figure 8-2. Likewise, an understanding of the cooling load demands on a building will require a reference to Figure 8-6 ("Annual Dry Bulk Degree Hours Above 85 F"). The use of the maps, when examined in relation to the Climatic Zone Form, is self-explanatory.

Figure 8-1 Engineering Data

Source: Climatic Atlas of the United States
U. S. Department of Commerce
June, 1968
Page 36

FIGURE 8-1 ANNUAL HEATING DEGREE DAYS (BASE 65°F)

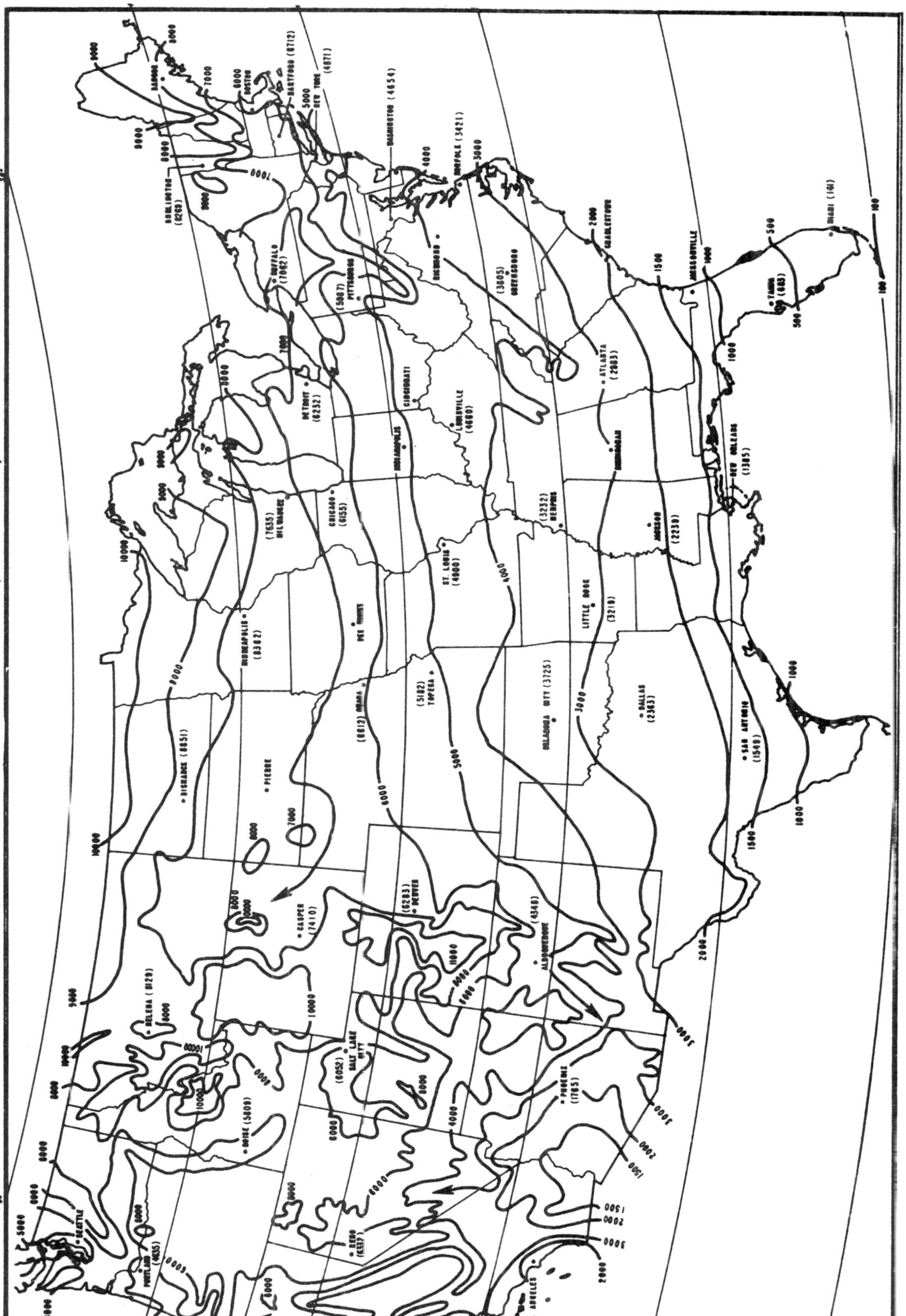

Figure 8-2 Engineering Data

Source: Climatic Atlas of the United States
U.S. Department of Commerce
June, 1968
Page 70

FIGURE 8-2 ANNUAL MEAN DAILY SOLAR RADIATION IN LANGLEYS

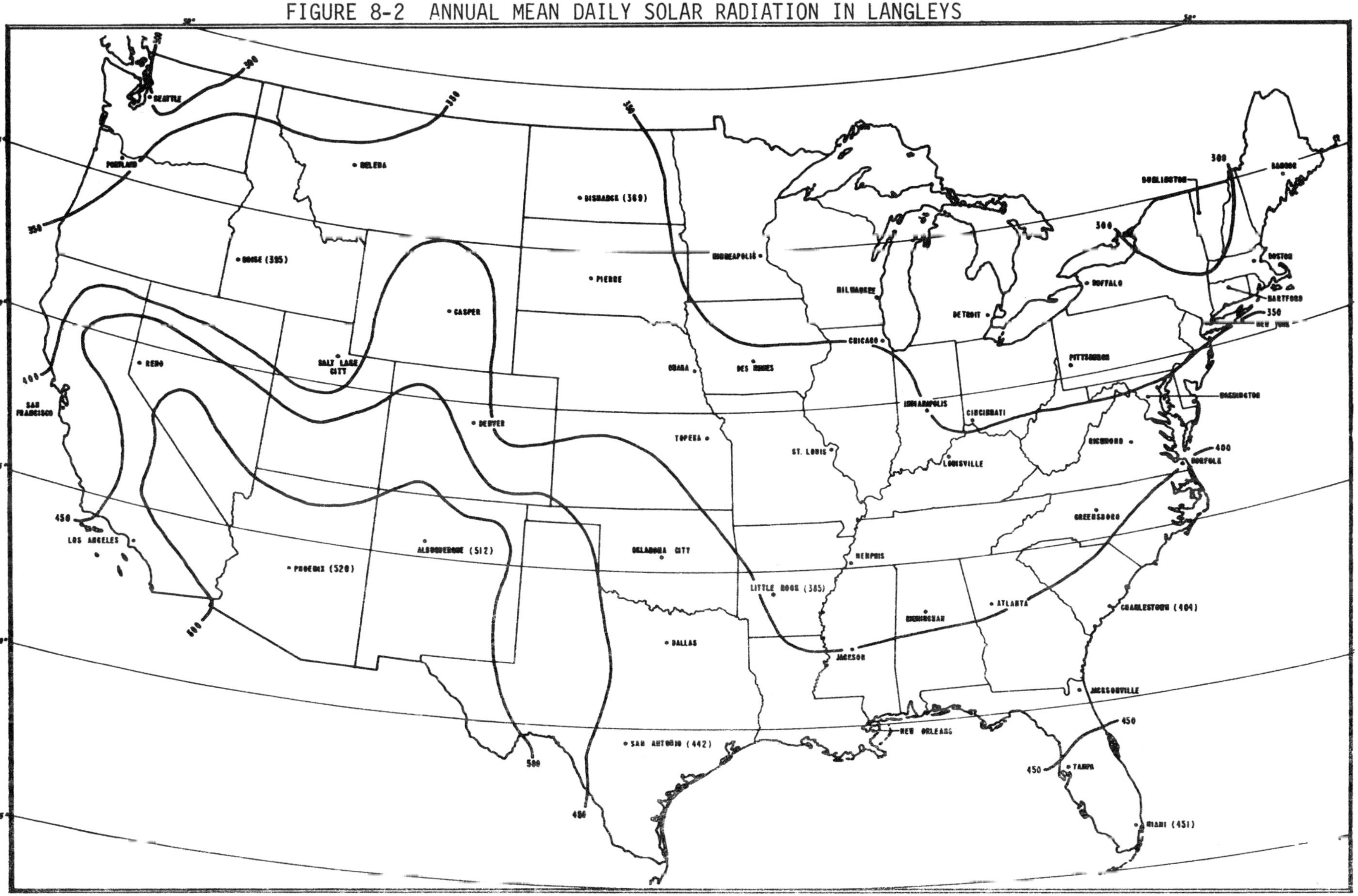

Figure 8-3 Engineering Data

Source: AFM 88-8
U.S. Government Printing Office
15 June 1967

Locations in United States-107

October through April-24 hrs/day

Dry bulb temperature less than 68°

Wet bulb temperature taken in 5° increments less than 54° F wet bulb.

Base wet bulb temperature taken at 54°F.

Wet bulb temperature difference taken from median of 5° range to 54°, i.e.: for 45 to 49 range, wet bulb difference = 54-47 or 7° F. Temperature difference multiplied by number of hours in each wet bulb temperature range, for October thruough April and summed for total. Locations with heating seasons significantly longer than October-April should be analyzed individually for maximum accuracy.

FIGURE 8-3 ANNUAL WET BULB DEGREE HOURS BELOW 54°F WET BULB TEMPERATURE AND 68°F DRY BULB

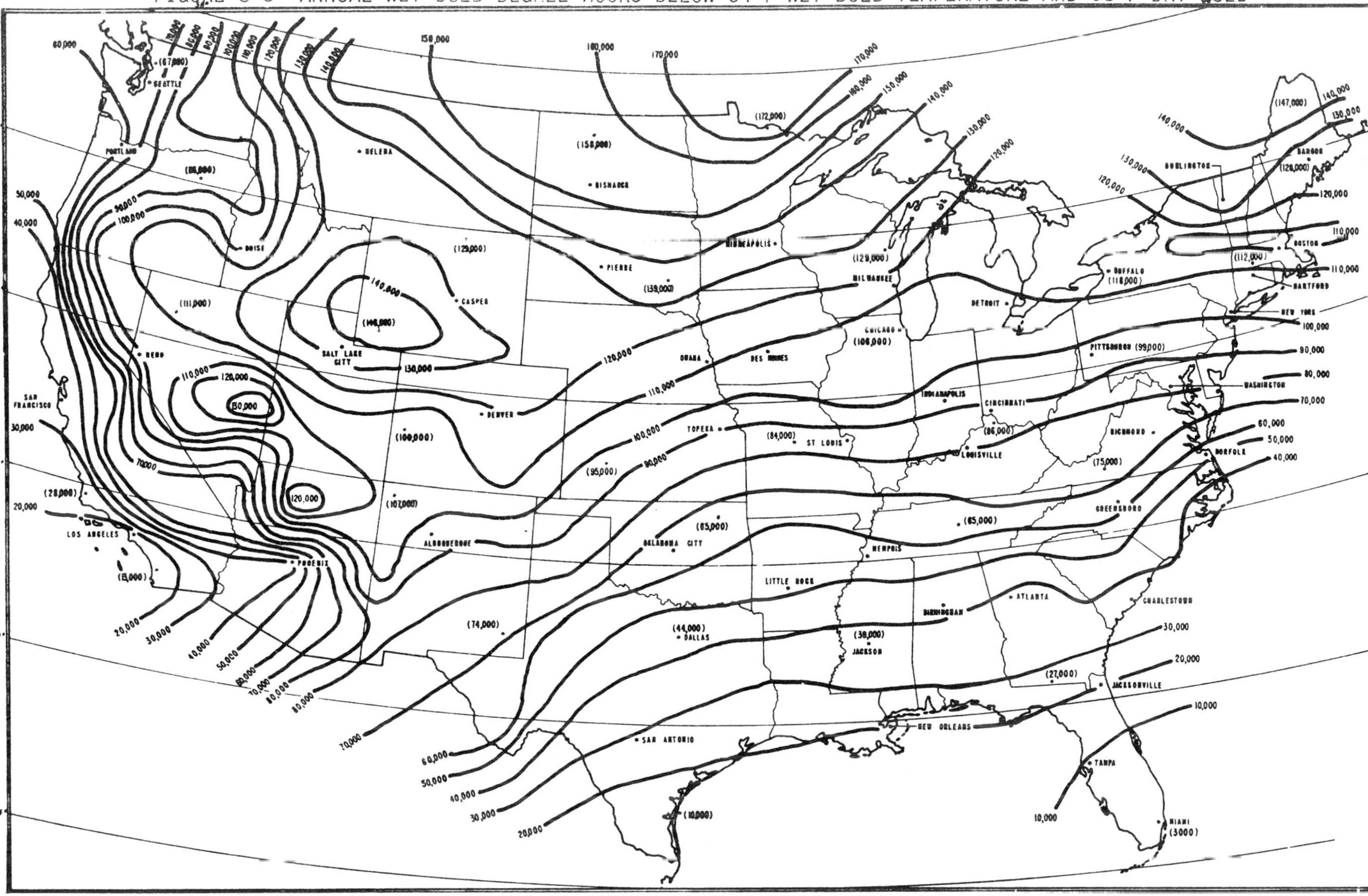

Figure 8-4 Engineering Data

Source: AFM 88-8
U.S. Government Printing Office
15 June 1967

Locations in U.S.-107

Twelve mos/yr, 8 hrs/day (0930-1730 hrs).

Dry bulb temperature taken in 5°F increments beginning with 80-84°F.

Dry bulb temperature difference taken from median of 5°F range to 78°F, i.e., for 85° to 89°F range, dry bulb Temperature difference = 87° - 78°F - 9°F. Temperature difference multiplied by number of hours in each dry bulb temperature range and summed for total.

FIGURE 8-4 ANNUAL DEGREE HOURS DRY BULB TEMPERATURE GREATER THAN 78°F

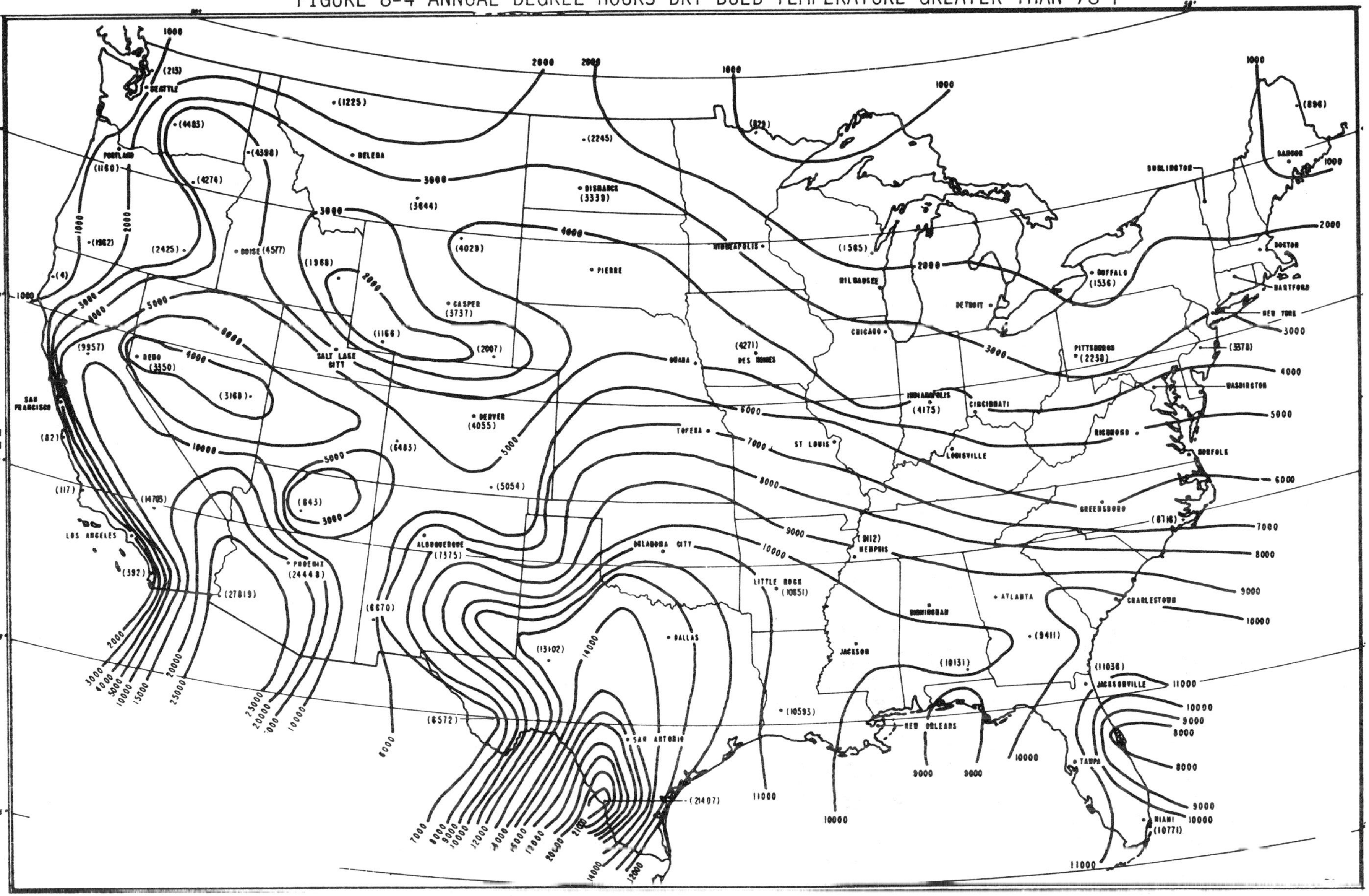

Figure 8-5 Engineering Data

Source: AFM 88-8
U.S. Government Printing Office
15 June 1967

Locations in the U.S.-107

Twelve mos/yrs, 8 hr/day (0930-1730).

Wet bulb temperature taken in 1°F increments beginning with 67°F.

Wet bulb temperature difference taken from 66° F.

Temperature difference multiplied by number of hours at each wet bulb temperature and summed for total.

FIGURE 8-5 ANNUAL DEGREE HOURS WET BULB TEMPERATURE GREATER THAN 66°F

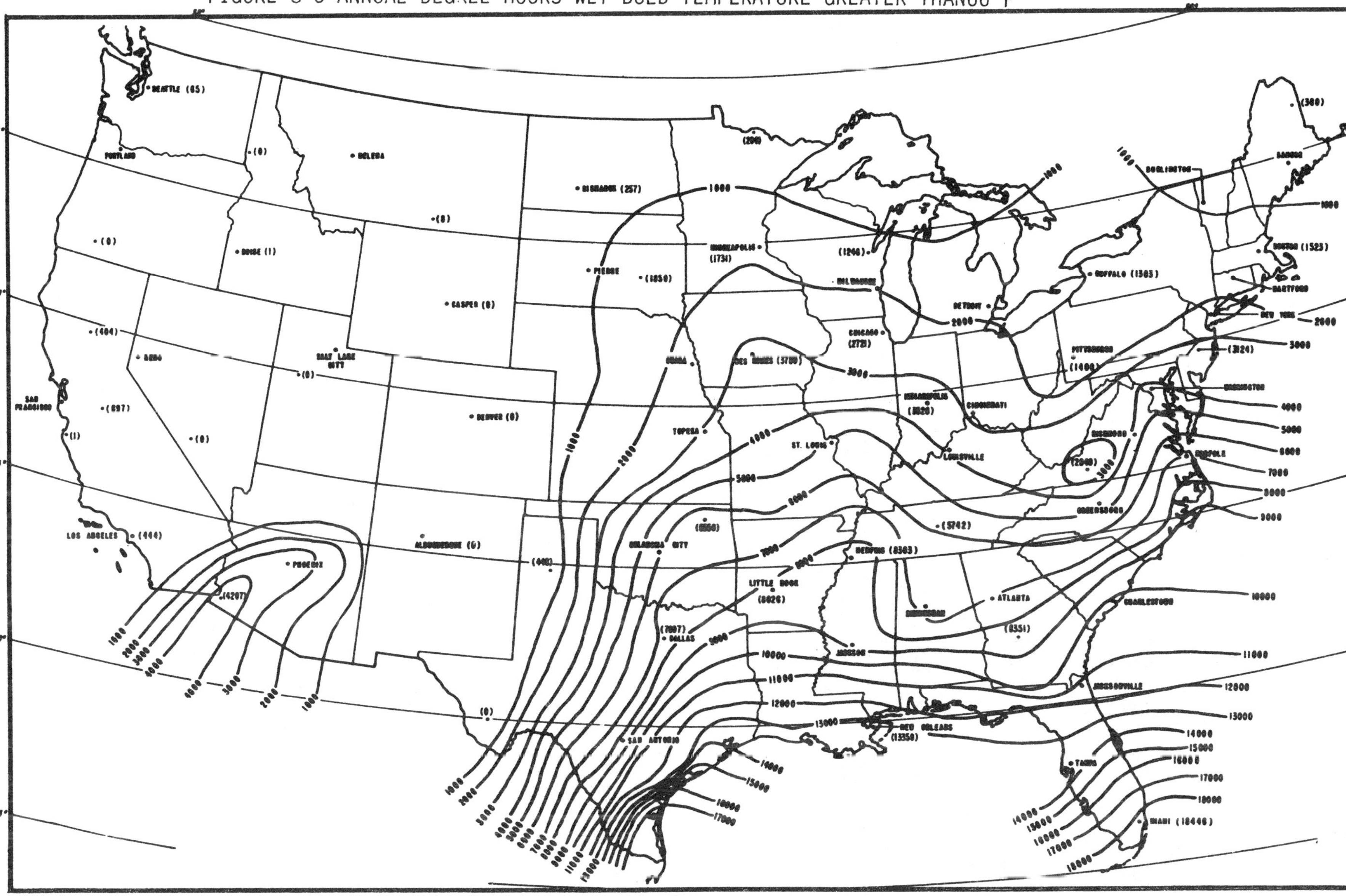

Figure 8-6 Engineering Data

Source: AFM 88-8
U.S. Government Printing Office
15 June 1967

Locations in U.S.-107

Twelve mos/yr, 8 hr/day (0930-1730).

Dry bulb temperature taken in 5°F increments beginning with 85-89°F.

Dry bulb temperature difference taken from median of 5°F range to 85°F, i.e., for 90 to 94°F range, dry bulb temperature difference = 92-85°F = 7°F.

Temperature difference multiplied by number of hours in each dry bulb temperature range and summed for total.

FIGURE 8-6 ANNUAL DEGREE HOURS DRY BULB GREATER THAN 85°F

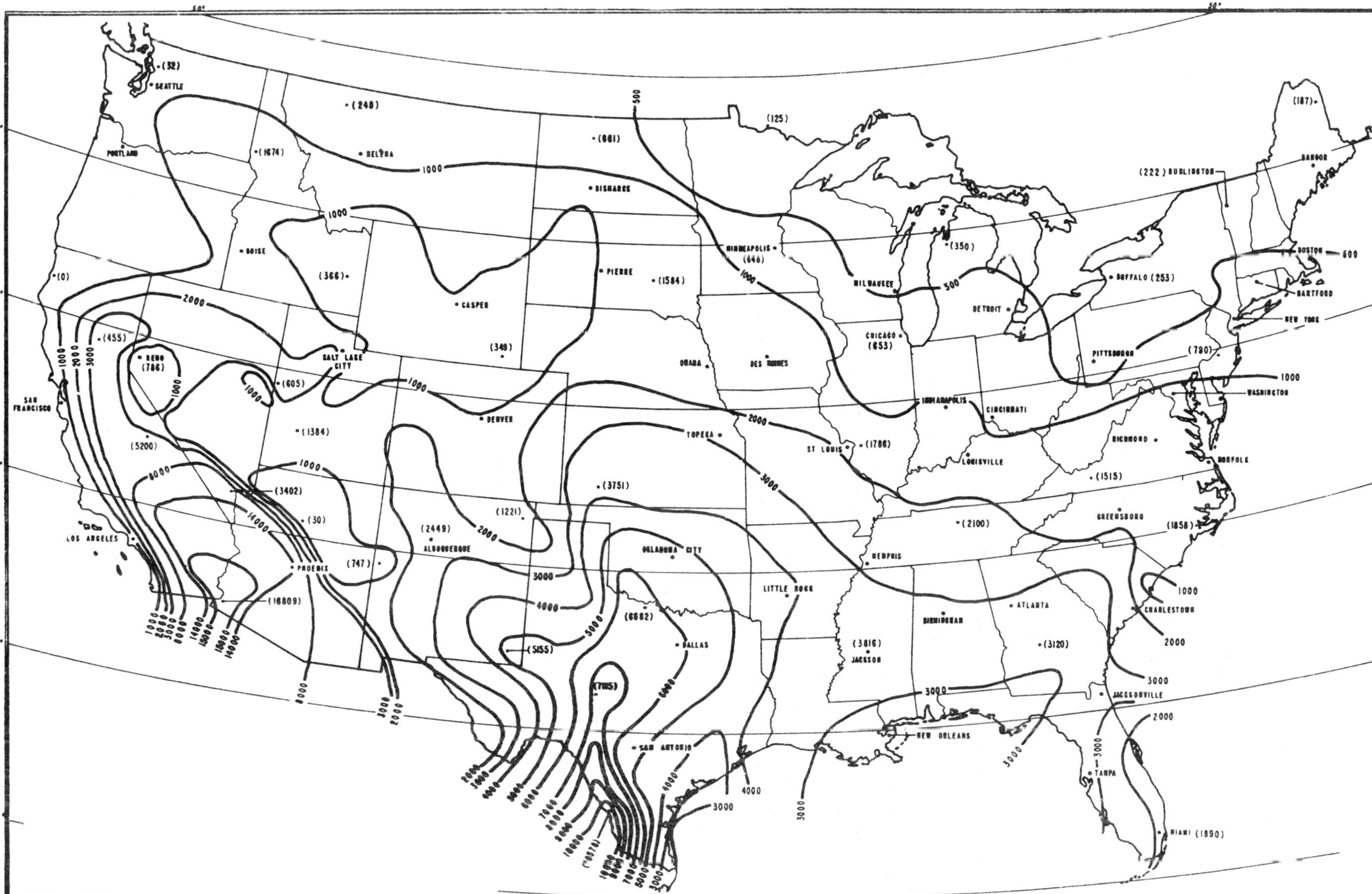

B. Survey and Tabulation Forms

The role of each of the following forms in a typical building survey has been discussed in one or another of the preceding chapters, particularly Chapters V, VI, and VII. What follow now are detailed instructions for the use of these forms in their various applications. Among those applicatons are: (A) surveying the building; (B) tabulating present energy use; (C) tabulating future energy use on the basis of proposed Energy Conservation Opportunities. In the latter application, the forms will often be used in conjunction with certain charts, tables, and nomographs (identified in the instructions for using each form).

The following forms are included along with descriptions of their use.

(A) Climatic Zone Form

(B) Utility Consumption/Cost Survey Form

(C) Building Load Survey Form

(D) Distribution Load Survey Form

(E) Annual Heating and Cooling Tabulation Form

(F) Energy Tabulation Form

(G) Project ECO Tabaulation Form

FIGURE 8-7 CLIMATIC ZONE FORM

CLIMATIC ZONE FORM

Job: 1 Date: 2

Design Temperature: Summer 3 Dbt 4 Wbt Winter 5 Dbt

Wind Velocity 6 Miles/Hr.

MAP	SOURCE	VALUE
Latitude	Fig 8-1	
Heating Degree Days	Fig 8-1	
Solar Radiation in Langleys	Fig 8-2	7
Degree Hours Less Than 54F Wbt/68F Dbt	Fig 8-3	
Degree Hours Greater Than 78F Dbt	Fig 8-4	
Degree Hours Greater Than 66F Wbt	Fig 8-5	
Degree Hours Greater Than 85F Dbt	Fig 8-6	

REMARKS

1. Name of the job or building surveyed.
2. Date of the survey.
3. Summer dry bulb temperature design for the area.*
4. Summer wet bulb temperature design for the area.*
5. Winter dry bulb temperature design for the area.*
6. Average coincident wind velocity for the area.*
7. The value for the area obtained from each map.

*Two sources for this data are: Department of Defense, Facility Design and Planning Engineering Weather Data; AFM 88-29, TM 5-785, NAVFAC P-89, July 1, 1978, or ASHRAE, Handbook of Fundamentals, New York, 1978 or current edition.

FIGURE 8-8 UTILITY CONSUMPTION/COST SURVEY FORM

UTILITY CONSUMPTION/COST SURVEY FORM

JOB: 1 DATE: 2

3 Mo.-Yr.	ELEC. 4 KWH	ELEC. DEMAND	ELEC. 4 $	FUEL	FUEL 4 $	FUEL	FUEL 4 $	DEGREE DAYS-HTG.
JAN.								
FEB.								
MAR.								
APR.								
MAY								
JUNE								
JULY								
AUG.								
SEPT.	5	6	7	8	9	8	9	10
OCT.								
NOV.								
DEC.								
JAN.								
FEB.								
MAR.								
APR.								
MAY								
JUNE								
JULY								
AUG.								
SEPT.								
OCT.								
NOV.								
DEC.								

1. Name of the job or building being surveyed.
2. Date of the survey.
3. Month and year of utility bill.
4. Multipliers - i.e., 1, 10, 100, 1000, etc.
5. Electrical consumption from utility bill.
6. Electrical demand from utility bill.
7. Total net dollars from utility bill.
8. Fuel consumption other than electric.
9. Dollars charged for fuel consumption.
10. Heating degrees-days for year. This must correspond to the base year elected. Record monthly degree-days if available for future monitoring.

FIGURE 8-9 BUILDING LOAD SURVEY FORM

BUILDING LOAD SURVEY FORM

JOB: 3 — AREA: 2 — WINTER: 5 — SUMMER: 6 — PECO NO: 1 — DATE: 4

GLASS					WALLS				ROOF OR FLOORS	
Facing	SqFt	Shading Ext	Shading Interior	Type U= 11	Facing	SqFt	Shade	Type U= 11	SqFt	Type U= 11
7	8	9	10	12	7	8	9	13	8	14
TOTAL					TOTAL					

LIGHTING Area	Type	OCCUPIED Kw	OCCUPIED Fc	OCCUPIED Hrs/Wk	OCCUPIED Kwh/Yr	CUSTODIAL Kw	CUSTODIAL Fc	CUSTODIAL Hrs/Wk	CUSTODIAL Kwh/Yr
15	16	19	20	21	22	19	20	21	22
	TOTAL			TOTAL				TOTAL	

PEOPLE Area	No	Hrs/Wk	Infiltration Item	Crack Usage	Figure	Factor	Cfm
23	24	25	26	27	28	29	30
						TOTAL	

DOMESTIC WATER Use	Temp Hot	Temp Cold	Gal/Meals or Person	People/Meals	Gallons Per Day	Btu/Yr E6	%Eff. & Conversion	39/Yr
31	3	33	34	35	36	37	38	40

DISTRIBUTION LOADS Item	Cfm	Gpm	Oa	Hp	Voltage	Amperage Rated	Amperage Opr	Kw	TOTAL Hrs/Wk	Kwh Per Year
41	4	43	44	45	46	47	48	49	50	51

1. No. of PECO being evaluated. Use No. "0" for the initial survey.

2. The functional area of the building covered by this particular survey sheet. It could be the entire building, but more likely is one of the several functional areas in the building.

3. Name of the job or building being surveyed.

4. Date of the survey.

5. Winter inside dry bulb temperature and relative jumidity if controlled.

6. Summer inside dry bulb temperature and relative humidity.

7. The orientation of the exposure.

8. The area of the exposure.

9. The percentage of exterior shading of the exposure by adjoining buildings, trees, etc. This will require consultants judgement as this must represent the percent shading effect during the entire year.

10. The interior shading, due either to the type of glass of internal shades such as blinds or drapes: the shading factor.

11. The coefficient of heat transfer.

12. The type of glass, whether single, double or triple glazed, whether double hung or metal casement, and the condition of the fit, whether tight, loose or average.

13. The type and color of wall construction.

14. The type and outside color of roof and floor construciton.

15. Areas where different lighting types, illumination levels, or operating hours occur.

16. The type of lighting in each of the areas.

17. The data for lighting which occurs during those hours when the area is in custodial, occupied or in normal use.

18. The data for lighting which occurs during those hours when the area is in custodial, unoccupied or other than normal use.

19. The total installed lighting in each area expressed in kilowatts.

20. The lighting level in each area, expressed in foot candles.

21. The number of hours per week the area is in use.

22. The total kilowatt hours per year for each area - a product of 21 x 22.

23. The particular part of the functional area where these people operate.

24. The number of people.

25. The number of hours per week the people are in the area.

26. The particular item where infiltration occurs, primarily windows and/or doors.

27. In the case of windows, the crack area around the window in lineal feet is needed for figuring infilatration. In the case of doors, the number of door openings per hour is needed.

28. The figure number in the manual covering infiltration. For windows use FIGURE 8-14 for doors FIGURE 8-15

29. The factor obtained from the figure in the manual.

30. The amount of air in cubic feet per minute for infiltration determined by the product of 27 x 29.

31. The type of domestic water use.

32. The hot water temperature required.

33. The yearly average cold water temperature entering the building.

34. Either the number of gallons per person per day in the case of handwashing or showering, or the gallons per meal in the case of dishwashing.

35. Either the number of people or the number of meals per day.

36. The number of gallons used per day determined by the product of 34 x 35.

37. The Btu per year determined by multiplying gallons/day (Item 31) x days/year x 8.34 to convert to pounds of water per year and multiplying this by hot water temperature minus cold water temperature (items 32-33).

38. The efficiency of the domestic water system and the conversion factor for Btu to fuel units.

39. The type of fuel unit.

40. The fuel units consumed in heating domestic water/year.

41. The particular item being surveyed; normally AH fans, pumps, etc.

42. The amount of air in cubic feet per minute.

43. The amount of water in gallons per minute.

44. The amount of outside air introduced into the area.

45. The name-plate horsepower of the motor driving the equipment.

46. The voltage furnished to the motor and also the phase.

47. The rated amperage draw of the motor.

48. The average operating draw of the motor.

49. The operating kw of the motor determined by multiplying item no. 45 x .746/motor eff. x item no. 48/item no. 47.

50. The number of hours per week the equipment operates.

51. The number of kilowatt hours per year the equipment consumes, determined by multiplying item 49 x item 50 x No. of weeks per year.

FIGURE 8-10 DISTRIBUTION LOAD SURVEY FORM

DISTRIBUTION LOAD SURVEY FORM

JOB: 2 AREA: 1 PECO NO: 3 DATE: 4

Area Served	Item	Cfm	Gpm	O A	Hp	Voltage	Amperage Rated	Amperage Opr.	Kw	Hrs/ Wk	Kwh/ Yr
5	6	7	8	9	10	11	12	13	14	15	16

(NOTE: Space for listing distribution loads that serve only the functional area being surveyed is provided on the building Load Survey Form. This form should be used for distribution loads that serve the entire building or more than one funcitonal area such as chilled water pumps, etc. This form can also be used for listing building equipment such as elevators, chillers, etc.)

1. The functional area being surveyed.

2. Name of the job being surveyed.

3. Number of PECO being evaluated. Use No. "0" for the initial survey.

4. Date of the survey.

5. This could be the same as item 1 above. However, if this is a sub-zone within the functional area, it should be identified at this point.

6. The particular item being surveyed; normally AH fans, pumps, etc.

7. The amount of air in cubic feet per minute.

8. The amount of water in gallons/minute.

9. The amount of outside air introduced into the area.

10. The name-plate horsepower of the motor driving equipment.

11. The voltage furnished to the motor and also the phase.

12. The rated amperage draw of the motor.

13. The average operating draw of the motor.

14. The operating kw of the motor determined by multiplying (item no. 10 x .746/motor efficiency) x (item no. 13/item no. 12).

15. The number of hours per week the equipment operates.

16. The number of kilowatt hours per year the equipment consumes, determined by multiplying item 14 x item 15 x number of weeks per year.

FIGURE 8-11 ANNUAL HEATING AND COOLING TABULATION FORM

ANNUAL HEATING AND COOLING TABULATION FORM

JOB: 2 AREA 1 PECO NO: 3 DATE: 4

ITEM	UNITS SqFt-Cfm	COOLING Fig No 8-	COOLING Factor	COOLING Shade Factor	COOLING Total Btu/Yr-E6	HEATING Fig No 8-	HEATING Factor	HEATING Total Btu/Yr-E6
North Glass		16 or 17				19 or 20		
South Glass	5	16 or 17				19 or 20		
East Glass		16 or 17		13		19 or 20		
West Glass		16 or 17				19 or 20		
Glass Conduction	6	18				--		
North Wall		22 or 23				25 or 26		
South Wall	7	22 or 23	12		14	25 or 26	12	14
East Wall		22 or 23				25 or 26		
West Wall		22 or 23				25 or 26		
Wall Conduction	8	24				--		
Roof Solar	9	27				--		
Roof Conduction		24				28		
Outside Air*	10	30				31		
Humidification*		--				32		
		11				11		
*Factor is Btu/1000Cfm		Sub-Total Cooling				Total Heating		

Item	Kwh/Yr 16	Conversion Factor	%Cooling Season/Yr	
Lighting 15	17	18	19	20
Fans				

People	No	Hrs/Yr	Btu/ Person	%Cooling Season/Yr	
21	22	23	24	25	26
TOTAL COOLING					

Summary	
Heating	
Conv. Factor	27
Efficiency	28
29 /Yr	30
Cooling	
Conv. Factor	31
COP	32
29 /Yr	30

All of the data to be listed on this form is derived from the Building Load Survey Form and the Distribution Load Survey Form. Space for listing distribution loads that serve only the functional area being surveyed is provided on the Building Load Survey Form.

1. The functional area covered on the Building Load Survey Form.
2. Name of the job or building being surveyed.
3. Number of PECO being evaluated. Use No."0" for the initial survey.
4. Date of the survey.
5. The square footage of the glass area for each exposure involved.
6. The total glass area for all exposures.
7. The wall area for each exposure involved.
8. The total of all wall areas.
9. The total roof area involved expressed in square feet.

10. Compare the outside air introduced through the HVAC Equipment (Item 44 on the Building Load Survey Form or Item 9 on the Distribution Load Survey Form) with the exhaust air (Item 42 on the Building Load Survey Form or Item 7 on the Distribution Load Survey Form) plus the infiltration (Item 30 on the Building Load Survey Form).

11. The figure number of the nomograph in the guide pertaining to the particular items.

12. The factor obtained from the nomograph.

12A. If the inside conditions are 78^{0} and 50% relative humidity during summer operation, use the factor obtained from Fig. 8-30 only. However, if conditions are lower than this, add to this factor an amount calculated in accordance with the method described in ECO No. 3, FIGURE 9-3.

13. The interior shade factor pertaining to the particular glass area where it applies.

14. The total Btu per year, obtained by multiplying the factor times the units involved times the shade factor.

15. The energy users in the functional area which cause a heat gain to be overcome by the cooling equipment.

16. The type of fuel units.

17. Total energy consumption per year for pertinent item.

18. Conversion factor to convert energy consumption to Btu.

19. The percentage of the total year which consititutes the cooling season.

20. The product of 17 x 18 x 19.

21. The area where the people are located.

22. The number of people.

23. The number of hours per year the people are in the area.

24. The number of Btu per hour per person.

25. The percentage of the total year which constitutes the cooling season.

26. The product of 22 x 23 x 24 x 25.

27. The conversion factor to convert from Btu to the heating fuel involved.

28. The efficiency of the heating system.

29. The type of fuel units.

30. The product of the heating or cooling Btu times the conversion factor divided by the efficiency or the COP to give the total fuel units per year consumed for either heating or cooling.

31. The conversion factor to convert from Btu to the cooling fuel involved.

FIGURE 8-12 ENERGY TABULATION FORM

ENERGY TABULATION FORM

JOB: 3 FUEL 1 UNITS 2 PECO NO 4 DATE: 5

ITEM \ AREA				6				TOTALS
7								8
SUB-TOTAL				9				
MISC.				10				
REHEAT				11				
TOTAL				12				

13 − (14 X 15) = 16

Utility Bill Energy Use − $\frac{\text{Base Year - Avg. Year}}{\text{Avg. Year Degree Days}}$ X Yearly Building Heating Use = Average Yearly Energy Use

1. The type of energy being tabulated, either electrical (consumption or demand), gas, oil, coal, etc.
2. The energy units involved, i.e.. kilowatt hours, Mcf, tons, etc.
3. Name of the job or building being surveyed.
4. Number of PECO being evaluated. Use No. "0" for the initial survey.
5. Date of the survey.
6. Each column will represent one of the functional areas of the building.
7. Individual items of energy-use found in the survey.
8. The total for each item listed in item 7.
9. Sub-totals of the energy-use for each functional area.
10. Miscellaneous energy-use, normally between two and five percent. This is at the discretion of the consultant.

11. The amount of reheat energy used in the facility. This must be determined for the total building, then prorated to each functional area by judgment of the consultant.

12. The total energy-use for each functional area and for the total building.

13. The total building energy use determined by the utility bill. If the fuel is <u>not</u> used for heating, use this same figure in item 16.

14. If the fuel is used for heating, enter the base year degree days minus average year degree-days and divide by average year degree-days.

15. The yearly total energy required to heat the building from item 30 under HEATING (Annual Heating and Cooling Tabulation Form).

16. The average Yearly Energy Use determined by item 13 - (item 14 x item 15).

FIGURE 8-13 PECO ENERGY TABULATION FORM

PECO ENERGY TABULATION FORM

JOB: 2 PECO NO: 1 DATE: 3

	AREA / ITEM				4				TOTALS
ELECTRICAL	Before ECO Energy Use				5				
	After ECO Energy Use				6				
	Kwh Savings				7				
	Before ECO Demand Use				9				
	After ECO Demand Use				10				
	Kw Savings				8				
FUEL 11	Before ECO Energy Use				5				
	After ECO Energy Use				6				
	12 Savings				13				

1. Number of PECO being tabulated.
2. Name of the job or building being surveyed.
3. Date of the survey.
4. Each column will represent one of the functional areas of the building.
5. The electrical energy consumption before the PECO examination.
6. The electrical energy consumption after the PECO examination.
7. The electrical energy savings in kw hours.
8. The electrical demand before the PECO examination.
9. The electrical demand after the PECO examination.
10. The demand savings in kw.
11. The type of fuel used other than electrical.
12. The fuel units.
13. The savings expressed in fuel units.

C. Charts and Nomographs

The following charts and nomographs have been designed to assist the energy consultant in three distinct phases of an energy study: (A) measuring the relative efficiency of various "energy-users" in a building; (B) calculating the present energy consumption of those users; (C) calculating the future energy consumption of users to be modified by one or more of the Energy Conservation Opportunities presented in Chapter IX.

With regard to calculating present and future energy use, some of the nomographs will be utilized in both phases (B) and (C), while others will be utilized only in phase (C). Those to be utilized in both phases are arranged at the front of the following section (Figures #8-14 through #8-32), while those to utilized primarily in phase (C) are arranged toward the back of the section (Figures #8-33 through #8-52). The use of nomographs in Phase (A), of course, might occur at various times under various circumstances according to the demands of a particular energy study.

Charts and nomographs intended primarily for calculating future energy use are referred to almost exclusively in the "ANALYSIS" portions of the Energy Conservation Opportunities presented in Chpater IX. It is in this portion of each ECO that instructions are given for analyzing the cost-benefit ratio of that ECO with regard to a particular building study, and often the nomographs are required to complete such an analysis.

As with the maps and forms on the pages above, the nomographs are here assembled in a single place for ease of reference in actual future building studies. They have been arranged as nearly as possible in the order in which they will usually be applied in such studies, rather than according to the various building systems to which they relate. Figure #8-15, for example, refers to "Heating," as does Figure #8-41 and other of the nomographs dispersed through the following pages. But Figure #8-15 would ordinarily be applied much earlier in a building study, while Figure #8-41 ("Effect of Insulation on 'U' Value") would be applied at a later stage involving a more substantial investment.

Instructions for using each nomograph are presented on the facing page, along with sources of data from which the nomograph was constructed.

The user might find it convenient to keep these nomographs in a separate small binder to permit easy reference while working through the ECO's in this manual. Some consultants have found it useful to use copies of the forms to mark their actual assumptions on as they move through each analysis. This copy is then filed with the other PECO calculations

The charts and nomographs are listed below:

LIST OF CHARTS AND NOMOGRAPHS

Present and Future Energy Consumption

Future Energy Consumption

FIGURE #8-14 Engineering Data

Source of Data: ASHRAE Handbook of Fundamentals, 1972, pgs. 333, 337,338

Instructions for Nomograph:

1. Enter chart on left-hand side at wind velocity in miles per hour as determined from design weather data for the area.
2. Proceed horizontally to the intersection of the curve that most closely fits existing conditions of building under survey.
3. Proceed vertically downward from this intersection to read the cfm per foot of crack on the bottom horizontal line.
4. This is the factor to be entered under "Infiltration" on the Building Load Survey Form.

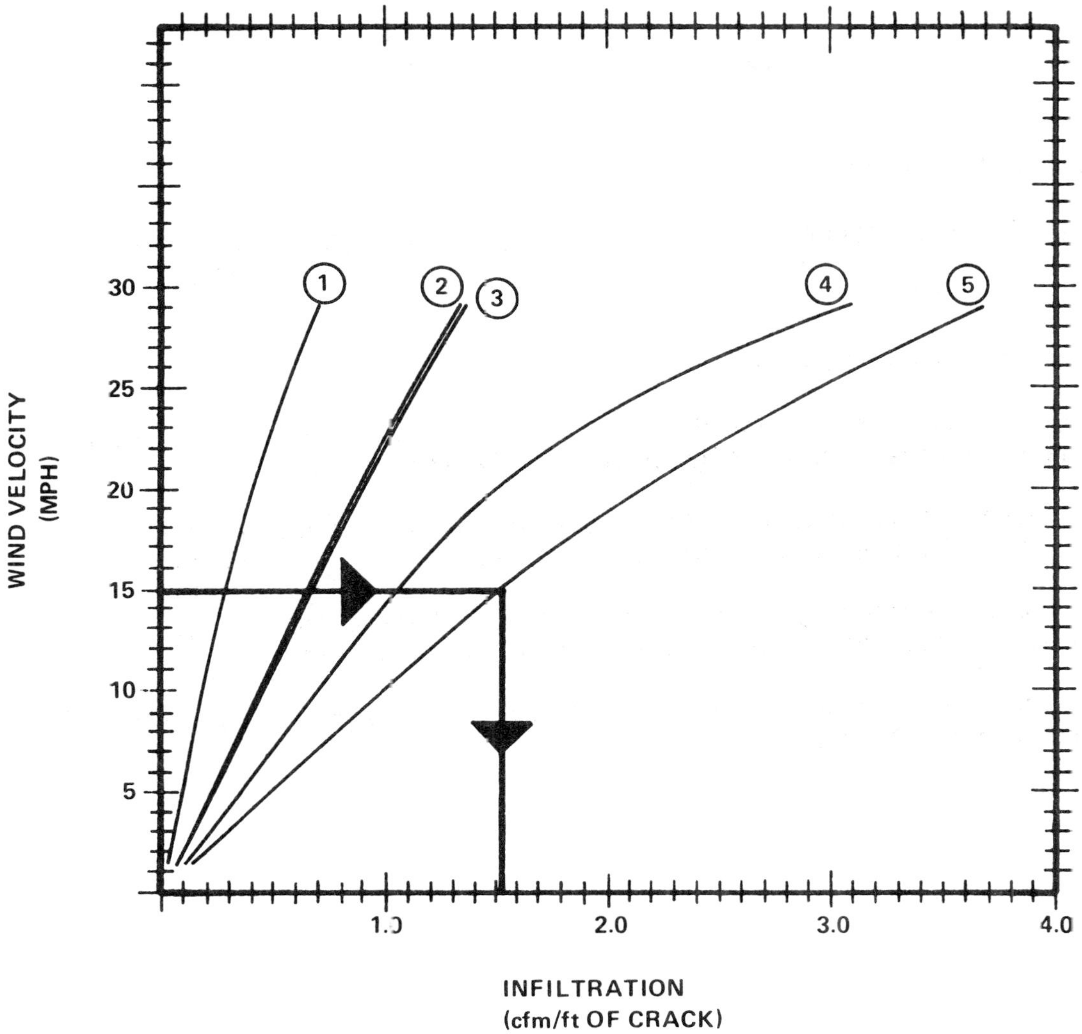

KEY:

NO.	TYPE	MATERIAL	WEATHER STRIPPED	FIT
1.	ALL	WOOD	YES	AVERAGE
	HINGED	METAL	YES	AVERAGE
2.	ALL	WOOD	NO	AVERAGE
	HINGED	METAL	NO	AVERAGE
	DOUBLE HUNG	STEEL	NO	AVERAGE
3.	ALL	WOOD	YES	LOOSE
	DOUBLE HUNG	STEEL	YES	AVERAGE
4.	CASEMENT/STEEL	STEEL	NO	AVERAGE
5.	ALL	WOOD	NO	LOOSE
	DOUBLE HUNG	STEEL	NO	AVERAGE

FIGURE 8-14. INFILTRATION THROUGH WINDOWS

Figure #8-15 Engineering Data

Source: Handbook of Air Conditioning, Heating and Ventilating 1965, Strock & Koral, 2nd Edition, Table #1, 2-105.

Instructions for Using Nomograph

1. Enter chart at the bottom horizontal line at the number of passages per hour through doors as estimated by job conditions.

2. Proceed vertically to the intersection of the curve or line that most closely fits the job conditions.

3. Proceed horizontally left from this intersection to read the infiltration in cfm.

4. This will be the factor entered under "Infiltration" on the Building Load Survery Form for doors.

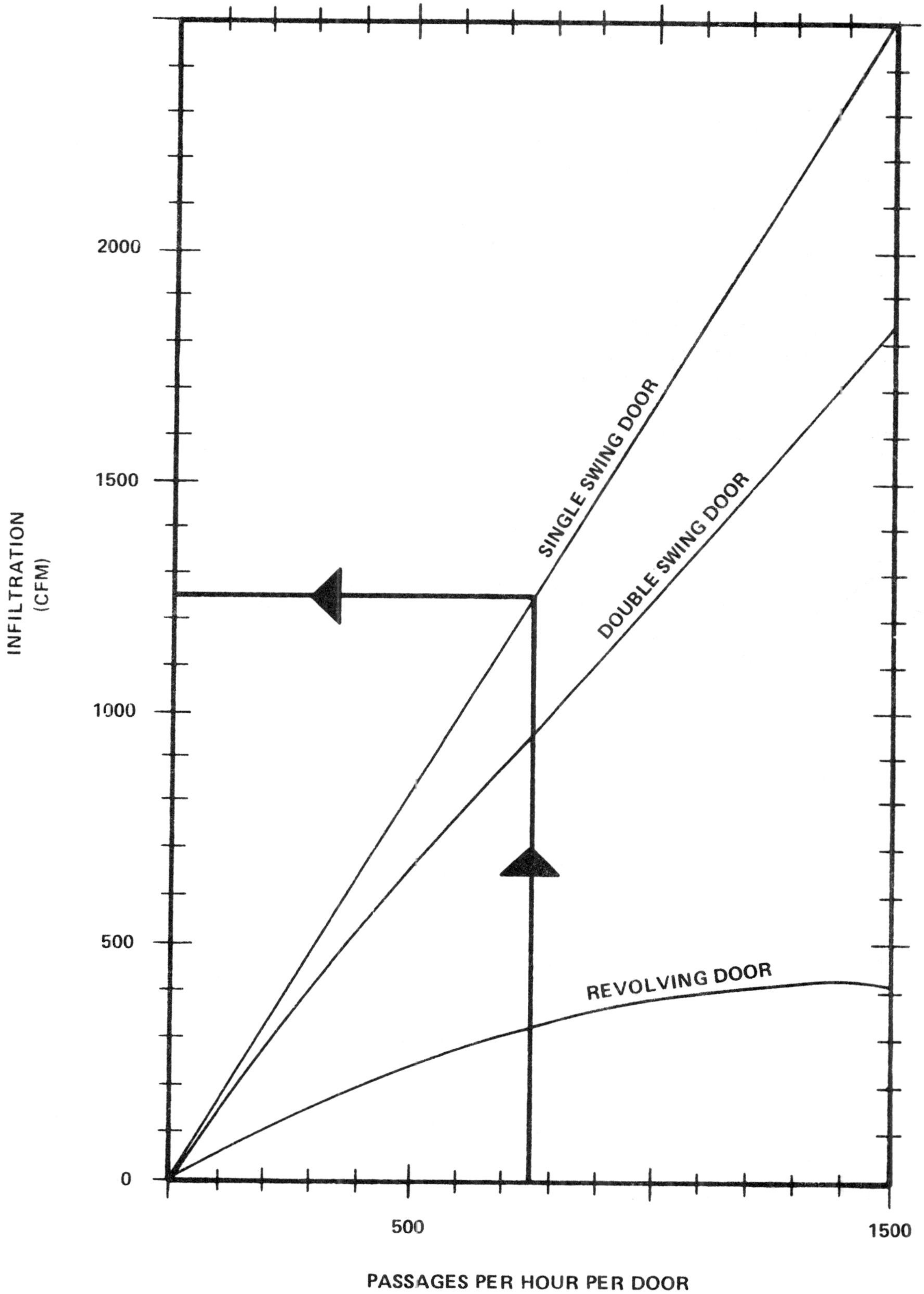

FIGURE 8-15. HEATING-INFILTRATION THROUGH EXTERNAL DOORS

Figure #8-16 Engineering Data

This figure is based on "Sunset" Computer program which was used to calculate solar effect on windows for 12 locations. The program calculates hourly solar angles and intensities for the 21st day of each month. Radiation intensity values were modified by the average percentage of cloud cover taken from weather records on an hourly basis. Heat gains are based on a 78^{0}F indoor temperature. During the cooling season, internal gains, ventilation, infiltration and conduction through the building can create a cooling load. The additional load caused by heat gain through the windows was calculated for each day. Daily totals were then summed for the number of days in each month to arrive at monthly heat gains. The length of the cooling season for each location considered was determined from weather data and characteristic operating periods. Yearly heat gains were derived by summing monthly totals for the length of the cooling season. These are summarized in Figure #8-56 for the 12 locations. Gains in the conduction heat gain component through windows read from Figure #8-18 were deducted from the total heat gains to derive the solar component. The solar component was then plotted and extrapolated to include the entire range of degree hours. Figure #8-16 was derived from locations with latitudes between 25^{0}N - 35^{0}N. The heat gains assume that the windows are subjected to direct sunshine. If shaded, gains should be read from the north exposure line. The accuracy of the graph diminishes for location with less than 5,000 degree hours.

Instructions for Use of Nomograph

1. Determine latitude from the Climatic Zone Form. Use either Figure 8-16 or Figure 8-17 depending on the latitude.

2. Enter the nomograph on the bottom left-hand vertical line at the annual dry bulb degree hours above 78 degrees F dry bulb as determined from the Climatic Zone Form.

3. Proceed horizontally right to the intersection of the appropriate number for Langleys as determined on the Climatic Zone Form.

4. Proceed vertically upward from this intersection to the orientation being investigated corresponding with existing glazing.

5. Proceed horizontally left from this intersection to read the yearly heat gain in Btu x 10^3 per square foot.

6. This is the factor to be entered under Cooling for Glass on the Annual Heating and Cooling Tabulation Form.

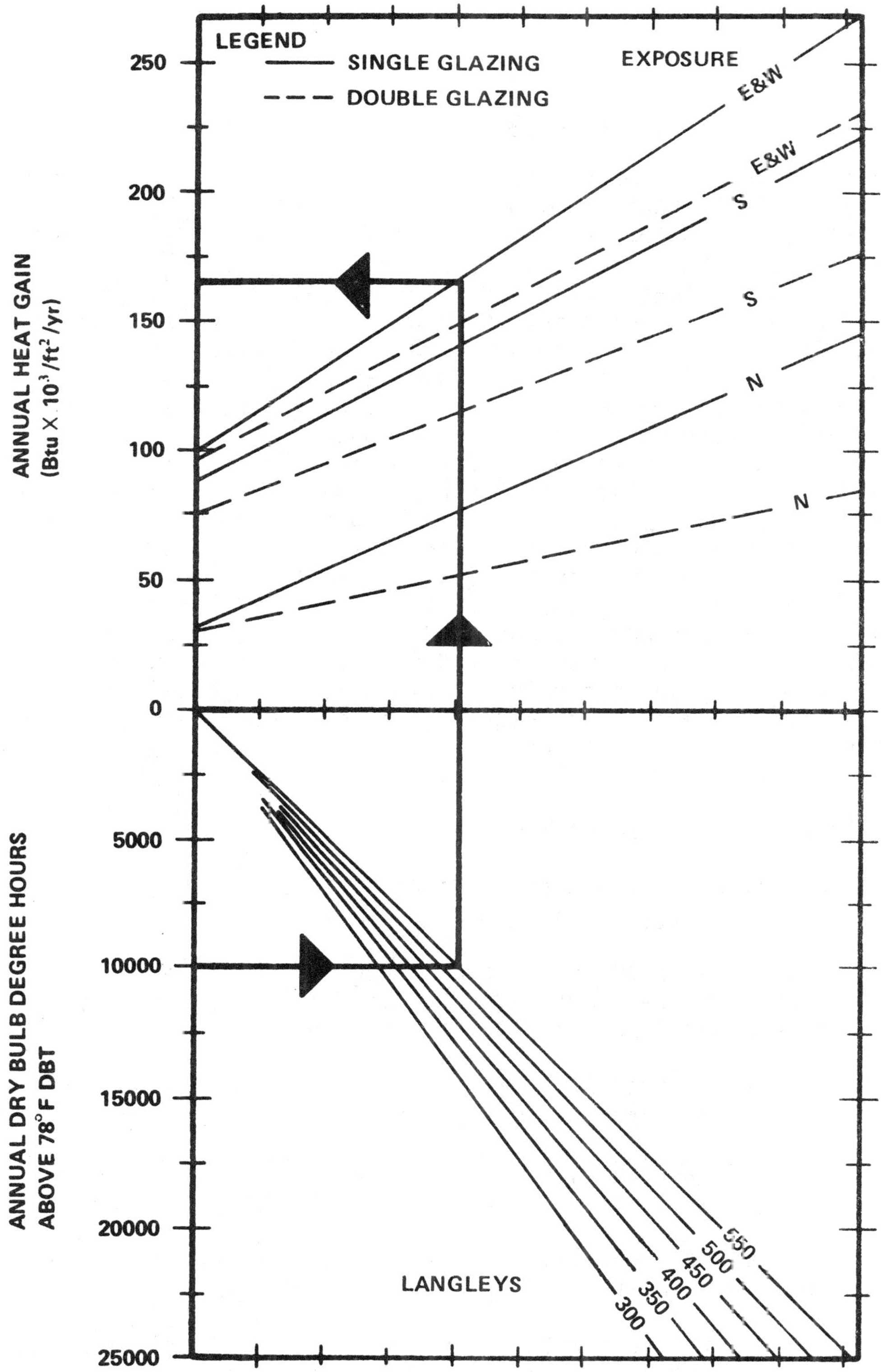

FIGURE 8-16. COOLING-ANNUAL SOLAR HEAT GAIN THROUGH WINDOWS LATITUDE 25°N–35°N

Figure #8-17 Engineering Data

This figure is based on "Sunset" Computer program which was used to calculate solar effect on windows for 12 locations. The program calculates hourly solar angles and intensities for the 21st day of each month. Radiation intensity values were modified by the average percentage of cloud cover taken from weather records on an hourly basis. Heat gains are based on a 78°F indoor temperature. During the cooling season, internal gains, ventilation, infiltration and conduction through the building can create a cooling load. The additional load caused by heat gain through the windows was calculated for each day. Daily totals were then summed for the number of days in each month to arrive at monthly heat gains. The length of the cooling season for each location considered was determined from weather data and characteristic operating periods. Yearly heat gains were derived by summing monthly totals for the length of the cooling season. These are summarized in Figure #8-56 for the 12 locations. Gains in the conduction heat gain component through windows read from Figure #8-18 were deducted from the total heat gains to derive the solar component. The solar component was then plotted and extrapolated to include the entire range of degree hours. Figure #8-16 was derived from locations with latitudes between 25°N - 35°N. The heat gains assume that the windows are subjected to direct sunshine. If shaded, gains should be read from the north exposure line. The accuracy of the graph diminishes for location with less than 5,000 degree hours.

Instructions for Use of Nomograph

1. Determine latitude from the Climatic Zone Form. Use either Figure 8-16 or Figure 8-17 depending on the latitude.

2. Enter the nomograph on the bottom left-hand vertical line at the annual dry bulb degree hours above 78 degrees F. dry bulb as determined from the Climatic Zone Form.

3. Proceed horizontally right to the intersection of the appropriate number for Langleys as determined on the Climatic Zone Form.

4. Proceed vertically upward from this intersection to the orientation being investigated corresponding with existing glazing.

5. Proceed horizontally left from this intersection to read the yearly heat gain in Btu x 10^3 per square foot.

6. This is the factor to be entered under Cooling for Glass on the Annual Heating and Cooling Tabulation Form.

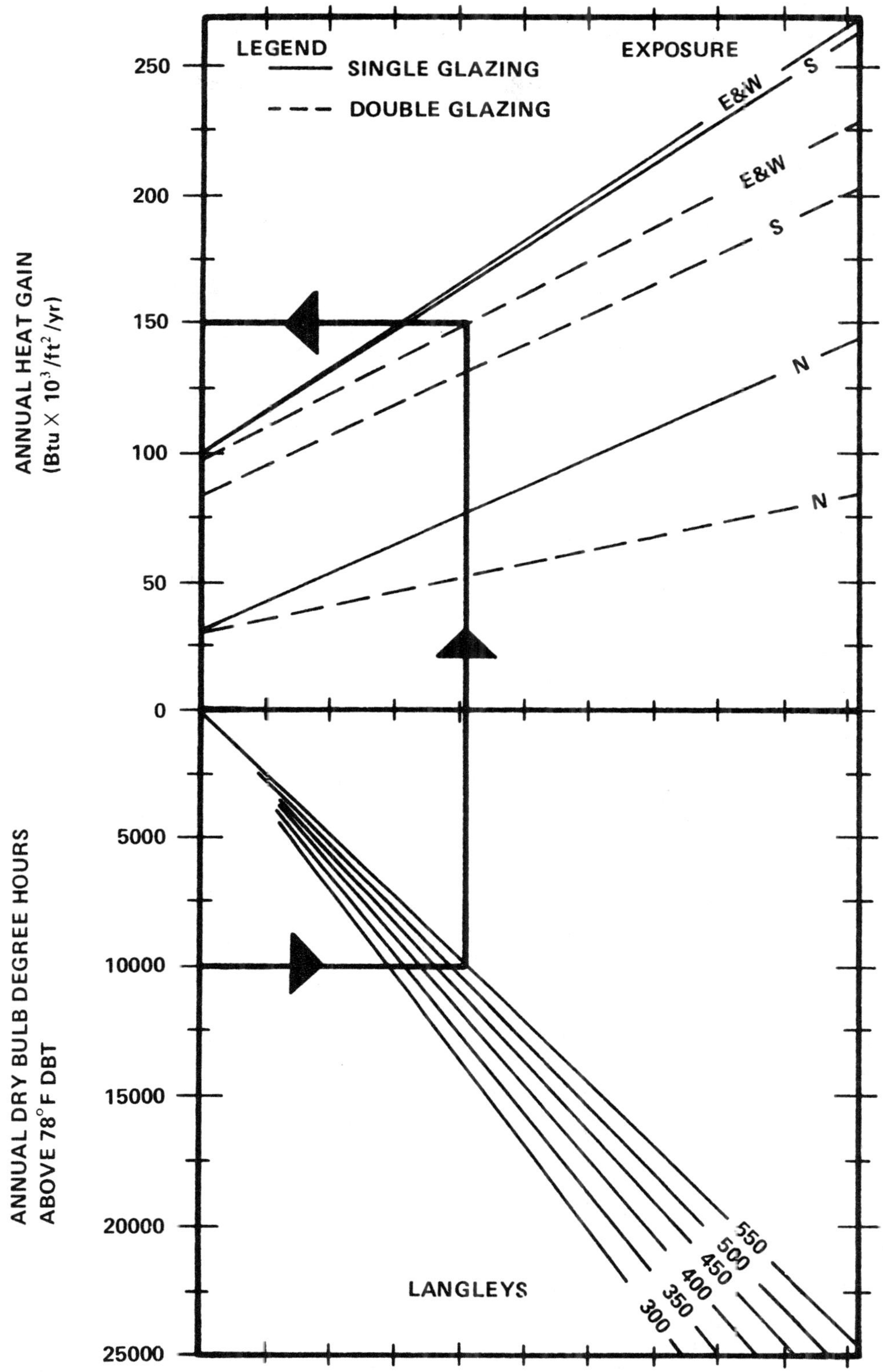

FIGURE 8-17. COOLING-ANNUAL SOLAR HEAT GAIN THROUGH WINDOWS LATITUDE 35°N–45°N

Figure #8-18 Engineering Data

This figure is based on degree hours read from Figure #8-20 which has a base of 56 hours/week. The figure is based on the formula: Q (heat gain)/yr = degree hours/yr x 'U' Value. 'U' Values assumed were 1.1 for single pane, 0.65 for double panes and 0.47 for triple pane. The major portion of degree hours occur between 10 a.m. and 3 p.m. Hence, for occupancies between 10 and 56 hours/week, the degree hour distribution can be assumed to be linear. However, for occupancies greater than 56 hours/week the degree hour distribution becomes non-linear, particularly in locations with greater then 15,000 degree hours. This is reflected by the curves for 72 and 96 hour/week occupancies.

Instructions for Use of Nomograph

1. Enter on lower left hand vertical line at the annual dry bulb degree hours above 78 degrees F. dry bulb as determined on Climatic Zone Form.

2. Proceed horizontally right to the number of occupied hours per week.

3. Proceed vertically upward from this intersection to the type of existing glazing.

4. Proceed horizontally left from this intersection to read the yearly heat gain in Btu x 10^3 per square foot.

5. This is the factor to be entered under Cooling for Glass conduction on the Annual Heating Annual Heating and Cooling Tabulation Form.

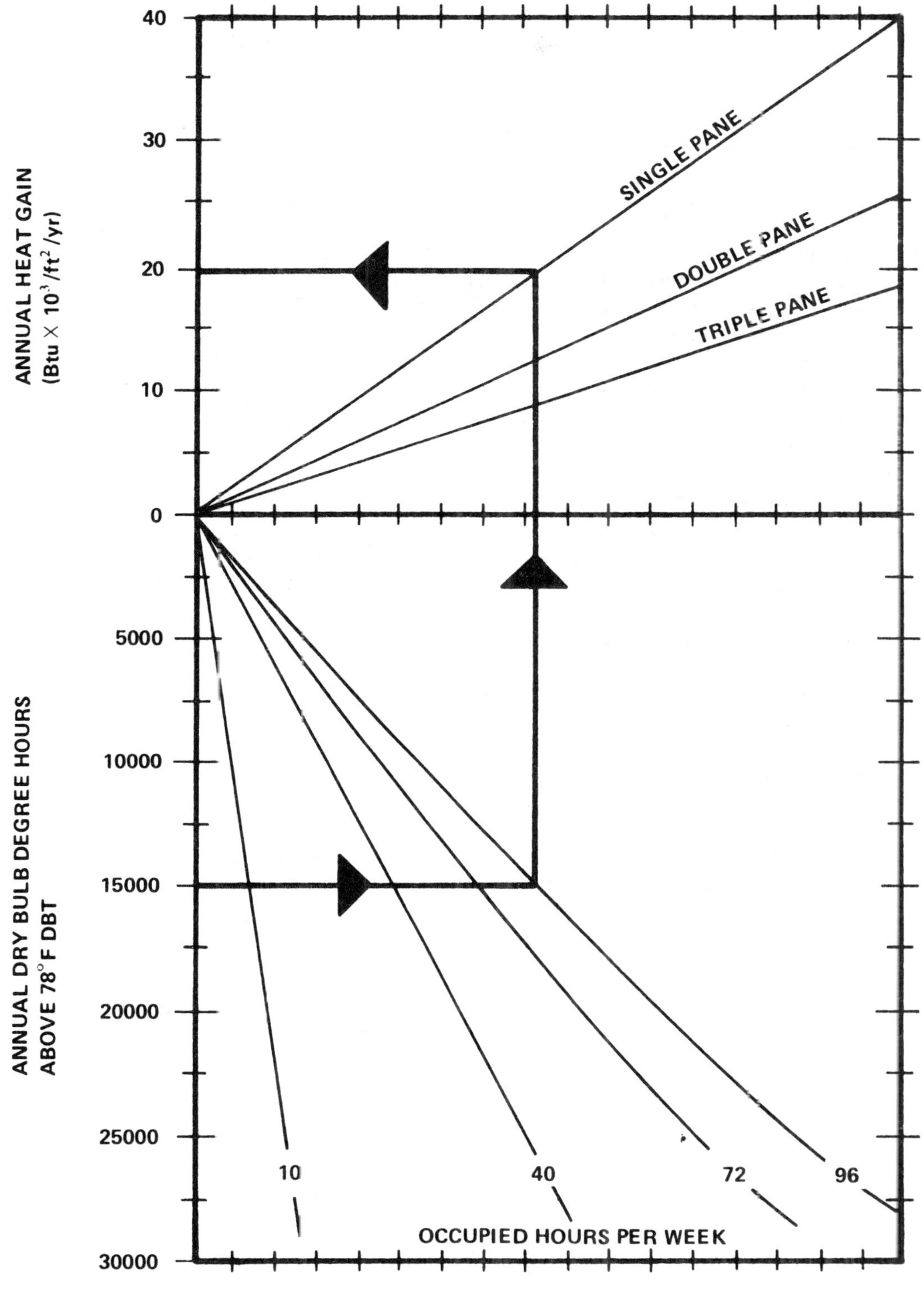

FIGURE 8-18. COOLING-ANNUAL CONDUCTION HEAT GAIN THROUGH WINDOWS

Figure #8-19 Engineering Data

This figure is based on "Sunset" Computer program which was used to calculate solar effect on windows for 12 locations. The program calculates hourly solar angles and intensities for the 21st day of each month. Radiation intensity values were modified by the average percentage of cloud cover taken from weather records on an hourly basis. Heat gains are based on a 68°F indoor temperature.

Additional assumptions were: 1) Total internal heat gain of 12 Btu/sq ft 2) Average outdoor air ventilation rate of 10%. 3) Infiltration rate of 1/2 air change per hour. Daily totals were then summed for the number of days in each month to arrive at monthly heat gains. The length of the heating season for each location considered was determined from weather data and characteristic operating periods. Yearly heat losses were derived by summing monthly totals for the length of the heating season. These are summarized in Figure #8-56 for the 12 locations. The data was then plotted and extrapolated to include the entire range of degree days. Figure #8-19 was derived from locations with latitudes between 25°N - 35°N.

Instructions for Use of Nomograph

1. Determine latitude from the Climatic Zone Form. Use either Figure 8-19 or Figure 8-20 depending on the latitude.

2. Enter the nomograph on the bottom left-hand vertical line at the degree days as determined from the Climatic Zone Form.

3. Proceed horizontally right to the number for Langleys as determined on the Climatic Zone Form.

4. Proceed vertically upward from this intersection to the orientation being investigated corresponding with existing glazing.

5. Proceed horizontally left from this intersection to read the yearly heat gain in Btu x 10^3 per square foot.

6. This is the factor to be entered under Cooling for Glass on the Annual Heating and Cooling Tabulation Form.

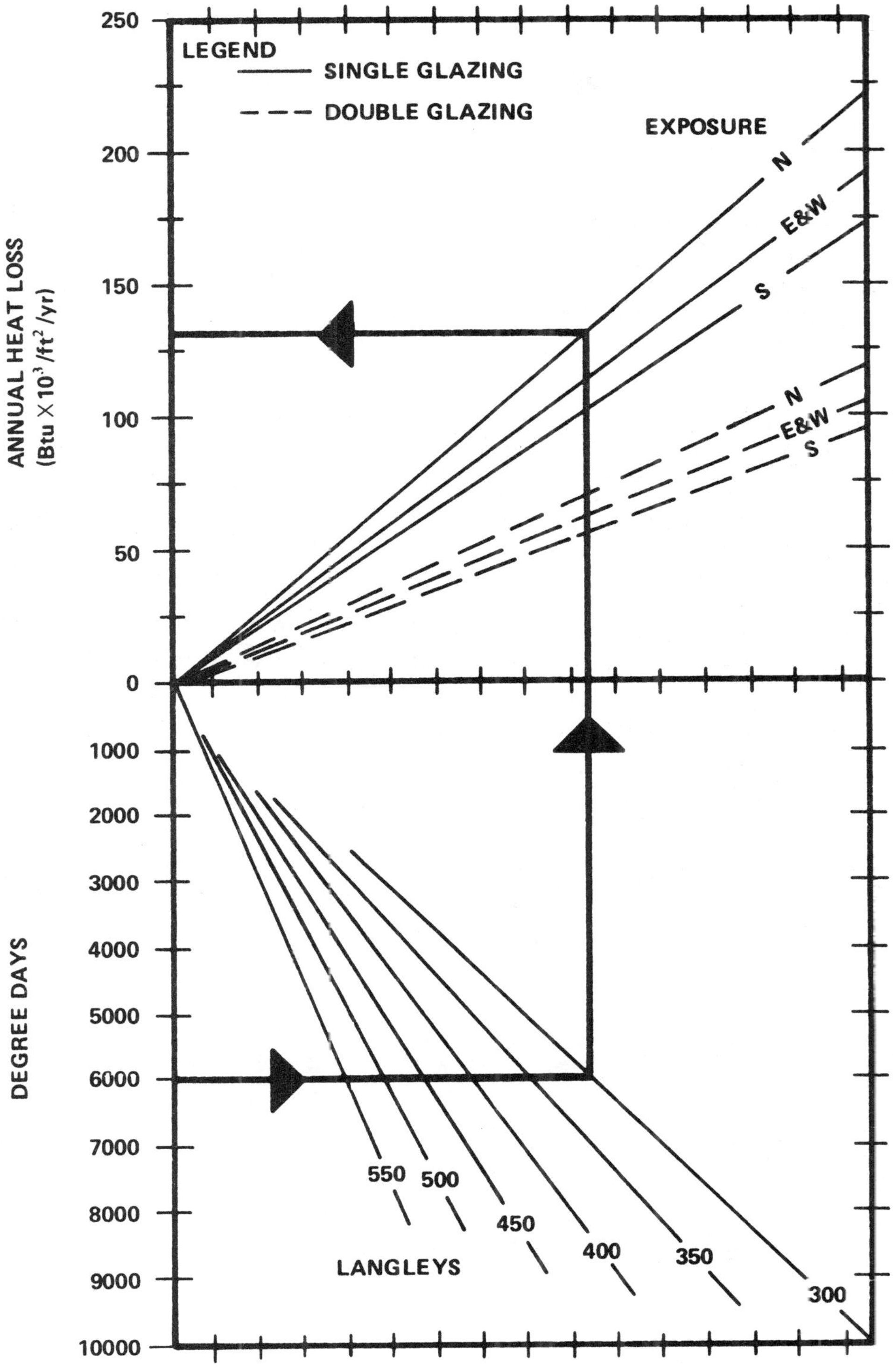

FIGURE 8-19. HEATING-ANNUAL HEAT LOSS THROUGH WINDOWS LATITUDE 25°N—35°N

Figure #8-20 Engineering Data

This figure is based on "Sunset" Computer program which was used to calculate solar effect on windows for 12 locations. The program calculates hourly solar angles and intensities for the 21st day of each month. Radiation intensity values were modified by the average percentage of cloud cover taken from weather records on an hourly basis. Heat gains are based on a 68°F indoor temperature.

Additional assumptions were 1) total internal heat gain of 12 Btu/sq ft 2) Average outdoor air ventilation rate of 10%. 3) Infiltration rate of 1/2 air change per hour. Daily totals were then summed for the number of days in each month to arrive at monthly heat losses. The length of the heating season for each location considered was determined from weather data and characteristic operating periods. Yearly heat losses were derived by summing monthly totals for the length of the heating season. These are summarized in Figure #8-56 for the 12 locations. The data was then plotted and extrapolated to include the entire range of degree days. Figure #8-20 was derived from locations with latitudes between 35°N - 35°N.

Instructions for Use of Nomograph

1. Determine latitude from the Climatic Zone Form. Use either Figure 8-19 or Figure 8-20 depending on the latitude.

2. Enter the nomograph on the bottom left-hand vertical line at the degree days as determined from the Climatic Zone Form.

3. Proceed horizontally right to the number for Langleys as determined on the Climatic Zone Form.

4. Proceed vertically upward from this intersection to the orientation being investigated corresponding with existing glazing.

5. Proceed horizontally left from this intersection to read the yearly heat gain in Btu x 10^3 per square foot.

6. This is the factor to be entered under Heating for Glass on the Annual Heating and Cooling Tabulation Form.

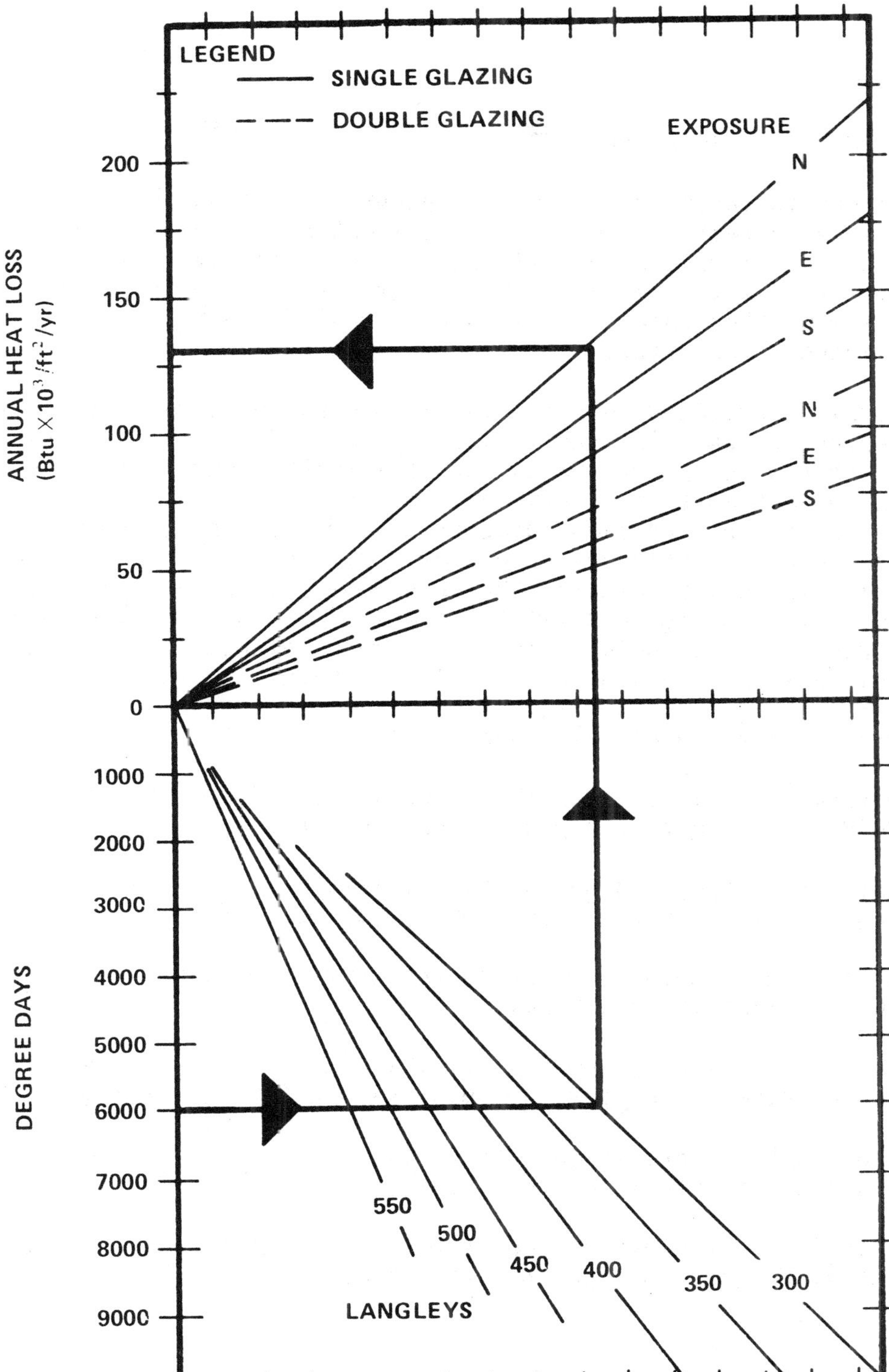

FIGURE 8-20. HEATING-ANNUAL HEAT LOSS THROUGH WINDOWS
LATITUDE 35°N–45°N

Figure #8-21

The development of this figure was based on the assumptions that

1) Thermal barriers are closed only when the building is unoccupied

2) The average degree day distribution is 25% during the daytime and 75% during nighttime.

The number of degree days occuring when the thermal barriers are closed were determined from the characteristic occupancy period shown in the figure. This can be expressed as a fraction of the total degree days by the relationship:

$$DD_A = 0.25\ DD_T \left[\frac{\text{unoccupied daytime hrs/wk}}{\text{total daytime hrs/wk}}\right] + 0.75\ DD_T \left[\frac{\text{unoccupied nightime hrs/wk}}{\text{total nighttime hrs/wk}}\right]$$

Where, DD_A = Adjusted degree days

DD_T = Total degree days

Instructions for Use of Nomograph

1. Enter the nomograph on the bottom left-hand vertical line at the degree days as determined from the Climatic Zone Form.

2. Proceed horizontally right to the intersection of the composite U-value as determined on the Building Load Survey Form.

3. Proceed vertically upward from this intersection to the number of occupied hours per week.

4. Proceed horizontally left to read the yearly heat loss in Btu x 10^3 per square foot.

5. This is the factor to be entered under Heating for Glass on the Annual Heating and Cooling Tabulation Form.

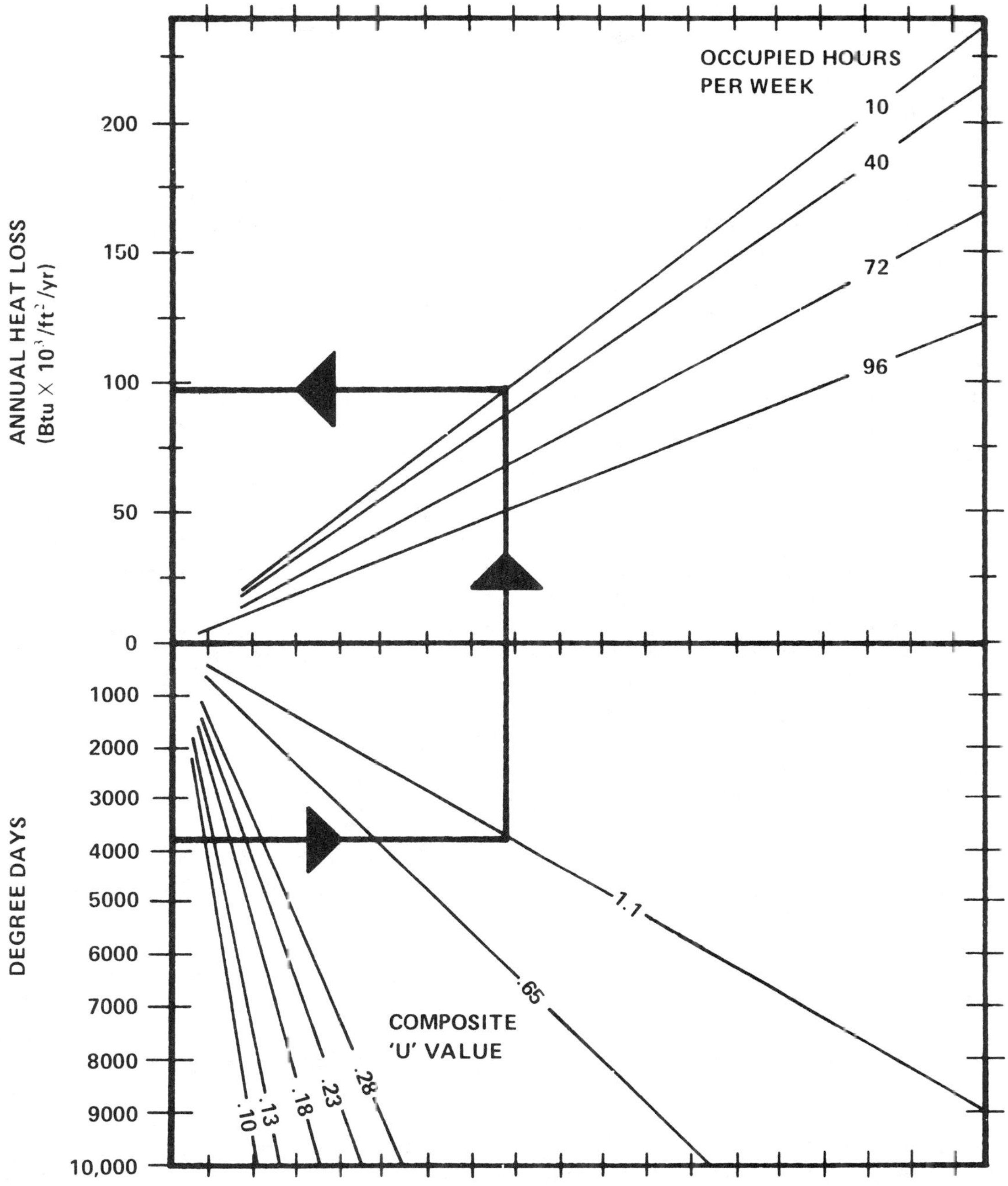

FIGURE 8-21. HEATING-ANNUAL HEAT LOSS FOR WINDOWS WITH THERMAL BARRIERS

Figure #8-22 Engineering Data

This figure is based on "Sunset" Computer program which was used to calculate solar effect on windows for 12 locations. The program calculates hourly solar angles and intensities for the 21st day of each month. Radiation intensity values were modified by the average percentage of cloud cover taken from weather records on an hourly basis. Heat gains are based on a $78^{o}F$ indoor temperature.

The solar effect was calculated using sol-air temperature and the heat entering or leaving a space was calculated with the equivalent temperature difference. Wall mass ranged from 50 - 60 lbs/sq ft and thermal lag averaged 4 1/2 hours. During this cooling season internal gains, ventilation, infiltration and conduction through the building skin can create a cooling load. The additional load caused by heat going through the walls was calculated for each day. Daily totals were then summed for the number of days in each month to arrive at monthly heat gains. The length of the cooling season for each location considered was determined from weather data and characteristic operating periods. Yearly heat gains were derived by summing monthly totals for the length of the cooling season.

Instructions for Use of Nomograph

1. Determine latitude from the Climatic Zone Form. Use either Figure 8-22 or 8-23 depending on the latitude.

2. Enter the nomograph on the bottom left-hand vertical line at the annual dry bulb degree hours above 78 degrees F. dry bulb as determined from the Climatic Zone Form.

3. Proceed horizontally right to the intersection of the appropriate number for Langleys as determined on the Climatic Zone Form.

4. Proceed vertically upward from this intersection of the orientation being investigated.

5. Proceed horizontally left from this intersection to read the U-value for the walls as determined and recorded on the Building Load Survey Form.

6. Proceed vertically downward from this intersection to the proper scale determined by the existing absorption coefficient indicated on the Build-Load Survey Form under type of walls (i.e., light or dark.)

7. Read the yearly heat gain in Btu x 10^3 per square foot.

8. Enter this factor under Cooling for Walls on the Annual Heating and Cooling Tabulation Form.

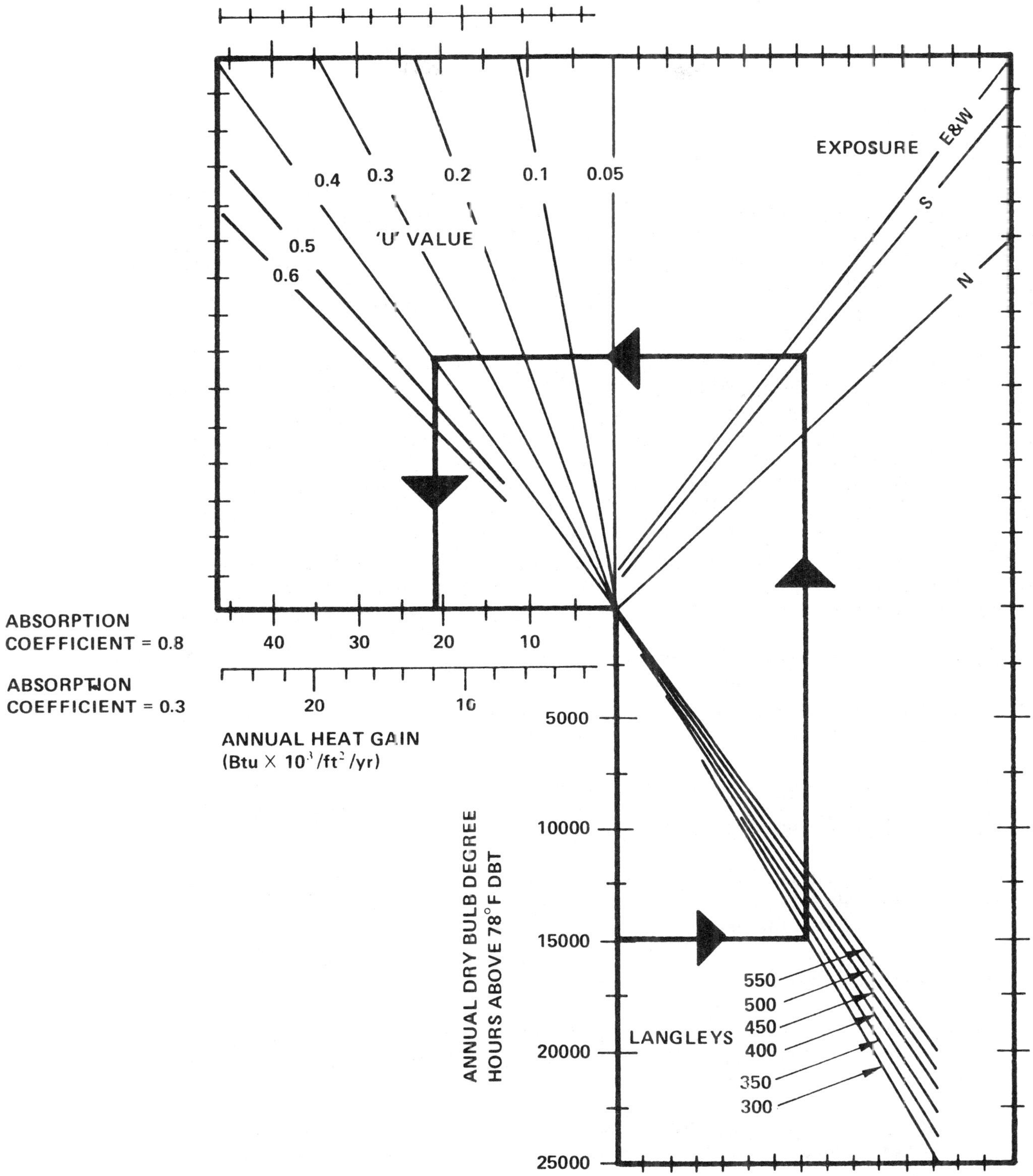

FIGURE 8-22. COOLING-ANNUAL SOLAR HEAT GAIN THROUGH WALLS LATITUDE 25°N–35°N

Figure #8-23 Engineering Data

This figure is based on "Sunset" Computer program which was used to calculate solar effect on windows for 12 locations. The program calculates hourly solar angles and intensities for the 21st day of each month. Radiation intensity values were modified by the average percentage of cloud cover taken from weather records on an hourly basis. Heat gains are based on a 78°F indoor temperature.

The solar effect on a wall was calculated using sol-air temperature and the heat entering of leaving a space was calculated with the equivalent temperature difference. Wall mass ranged from 50-60 lbs/sq ft and thermal lag averaged 4 1/2 hours. During the cooling season, internal gains, ventilation infiltration and conduction through the building skin can create a cooling load. The additional load caused by heat gain through the walls was calculated for each day. Daily totals were then summed for the number of days in each month to arrive at monthly heat gains. The length of the cooling season for each location considered was determined from weather data and characteristic operating periods. Yearly heat gains were derived by summing monthly totals for the length of the cooling season. (The heat losses assume that the walls are subjected to direct sunlight. If shaded, losses should be read from the North exposure line.)

Instructions for Use of Nomograph

1. Determine latitude from the Climatic Zone Form. Use either Figure 8-22 or 8-23 depending on the latitude.

2. Enter the lower left-hand vertical line at the annual dry bulb degree hours above 78 degrees F. dry bulb as determined from the Climatic Zone Form.

3. Proceed horizontally right to the number for Langleys as determined on the Climatic Zone Form.

4. Proceed vertically upward from this intersection to the orientation being investigated.

5. Proceed horizontally left from this intersection to the U-value for the walls as determined and recorded on the Building Load Survey Forms

6. Proceed vertically downward from this intersection to the proper scale determined btu the existing absorption coefficient indicated on the Building Load Survey Form under type of walls (i.e. light or dark).

7. Read the yearly heat gain in Btu x 10^3 per square foot.

8. Enter this factor under Cooling for Walls on the Annual Heating and Cooling Tabulation Form.

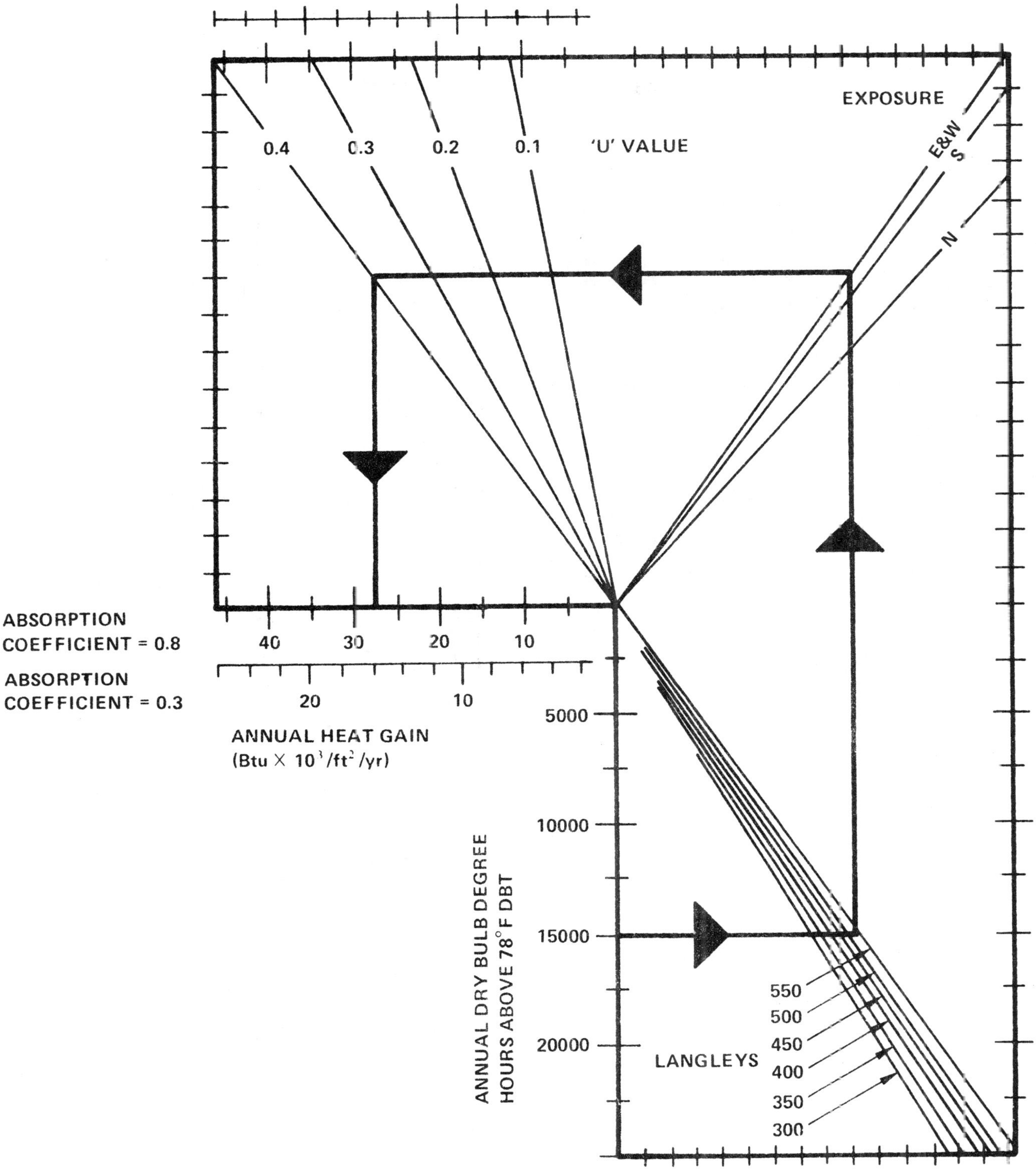

FIGURE 8-23. COOLING-ANNUAL SOLAR HEAT GAIN THROUGH WALLS LATITUDE 35°N 45°N

Figure #8-24 Engineering Data

This figure is based on degree hours read from Figure #8-4 which has a base of 56 hours/week. The figure is based on the formula: Q (Heat Gain)/yr = Degree Hours/yr x 'U' Value. The major portion of degree hours occur between 10 a.m. and 3 p.m. Hence for occupancies between 10 and 56 hrs./week, the degree hour distribution can be assumed to be linear. However, for occupancies greater than 54 hrs/week the degree hour distribution becomes non-linear, particularly in locations with greater than 15,000 degree hours. This is reflected by the curves for 72 and 96 hour/week occupancies

Instructions for Use of Nomograph

1. Enter the nomograph on the lower left-hand vertical line at the annual dry bulb degree above 78 degrees F. dry bulb as determined on the Climatic Zone Form.

2. Proceed horizontally right to the number of occupied hours per week.

3. At this intersection, proceed vertically upward to the U-value as determined and recorded on the Building Load Survey Form.

4. Proceed horizontally left at this intersection to read the yearly heat gain in Btu x 10^3 per square foot.

5. This is the factor to be entered under Cooling for Conduction of Heat through Walls, Roofs and Floors on the Annual Heating and Cooling Tabulation Form.

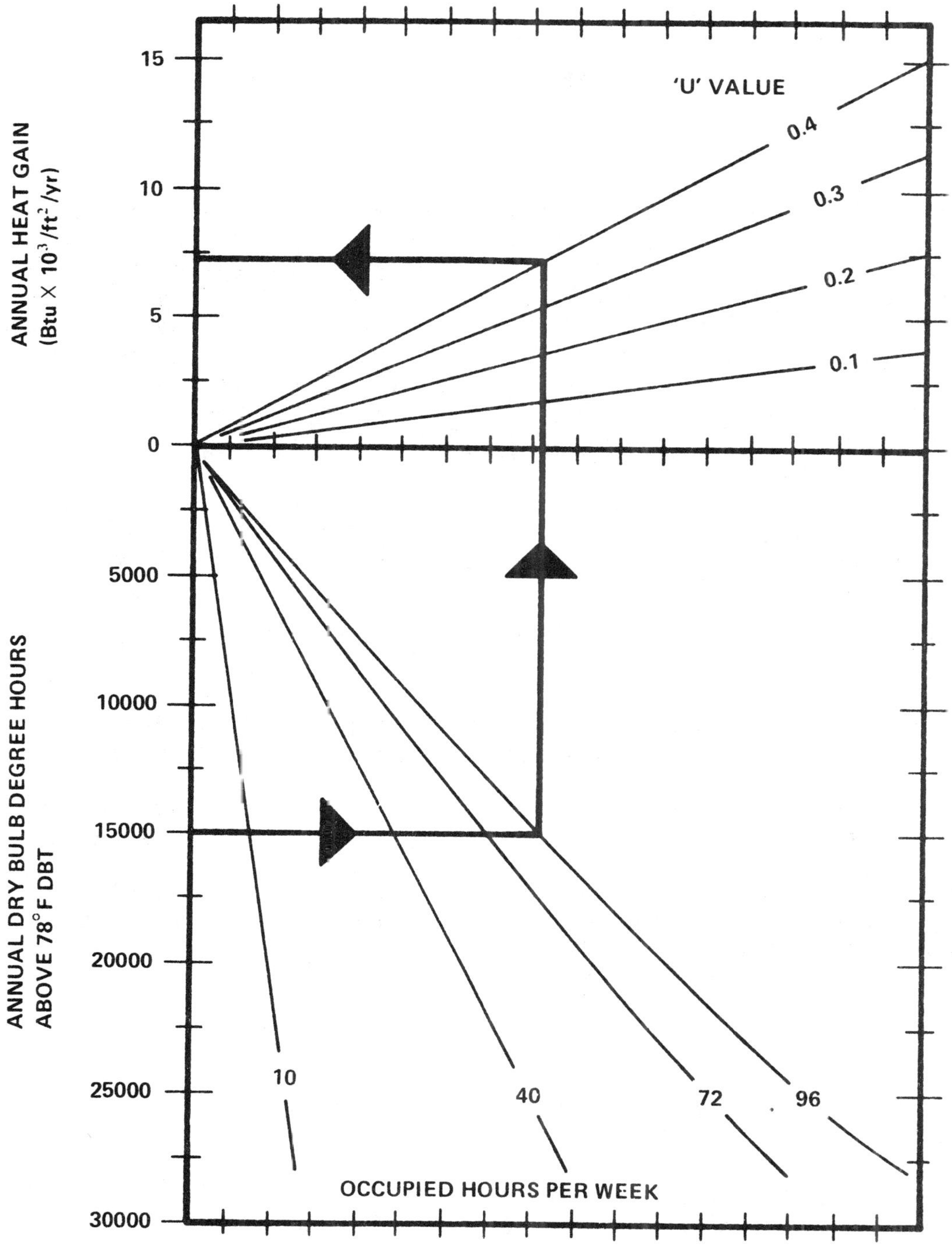

FIGURE 8-24. COOLING-ANNUAL CONDUCTION HEAT GAIN THROUGH WALLS, ROOFS, AND FLOORS

Figure #8-25 Engineering Data

This figure is based on "Sunset" Computer program which was used to calculate solar effect on windows for 12 locations. The program calculates hourly solar angles and intensities for the 21st day of each month. Radiation intensity values were modified by the average percentage of cloud cover taken from weather records on an hourly basis. Heat losses are based on a 68°F indoor temperature.

The solar effect on a wall was calculated by the computer using sol-air temperature and entering or leaving a space was calculated with the equivalent temperature difference. Wall mass ranged form 50-60 lbs/sq ft and thermal lag averaged 4 1/2 hours. Additional assumptions were: 1) Total internal heat gain of 12 Btu/sq ft. 2) Average outdoor air ventilation rate of 10%. 3) Infiltration rate of 1/2 air change per hour. Daily totals were then summed for the number of days in each month to arrive at monthly heat losses. The length of the heating season for each location considered was determined from weather data and characteristic operating periods. Yearly heat losses were derived by summing monthly totals for the length of the heating season. (The heat losses assume that the walls are subjected to direct sunlight. If shaded, losses should be read from the North exposure line.)

Instructions for Use of Nomograph

1. Determine latitude from the Climatic Zone Form. Use either Figure 8-25 or 8-26, depending on the latitude.

2. Enter the nomograph right on the bottom left-hand vertical line at the degree days determined from the Climatic Zone Form.

3. Proceed horizontally right to the number of Langleys as determined on the Climatic Zone Form.

4. Proceed vertically upward from this intersection to the orientation being investigated.

5. Proceed horizontally left from this intersection to the U-Value as determined and recorded on the Building Load Survey Form.

6. Proceed vertically downward at this intersection to the proper scale for the absorption coefficient as indicated by the type of wall on the Building Load Survey Form (i.e. dark or light). Read the yearly loss in Btu x 10^3 per square foot.

7. This factor is to be entered under Heating for Walls on the Annual Heating and Cooling Tabulation Form.

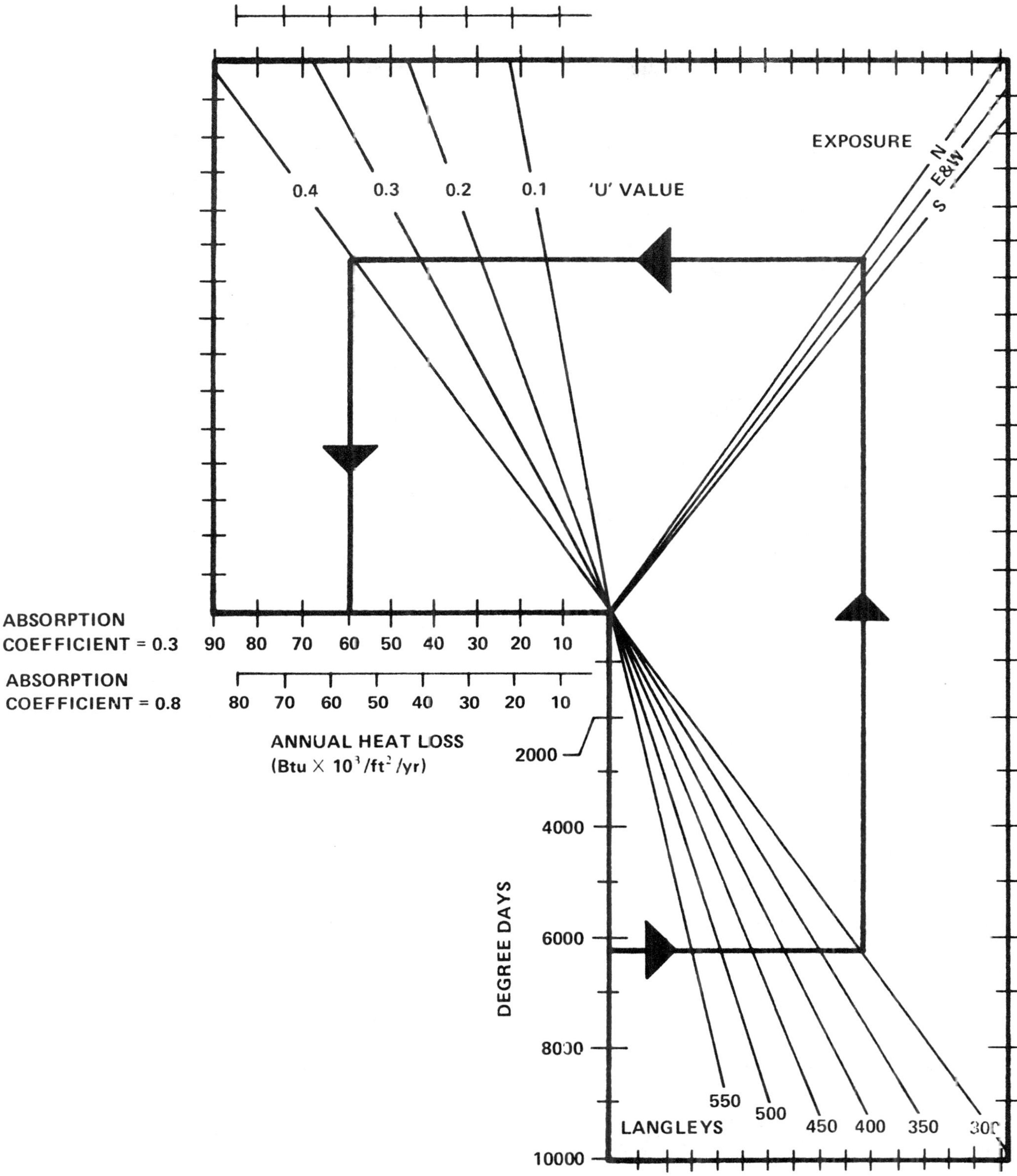

FIGURE 8-25. HEATING-ANNUAL HEAT LOSS THROUGH WALLS
LATITUDE 25°N–35°N

Figure #8-26 Engineering Data

This figure is based on "Sunset" Computer program which was used to calculate solar effect on windows for 12 locations. The program calculates hourly solar angles and intensities for the 21st day of each month. Radiation intensity values were modified by the average percentage of cloud cover taken from weather records on an hourly basis. Heat losses are based on a 68°F indoor temperature.

The solar effect on a wall was calculated by the computer using sol-air temperature and entering or leaving a space was calculated with the equivalent temperature difference. Wall mass ranged form 50-60 lbs/sq ft and thermal lag averaged 4 1/2 hours. Additional assumptions were: 1) Total internal heat gain of 12 Btu/sq ft 2) Average outdoor air ventilation rate of 10%. 3) Infiltration rate of 1/2 air change per hour. Daily totals were then summed for the number of days in each month to arrive at monthly heat losses. The length of the heating season for each location considered was determined from weather data and characteristic operating periods. Yearly heat losses were derived by summing monthly totals for the length of the heating season. (The heat gains assume that the walls are subjected to direct sunlight. If shaded, losses should be read from the North exposure line.)

Instructions for Use of Nomograph

1. Determine latitude from the Climatic Zone Form. Use either Figure 8-25 or 8-26, depending on the latitude.

2. Enter the nomograph on the bottom left-hand vertical line at the degree days determined from the Climatic Zone Form.

3. Proceed horizontally right to the number of Langleys as determined on the Climatic Zone Form.

4. Proceed vertically upward from this intersection to the orientation being investigated.

5. Proceed horizontally left from this intersection to the U-Value as determined and recorded on the Building Load Survey Form.

6. Proceed vertically downward at this intersection to the proper scale for the absorption coefficient as indicated by the type of wall on the Building Load Survey Form (i.e. dark or light). Read the yearly loss in Btu x 10^3 per square foot.

7. This factor is to be entered under Heating for Walls on the Annual Heating and Cooling Tabulation Form.

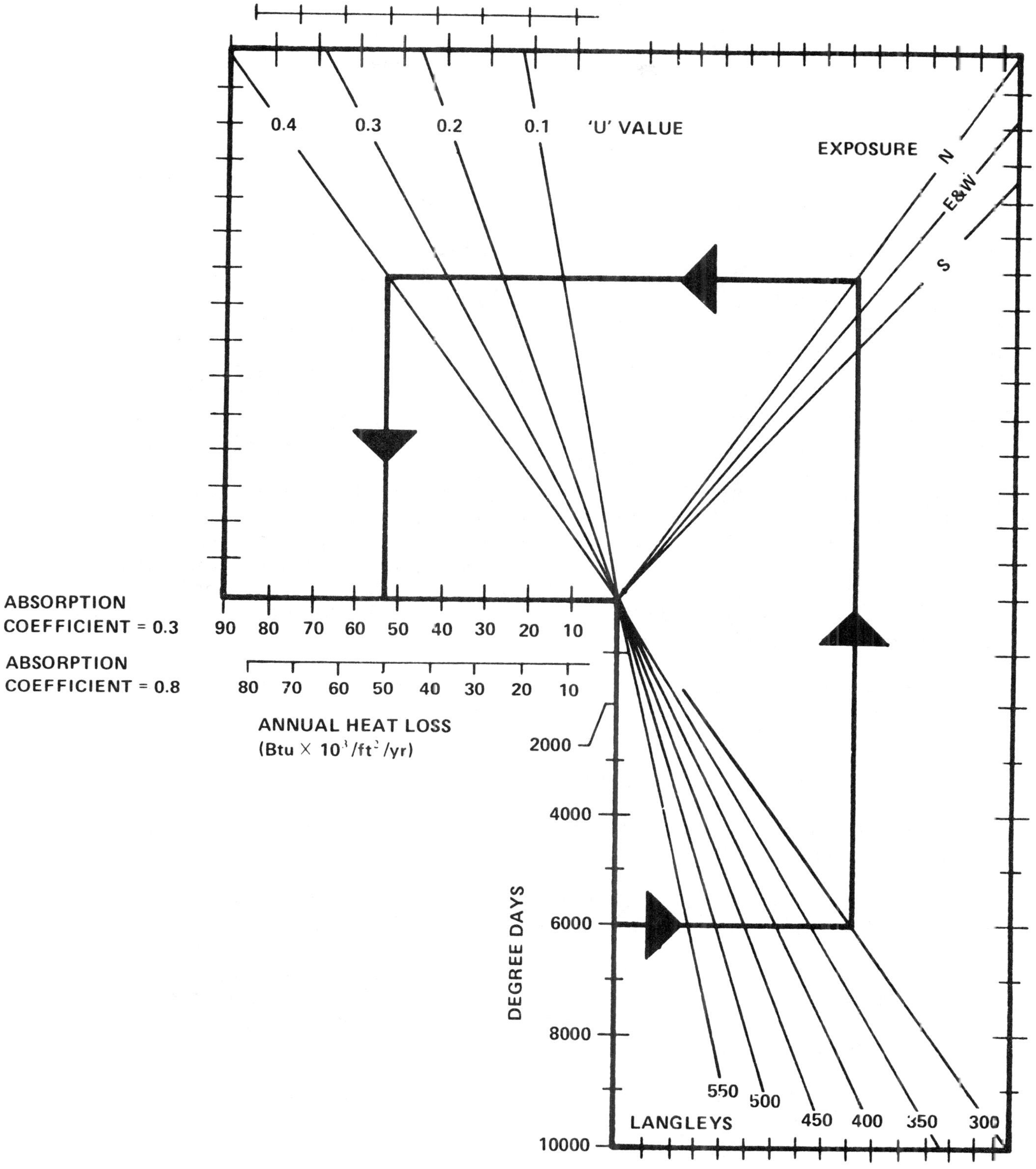

FIGURE 8-26. HEATING-ANNUAL HEAT LOSS THROUGH WALLS
LATITUDE 35°N–45°N

Figure #8-27 Engineering Data

This figure is based on "Sunset" Computer program which was used to calculate solar effect on windows for 12 locations. The program calculates hourly solar angles and intensities for the 21st day of each month. Radiation intensity values were modified by the average percentage of cloud cover taken from weather records on an hourly basis. Heat losses are based on a $78^{O}F$ indoor temperature.

The solar effect on a wall was calculated by the computer using sol-air temperature and entering or leaving a space was calculated with the equivalent temperature difference. Roof mass ranged from 25 - 35 lbs/sq ft and thermal lag averaged 3 1/2 hours. During the cooling season, internal gains, ventilation, infiltration and conduction through the building skin can create a cooling load. The additional load caused by heat gain through the roof was calculated for each day. Daily totals were then summed for the number of days in each month to arrive at monthly heat gains. The length of the cooling season for each location considered was determined from weather data and characteristic operating periods. Yearly heat gains were derived by summing monthly totals for the length of the cooling season.

Instructions for Use of Nomograph

1. Enter the nomograph on the bottom left-hand vertical line at the annual dry bulb degree hours above 78 degrees F. dry bulb as determined on the Climatic Zone Form.

2. Proceed horizontally right to the number of Langleys as determined on the Climatic Zone Form.

3. Proceed vertically upward from this intersection to the U-Value as determined adn recorded on the Building Load Survey.

4. Proceed horizontally left at this intersection to the proper scale as determined by the absorption coefficient as indicated on the Building Load Survey Form under type of roof (i.e. dark or light). Read the yearly heat gain in Btu x 10^3 per square foot.

5. This factor is to be entered under Cooling for Roof, Solar, on the Annual Heating and Cooling Tabulation Form.

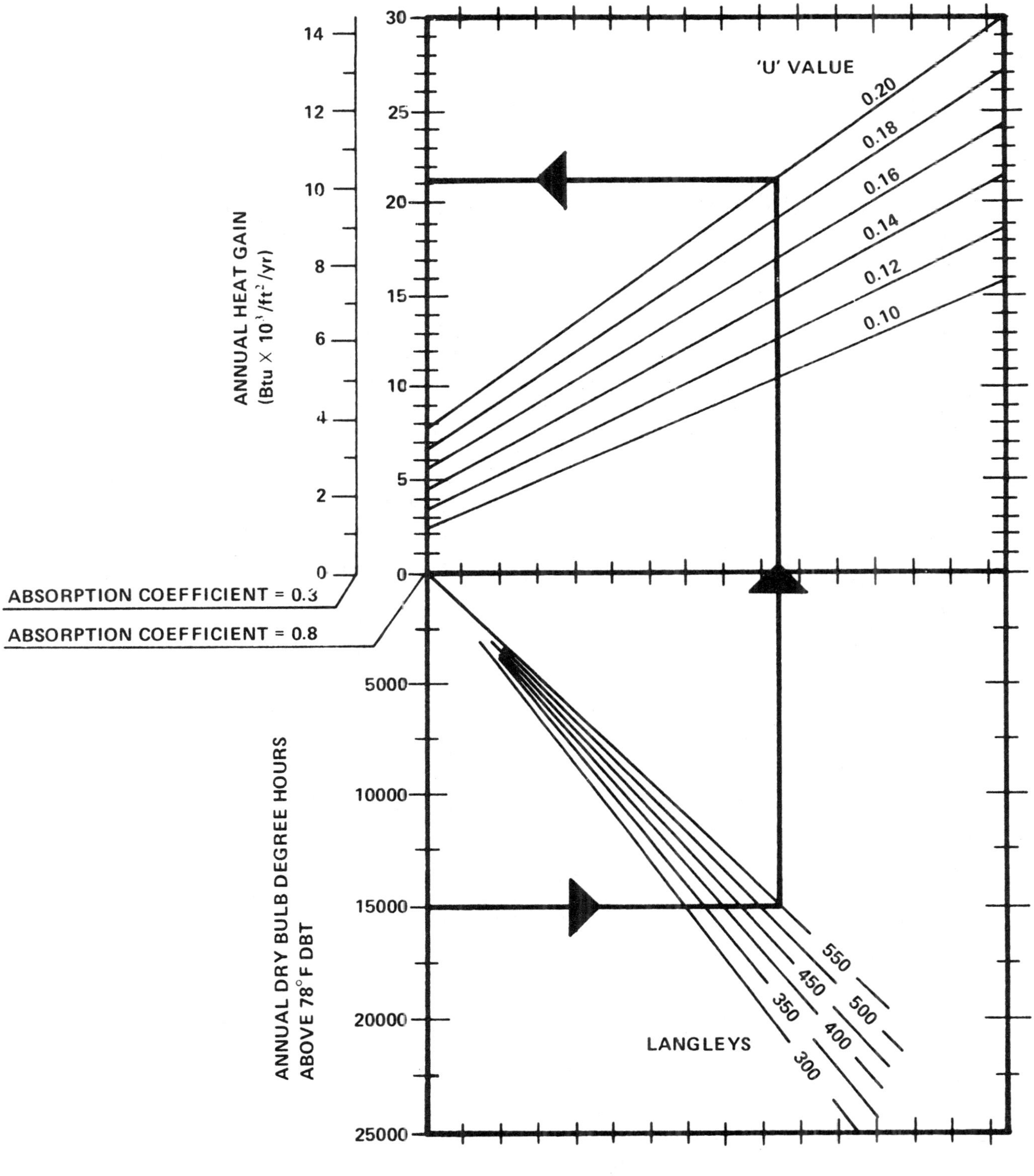

FIGURE 8-27. COOLING-ANNUAL SOLAR HEAT GAIN THROUGH ROOF

Figure #8-28 Engineering Data

This figure is based on "Sunset" Computer program which was used to calculate solar effect on windows for 12 locations. The program calculates hourly solar angles and intensities for the 21st day of each month. Radiation intensity values were modified by the average percentage of cloud cover taken from weather records on an hourly basis. Heat losses are based on a 68°F indoor temperature.

The solar effect on a wall was calculated by the computer using sol-air temperature and entering or leaving a space was calculated with the equivalent temperature difference. Roof mass ranged from 25 - 35 lbs/sq ft and thermal lag averaged 3 1/2 hours. Additional assumptions were: 1) total internal heat gain of 12 Btu/sq. ft. 2) Outdoor air ventilation rate of 10% 3) Infiltration rater of 1/2 air change per hour. Daily totals were then summed for the number of days in each month to arrive at monthly heat losses. The length of the heating season for each location considered was determined from weather data and characteristic operating periods. Yearly heat losses were derived by summing monthly totals for the length of the cooling season.

Absorption coefficients and U values were varied and summarized for the 12 locations as shown in Figure #8-58. The data was then plotted and extrapolated to include the entire range of degree days.

Instructions for Use of Nomograph

1. Enter the nomograph on the bottom left-hand vertical line at the degree days as determined on the Climatic Zone Form.

2. Proceed horizontally right to the number of Langleys as determined on the Climatic Zone Form.

3. Proceed vertically upward from this intersection as determined and recorded on the Building Load Survey.

4. Proceed horizontally left at this intersection to the proper scale as determined by the absorption coefficient as indicated on the Building Load Survey Form under type of roof (i.e. dark or light). Read the yearly heat loss in Btu x 10^3 per square foot.

5. This factor is to be entered under Heating for Roof conduction on the Annual Heating and Cooling Tabulation Form.

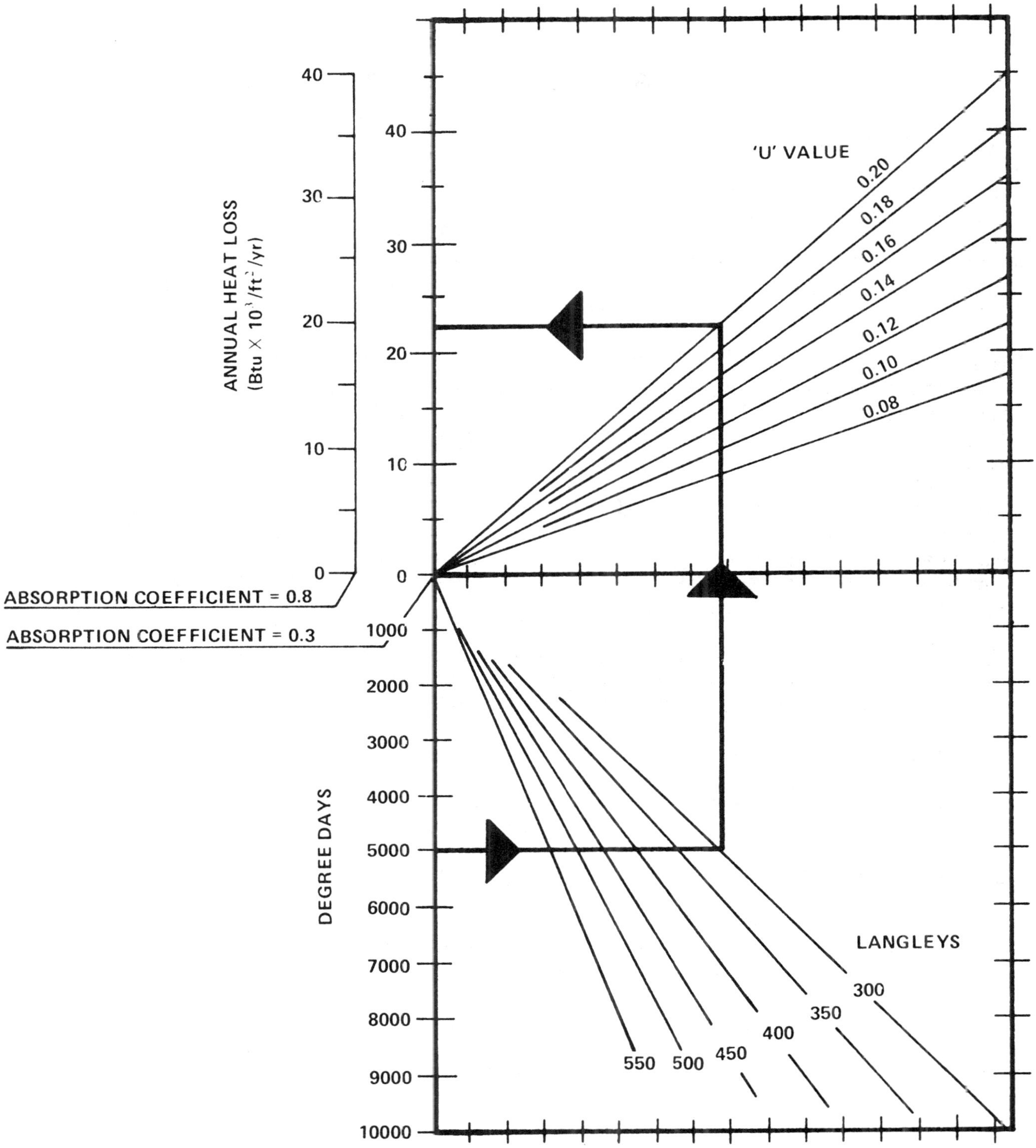

FIGURE 8-28. HEATING-ANNUAL HEAT LOSS THROUGH ROOF

Figure #8-29 Engineering Data

Heat losses determined from this figure are based on the assumption that the floor is over an unheated space which is at outdoor ambient air temperature. Since the load on the heating system during occupied hours is generally a small percentage of the total annual heating load, the figure gives heat loss during unoccupied times only with 10 hours occupied time per week being the maximum (158 hours unoccupied). It was also assumed that the temperature during unoccupied time is set back to 55°F and since drgree days are based on 65°F, all losses were multiplied by a factor of 55/65, thus the formula the figure is based on is : Q (heat loss) = Degree Days x 24 x 'U' Value x 55/65 x $\frac{\text{unoccupied hours}}{168}$

Instructions for Use of Nomograph

1. Enter the nomograph on lower left-hand vertical line at the degree days as determined on the Climatic Zone Form.

2. Proceed horizontally right to the intersection of the U-Value as determined and recorded on Building Load Survey Form.

3. Proceed vertically upward from this intersection to the number of occupied hours per week.

4. Proceed horizontally left from this intersection to the read the yearly loss in Btu x 10^3 per square foot.

5. This factor is to be entered under Heating for Walls on the Annual Heating and Cooling Tabulation Form.

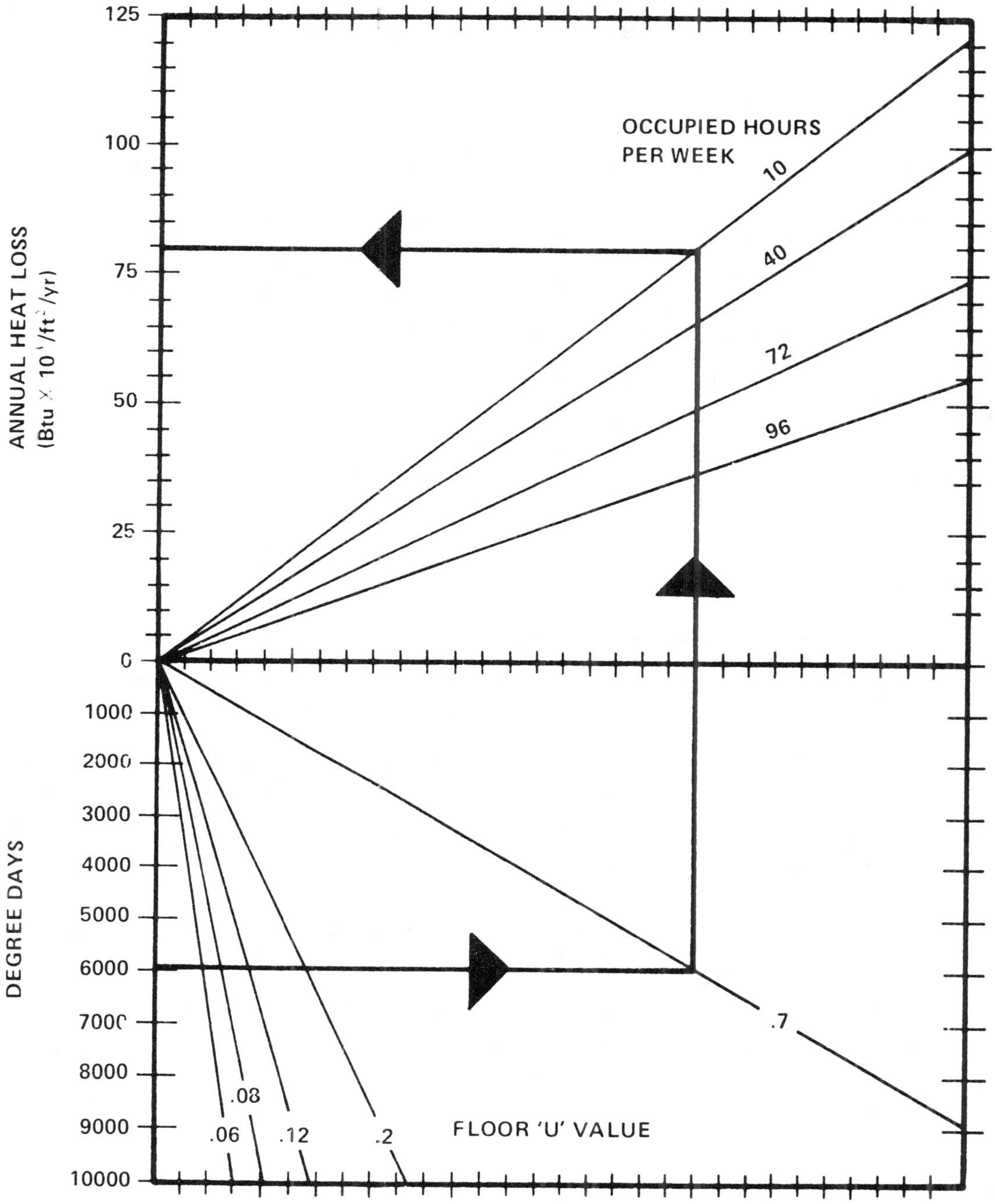

FIGURE 8-29. HEATING-ANNUAL HEAT LOSS THROUGH FLOORS EXPOSED TO OUTDOOR TEMPERATURES

Figure 8-30 Engineering Data

The wet bulb degree hours are based on 12 mos/yr, and 8 hr/day.

Energy used is a function of the wb degree hours above the base of 66° F, the rh maintained the number of hours of controlled humidity. The base re is 50% which is approximately 78°F dbt, 66°F wbt. The figure expresses the energy used per 1000 cfm of air conditioned or dehumidified. An analysis of the total heat content of air in the range under consideration indicates an average total heat variation of 0.93 Btu/lb for each degree wbt change at constant db temperature and that the total heat varies nearly directly with rh. One thousand cfm is equal to 4286 x 0.93 or 3986 Btu. The lower section of the figure shows the direct relationship from the base of 50% rh and the upper section proportions the hours of system operation with 56 hr/wk being 100%.

Instructions for use of Nomograph

1. Enter the nomograph on the lower left hand vertical line at the annual wet bulb degree hours above 66 degrees F wet bulb determined on the Climatic Zone Form.

2. Proceed horizontally right to the intersection of the indoor relative humidity as determined and recorded on the Building Load Survey Form.

3. Proceed horizontally upward from this intersection to the number of hours of operation per week.

4. Proceed horizontally left from this intersection to read the energy used in Btu x 10^6 per year per 1,000 cfm.

5. This is the factor to be entered under Cooling for Outside Air on the Annual Heating and Cooling Tabulation Form.

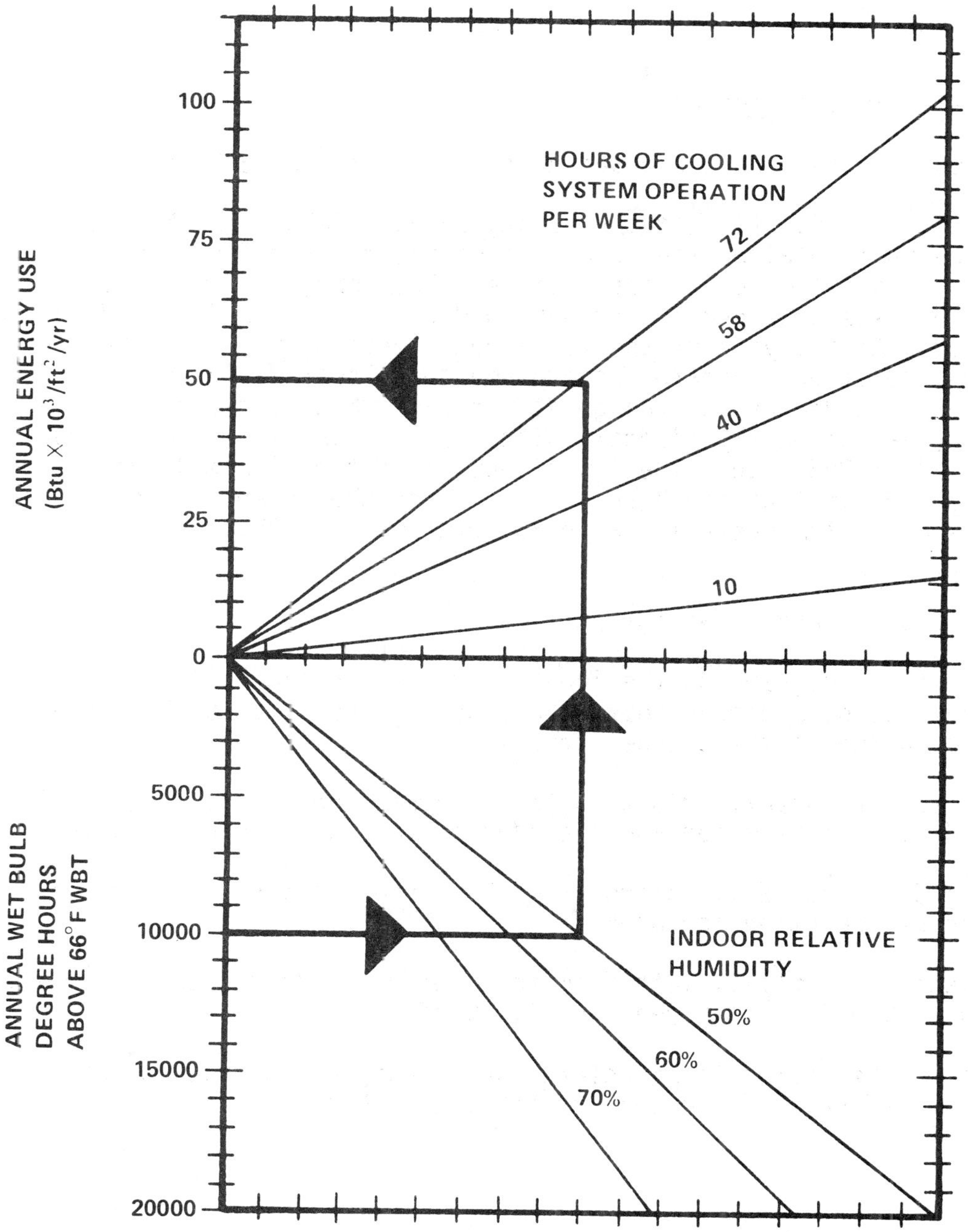

FIGURE 8-30. COOLING-ANNUAL ENERGY USED PER 1000 cfm TO MAINTAIN VARIOUS HUMIDITY CONDITIONS

Figure #8-31 Engineering Data

Energy used is a function of the number of degree days, indoor temperature and the number of hours that temperature is maintained and is expressed as the energy used per 1000 cfm of air conditioned.

The energy used per year was determined as follows:

Btu/yr. = (1000 cfm)x(degree days/yr.)x(24 hr./day)x(1.08) *

Since degree days are base $65^{O}F$, the other temperatures in the lower section of the figure are directly proportional to the $65^{O}F$ line. The upper section proportions the hours of system operation with 168 hr./week being 100%.

*1.08 is a factor which incorporates specific heat, specific volume, and time.

Instructions for Use of Nomograph

1. Enter the nomograph on the lower left hand vertical line at the degree days determined on the Climatic Zone Form.

2. Proceed horizontally right to the indoor temperature line.

3. Proceed vertically upward from the intersection of this line to the number of hours of occupancy per week.

4. Proceed horizontally left from the intersection of this line to determine the energy used in Btu x 10^6 per year per 1,000 cfm.

5. This is the factor to be entered under Heating for Outside Air on the Annual Heating and Cooling Tabulation Form.

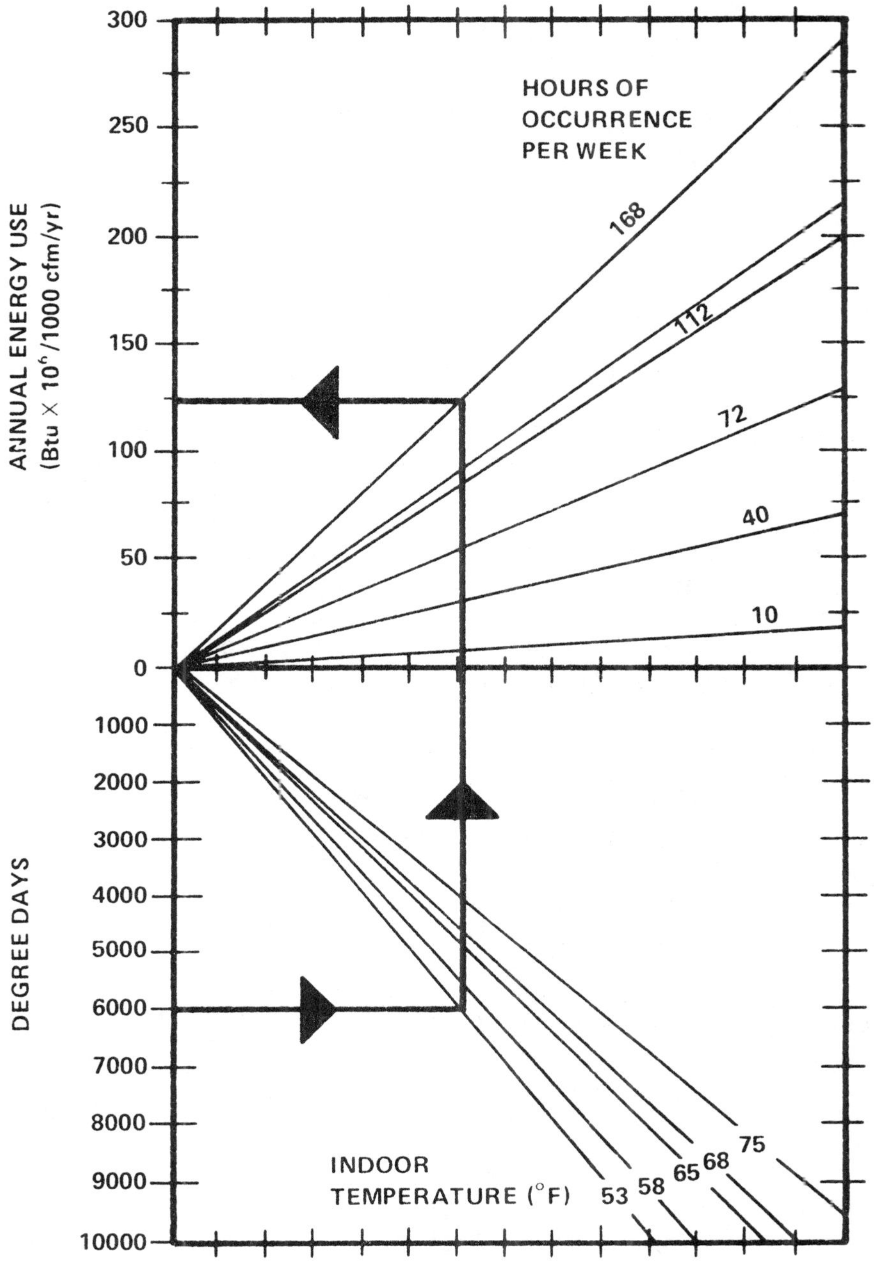

FIGURE 8-31. HEATING-ANNUAL ENERGY USED PER 1000 cfm OUTDOOR AIR

Figure #8-32 Engineering Data

The wet bulb degree hours are based on 24 hours/day, October through April.

Base indoor condition for figure is dbt = 68°F, wbt = 54°F. rh = 40%.

Energy used is a function of the wet bulb degree hours below the base conditions, the RH maintained and the number of hours of controlled humidity. The figure expresses the energy used per 1000 cfm of air conditioned or humidified.

An analysis of the total heat content of air in the range under consideration indicates an average total heat variation of 0.522 Btu/lb for each degree wb change. Utilizing the specific heat of air, this can be further broken down to 0.24 Btu/lb sensible heat and 0.282 Btu/lb latent heat. One thousand cfm is equal to 4286 x 0.282 or 1208 Btu. Further investigation of the relationship between wb temperature, db temperature, and total heat shows that latent heat varies directly with rh at constant db temperature. The lower section of the figure shows this proportional relationship around the base of 40% rh. The upper section proportions the hours of system operation with 168 hr/wk being 100%.

Instructions for Use of Nomograph

1. Enter the nomograph on the lower left hand vertical line at the annual wet bulb degree hours below 54 degrees F wet bulb and 68 degrees F dry bulb as determined on the Climatic Zone Form.

2. Proceed horizontally right to the intersection of the indoor relative humidity as determined and recorded on the Building Load Survey Form.

3. Proceed vertically upward at this intersection to the number of hours of controlled humidity per week.

4. Proceed horizontally left from this intersection to read the energy used in Btu x 10^6 per year per 1,000 cfm.

5. This factor is entered under Heating for Humidification on the Annual Heating and Cooling Tabulation Form.

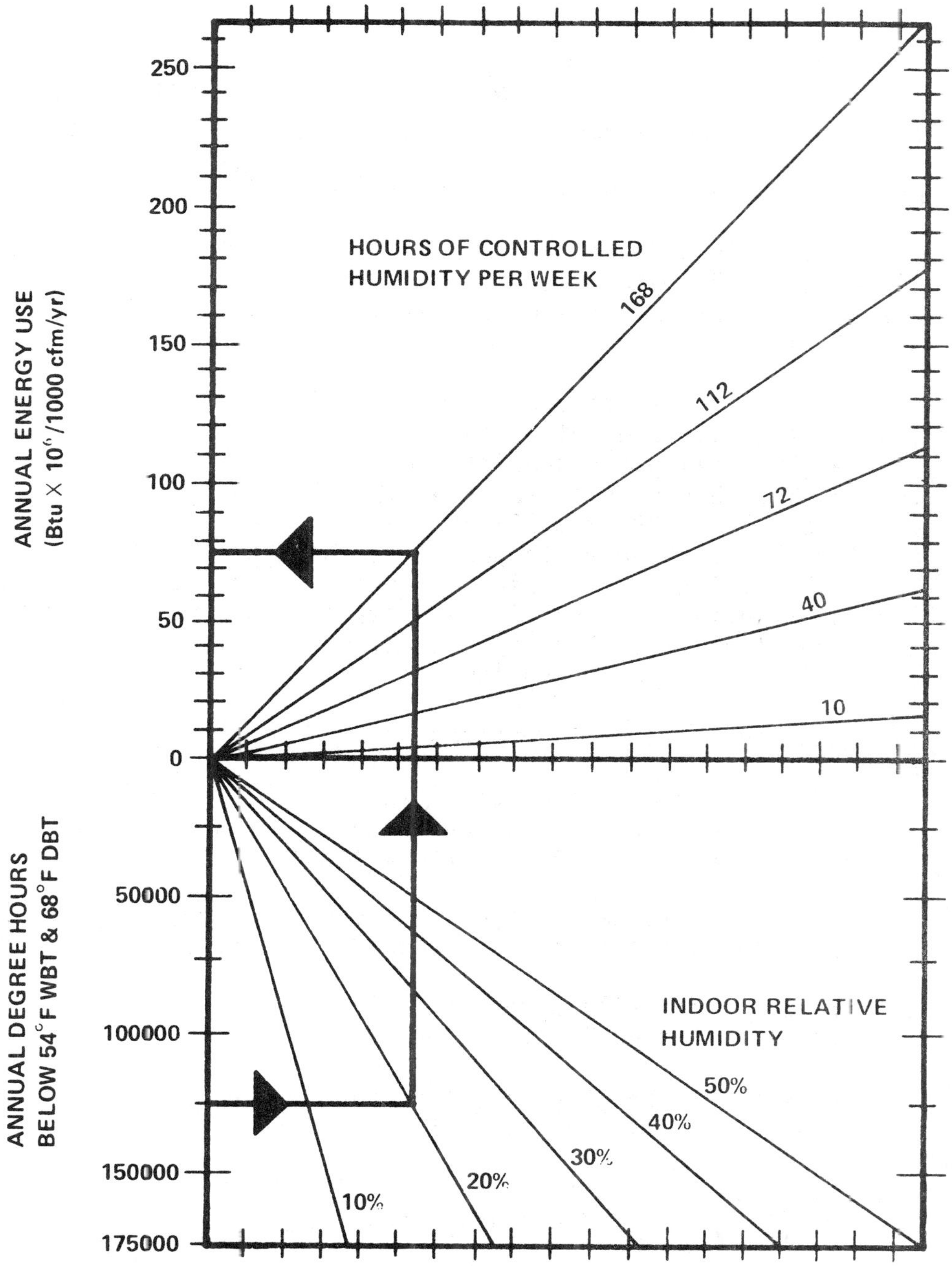

FIGURE 8-32. HEATING-ANNUAL ENERGY USED PER 1000 cfm TO MAINTAIN VARIOUS HUMIDITY CONDITIONS

Figure #8-33 Engineering Data

References: AFM 88-8
U. S. Government Printing Office
15 June 1967

Climatic Atlas of the United States
U. S. Department of Commerce
June 1968

Five locations with heating degree day totals ranging from 1400-8400 were analyzed regarding time temperature distribution below 68°F DB. The percentage temperature distribution for each of the 5°F ranges which includes the setback from 68°F shown in the lower half of the figure related to the total number of degree hours below 68°F was determined for 24 hr/day, 365 days/yr. This percentage was then plotted against annual degree days and is shown as the lower half of the Figure.

The upper half of the Figure represents the range of heating energy consumed/sq ft for various buildings over the range of 1000 to 10,000 degree days. The extreme right hand line in the upper half represents 90% of the present consumption when projected vertically. Analysis of energy usage for heating by various buildings in several locations showed that it can be safely assumed that approximately 10% of the total heating energy consumption is during occupied hours. Therefore, savings by night setback are applicable to only 90% of the total heating energy consumption. The remainder of the upper half of the Figure simply proportions the energy saved based on the point of entry from the lower section.

Instructions for use of nomograph

1. Enter the nomograph on the lower left hand vertical line at the "degree days" from Climatic Zone Form.

2. Proceed horizontally right to the number of degrees of night setback.

3. Proceed vertically upward at the intersection of this line to the horizontal "300" line and then proceed parallel to the closest plotted line.

4. Enter the top left hand vertical line at the present heating energy consumption ("Annual Heating and Cooling Load Form") divided by the square footage of the building.

5. Proceed horizontally right to intersect the line mentioned in 3 above.

6. Proceed upward vertically at this intersection and read the savings in Btu x 10 cubed per square feet per year in the same selected order of magnitude used to enter at step 4 above.

7. Multiply this by the square footage of the building to determine the total savings per year.

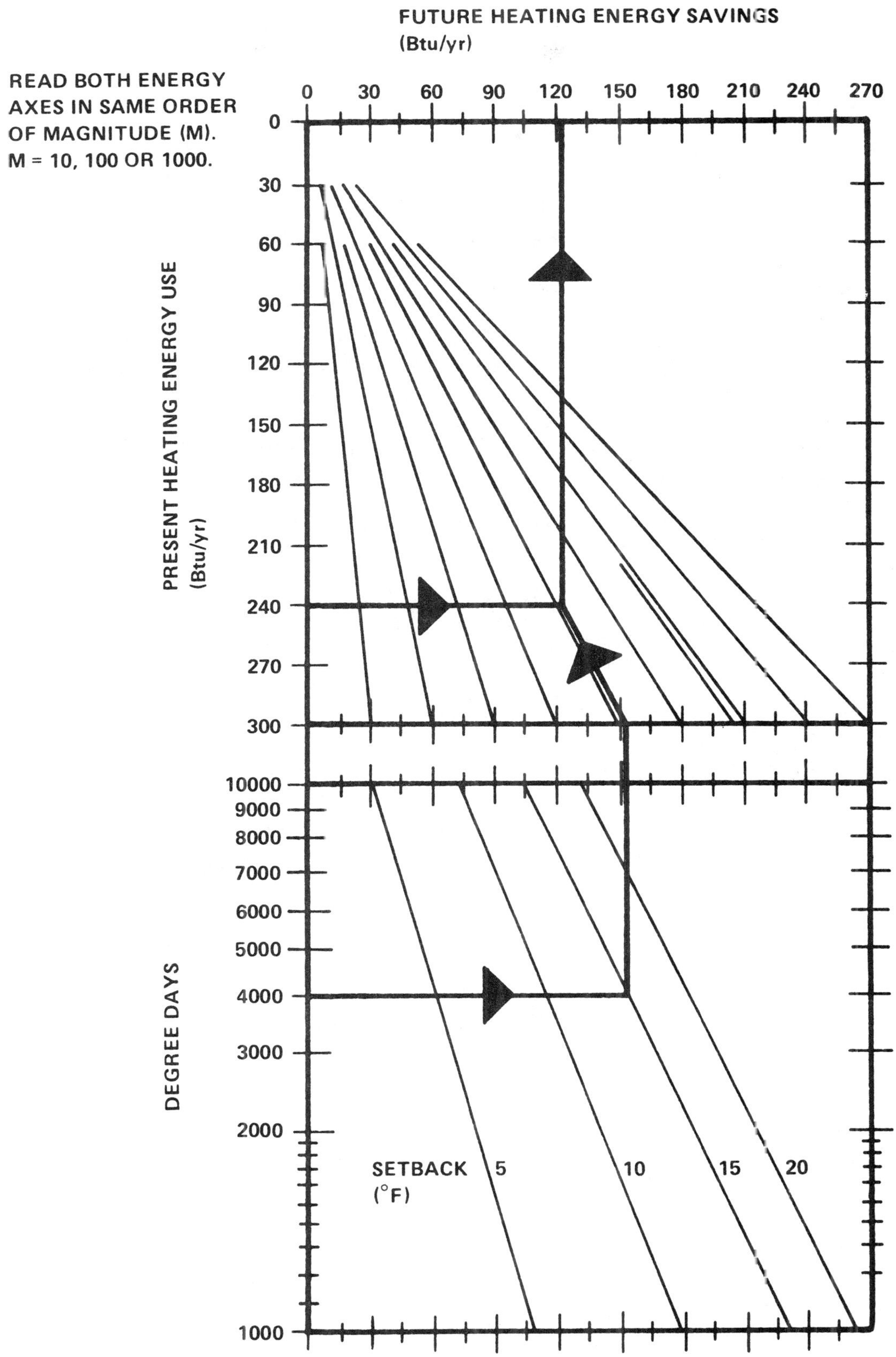

FIGURE 8-33. HEATING-ENERGY SAVED BY NIGHT SETBACK

Figure #8-34 Engineering Data

Sources: Data from various boiler and burner manufacturers was analyzed and integrated to make up one graph.

Instructions for use of nomograph

1. Enter the nomograph at the lower horizontal line at the percentage of CO_2 in the flue gas for the fuel being used.

2. Enter the lower left hand vertical part of the nomograph at the percentage O_2 in the flue gas and proceed horizontally right to the intersection of the plotted curved line.

3. Proceed vertically upward at this intersection to the stack temperature line.

4. Proceed horizontally left at this intersection and read the boiler efficiency corresponding to the fuel used.

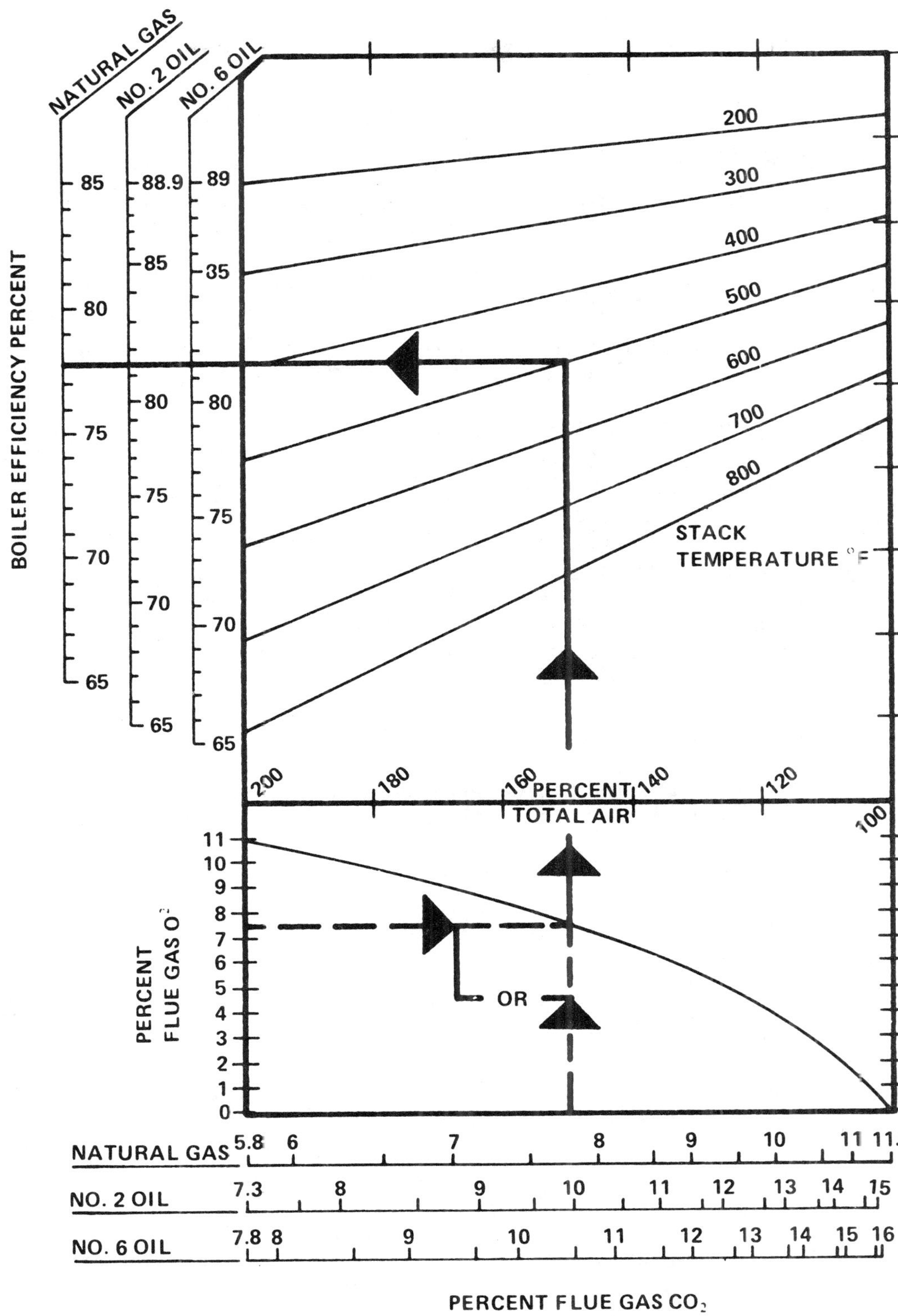

FIGURE 8-34. HEATING-EFFECT OF FLUE GAS COMPOSITION AND TEMPERATURE ON BOILER EFFICIENCY

FIGURE 8-35 ENGINEERING DATA

This graph is derived from calculations based on manufacturers data and should be used only to obtain order of magnitude savings that may be available. It is based upon the following assumptions:

$$\text{Footcandles} = \frac{\text{No. Lamps x Lamp Lumens x No. Fixtures x CU x MF}}{\text{Area}}$$

$$= \text{No. Lamps x} \left[\frac{\text{Lamp Lumens x No. Fixtures x CU x MF}}{\text{Area}}\right]$$

where

CU = Coefficient of Utilization
MF - Maintenance Factor

Since the coefficient of utilization varies by only about 10%, for the purposes of this approximation it can be neglected. Then, by varying only the number of lamps,

Footcandles = No. Lamps x Constant (for any specific area)

Therefore, footcandles vary directly with the number of lamps.

Since wattage consumption depends on the number of lamps and ballasts used, wattage consumption varies directly with the number of lamps.

Instructions for Use of Nomograph

1. Enter the chart on the vertical left hand line at the initial foot candles.

2. Proceed horizontally right to the curve representing the planned modification.

3. Proceed vertically upward from this intersection to read the result in foot candles.

4. Proceed vertically downward from the intersection in item 2 above to read the resultant watts per square foot and Btu per square foot.

5. Proceed horizontally across the chart from the initial foot candles and read the initial watts per square foot and Btu per square foot.

6. This chart may be read starting with the vertical right hand line if initial watts per square foot is known.

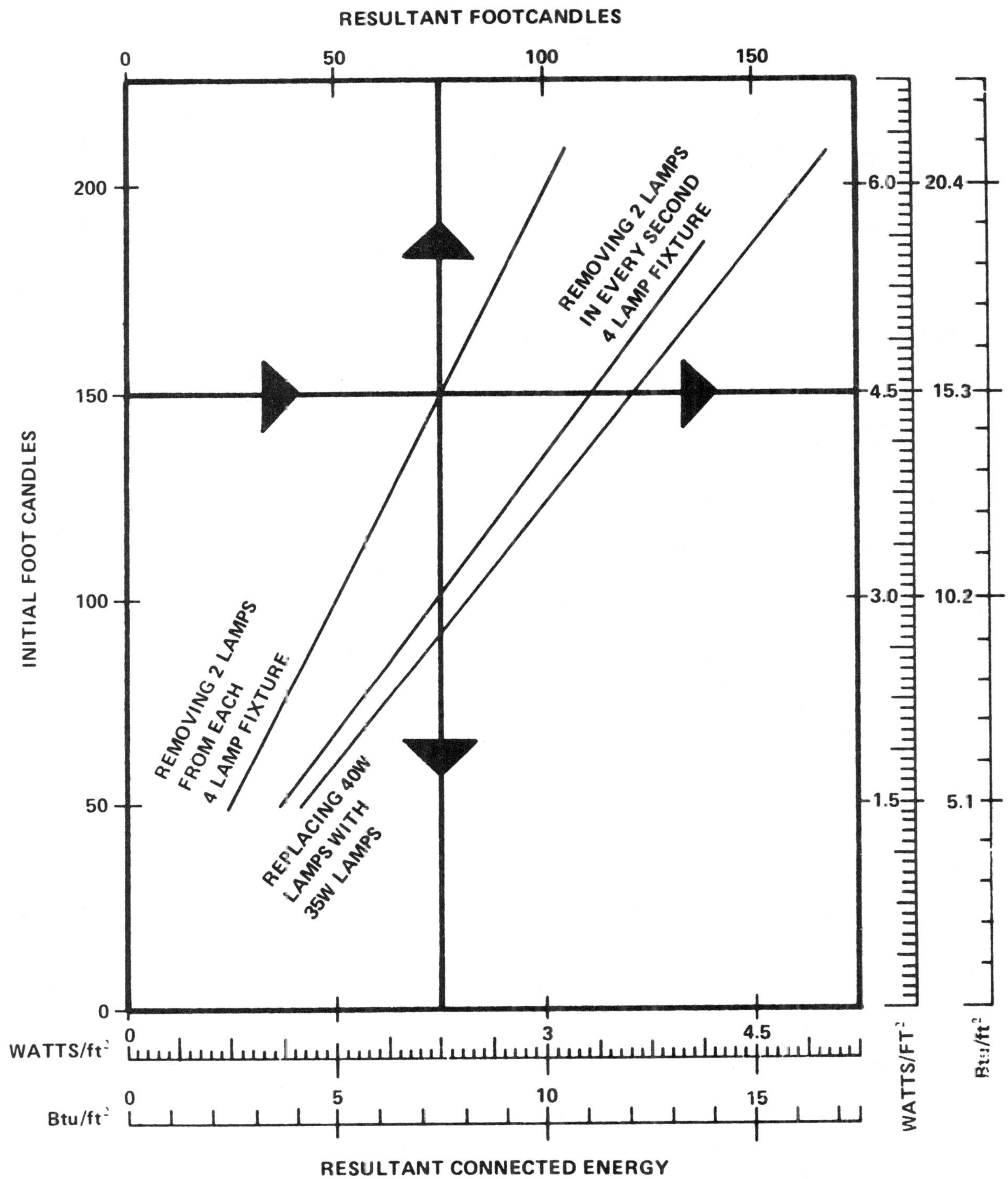

FIGURE 8-35. LIGHTING-RESULTANT FOOTCANDLES AND WATTS/SQ FT FOR VARIOUS LAMP CHANGES

Figure 8-36 Engineering Data

Reference: I.E.S. Lighting Handbook, 4th edition 1968, page 9-17, figure 9-7 Illuminating Engineering Society 345 East 47th Street, New York, N.Y.

The data is computed from actual measurements of lumen output from lamps and fixtures exposed to various environments for different lengths of time.

Instructions for use of nomograph

1. Determine whether to use the top or the bottom of the chart depending on the type of fixture involved.

2. Enter the chart on the lower horizontal line at the number of months between cleaning.

3. Proceed vertically upward to intersect the curve which most accurately suits the dirt conditions involved in space.

4. Proceed horizontally left from this intersection and read the percent of lumen output.

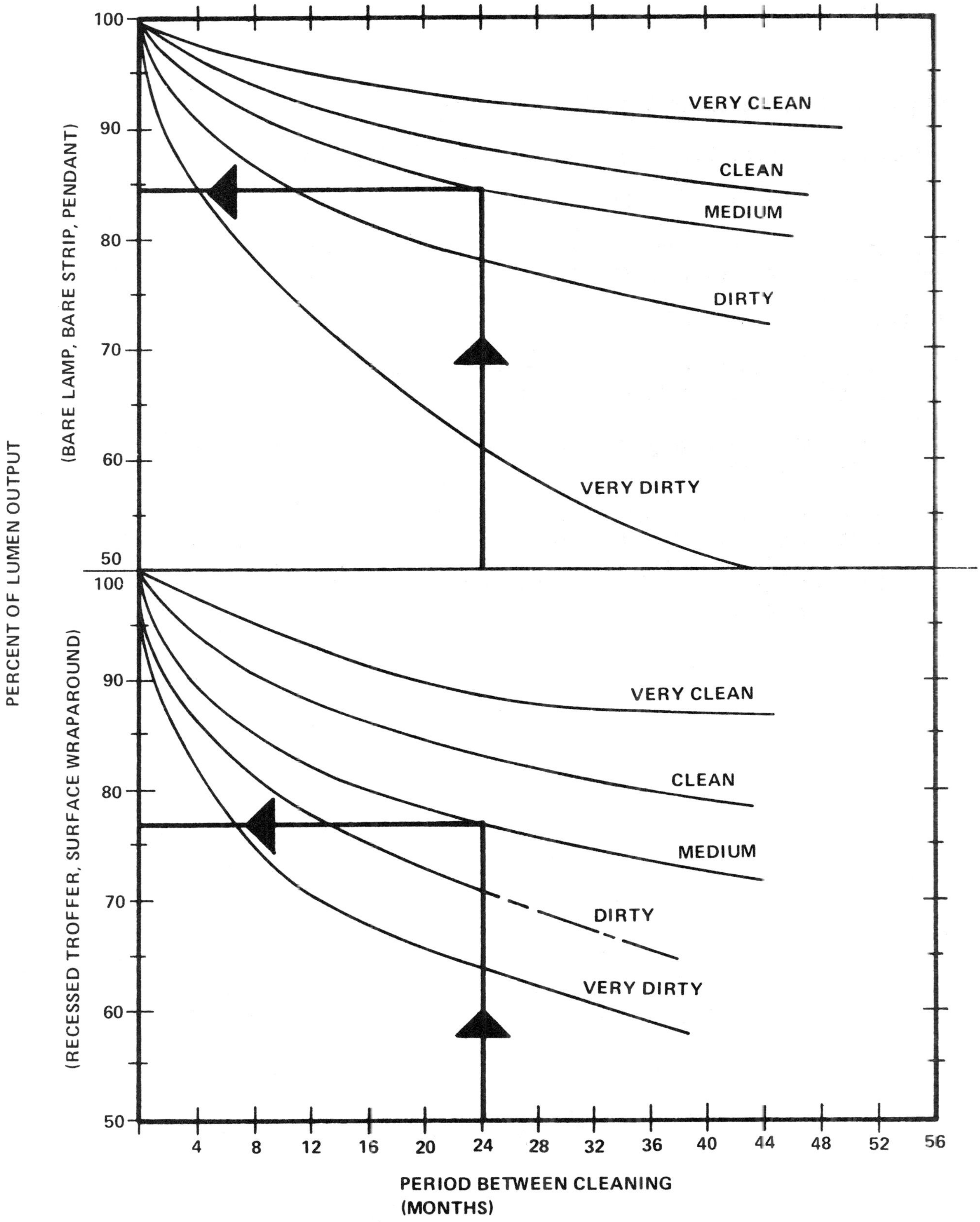

FIGURE 8-36. LIGHTING-FIXTURE CLEANING CYCLE

Figure #8-37 Engineering Data

Graph based on manufacture's information of lamp performance, typical for the types listed.

Instructions for use of Nomograph

1. Enter the chart on lower horizontal line at the number of hours of operation.

2. Proceed vertically upward to the intersection of curve for the type of lamps in use.

3. Proceed horizontally left from this intersection to read the percentage of relative light output.

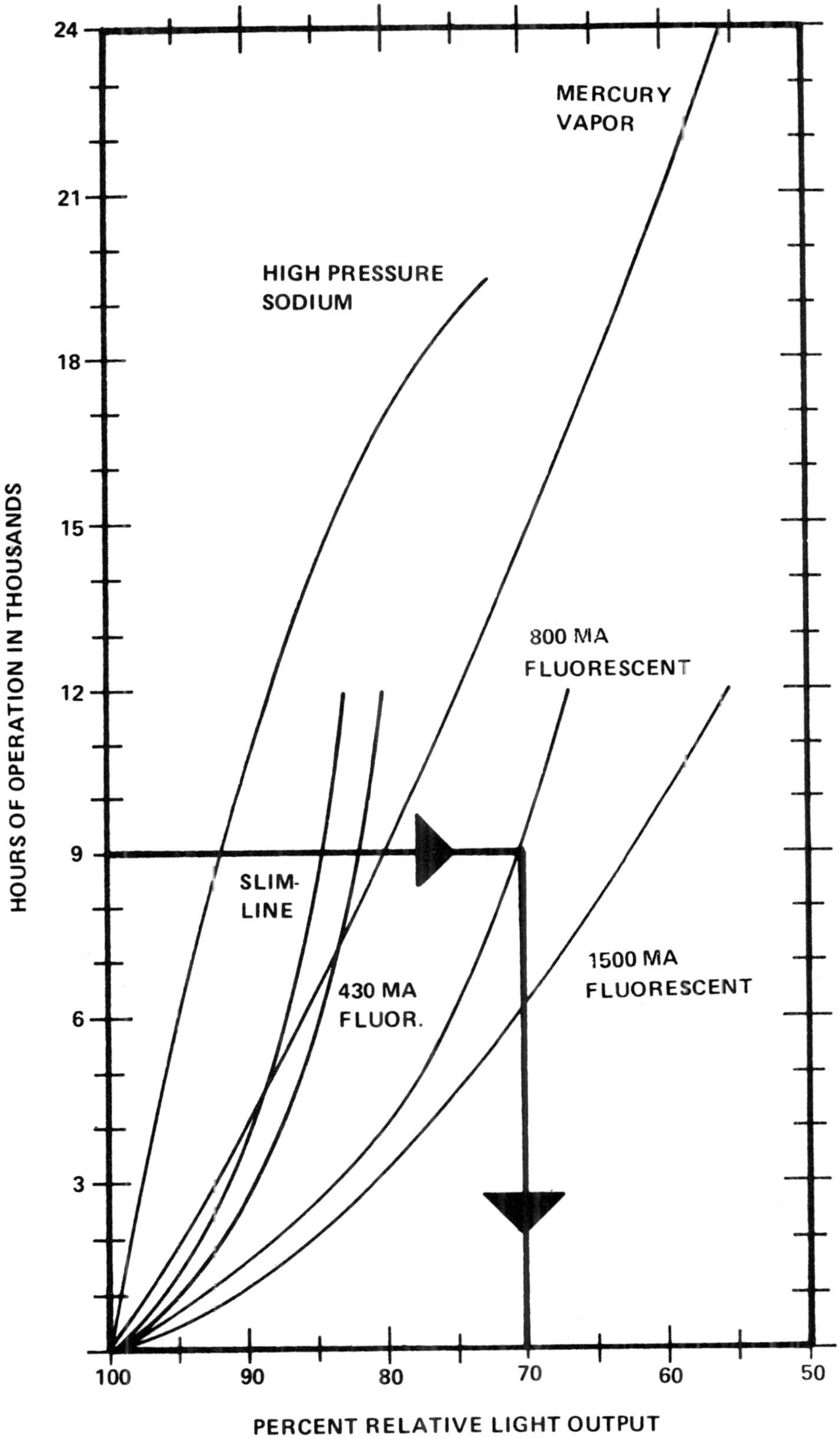

FIGURE 8-37. LIGHTING-LAMP LUMEN DEPRECIATION OF STANDARD LAMPS

Figure #8-38 Engineering Data

This figure is based upon the following assumptions:

$$\text{Footcandles} = \frac{\text{No . Lamps X Lamp Lumens X No. Fixtures X CU X MF}}{\text{Area}}$$

$$= \text{MF X} \left[\frac{\text{No Lamps X Lamp Lumens X No Fixtures X CU}}{\text{Area}} \right]$$

Where

CU = Coefficient of Utilization
MF = Maintenance Factor

By varying <u>only</u> the maintenance factor (MF)
Footcandles = MF X Constant (for any specific area)

Therefore, footcandles vary directly with the maintenance factor <u>where other factors are maintained constant</u>.

Instructions for use of Nomograph:

1. Enter the chart on the lower horizontal line at the appropriate maintenance factor.*

2. Proceed vertically upward to the intersection of the plotted curve.

3. Proceed horizontally left at this intersection and read the percentage decrease in foot candles.

*The maintenance factor is determined by multiplying the results of Figure 8-36 by Figure 8-37.

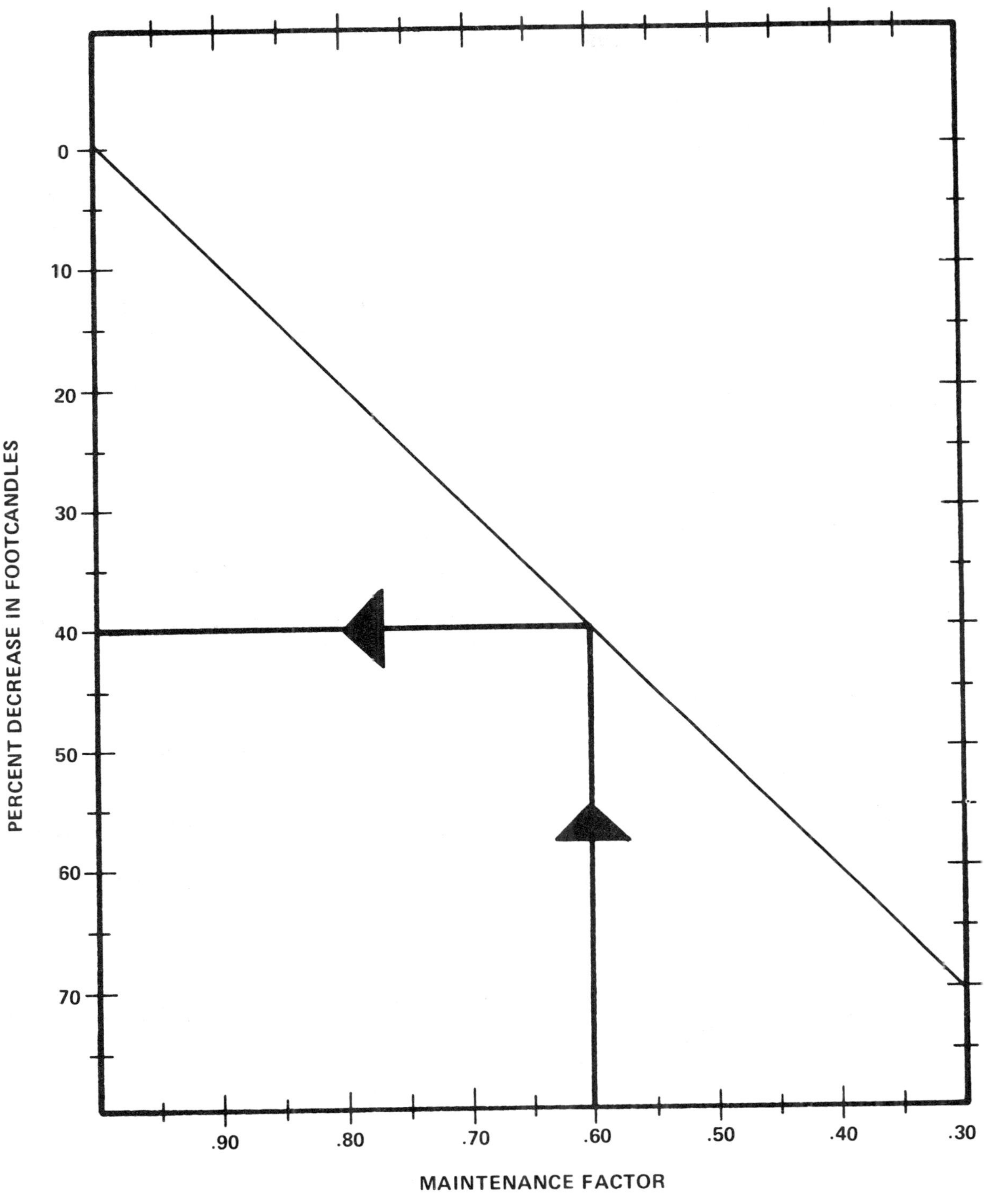

FIGURE 8-38. LIGHTING-EXAMPLE OF MAINTENANCE FACTOR EFFECT ON FOOTCANDLES

Figure #8-39 Engineering Data

Graph based on manufacturer's information of statistics of flourescent lamp life.

Instructions for using Nomograph

1. Enter the chart on the lower horizontal line at the percent of rated life.

2. Proceed vertically upward to the intersection of the curve.

3. Proceed horizontally left at this intersection and read the percent of initial lamp survival.

4. This chart should be used in reverse, starting with the percent of initial lamp survival, to determine the percentage of rated life.

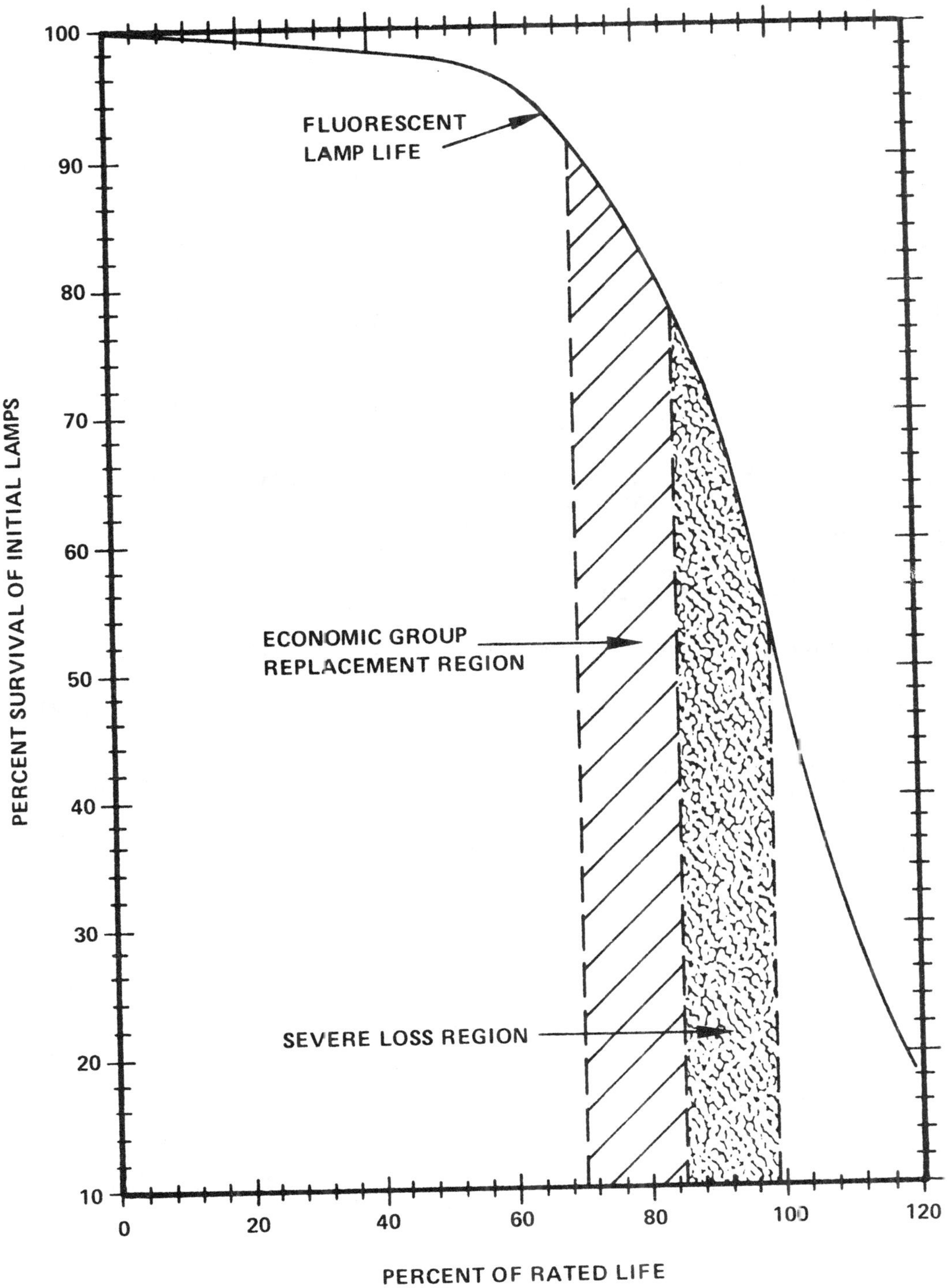

FIGURE 8-39. LIGHTING-FLUORESCENT GROUP RELAMPING

Figure #8-40 Engineering Data

This graph is developed from manufacturer's information and is typical for lamp generic types.

Based on the following lamps

High pressure sodium - 100, 150, 250, 400 & 1000 watt size.
Metal halide - 400, 1000 & 1500 watt size, standard line lamps.
Fluorescent
430 ma - 4-90 watt size, cool white color
800 ma - 20-110 watt size, cool white
1500 ma - 110-220 watt size, cool white
Mercury vapor - 40-1500 watt sizes, dulex white color
Tungsten halogen - 43-1250 watt size
Incandescent - 3-1500 watt size, inside frosted.

Instructions for use of Nomograph

1. Enter the chart on the lower horizontal line at the lamp wattage.

2. Proceed vertically upward to intersect curve corresponding to the type of lamp

3. Proceed horizontally left at this intersection to read the efficacy in lumens per watt.

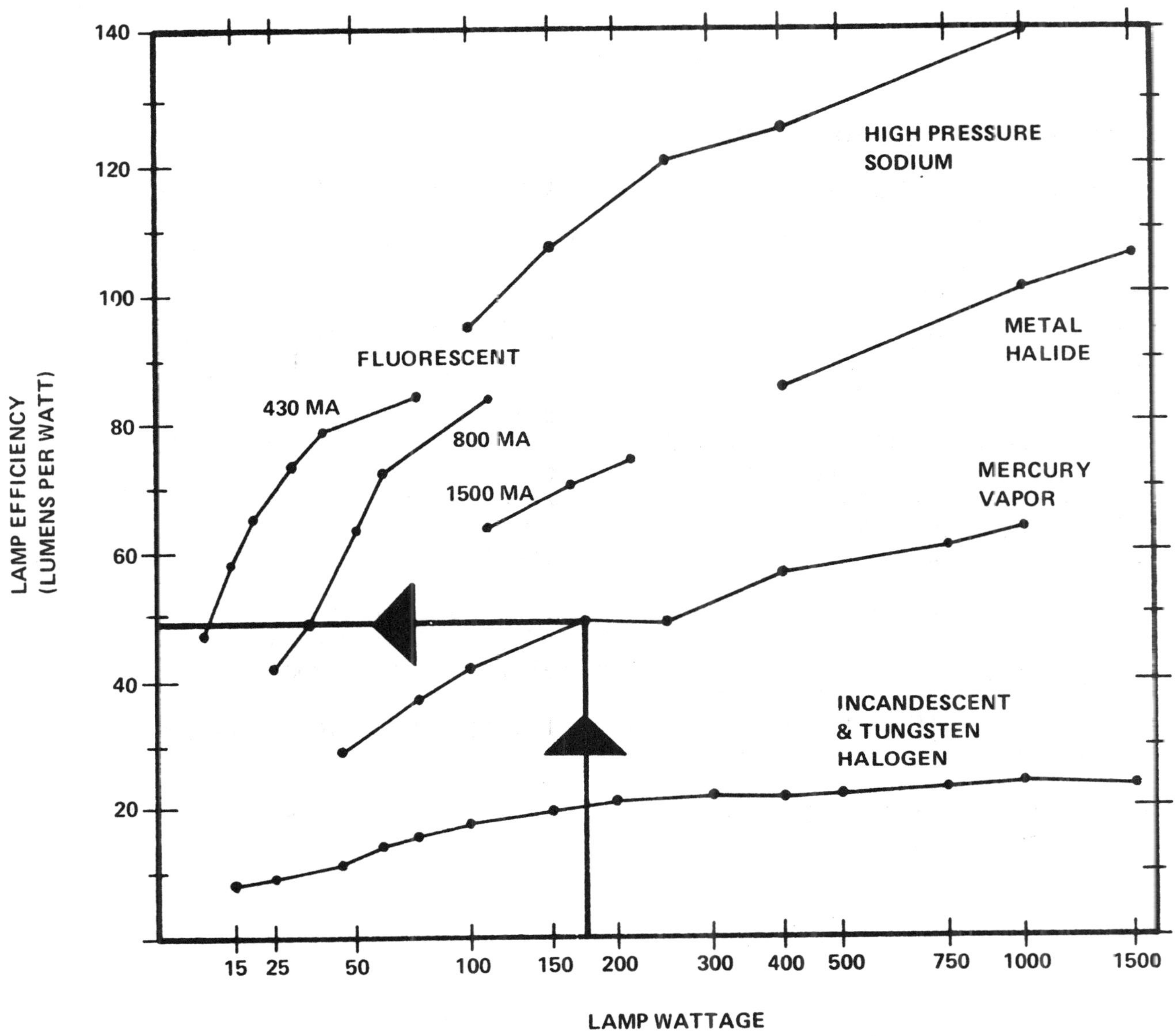

FIGURE 8-40. LIGHTING EFFICIENCY OF DIFFERENT LAMP TYPE AND WATTAGE

Figure #8-41 Engineering Data

This figure is based on the addition of wall insulation with an R Value of 4.0. The calculations to determine the U value after the addition of insulation assumed a combined inside/outside film coefficient of R=1. The standard formula U=1/R was used

Instructions for use of Nomograph

1. Enter the chart on the left hand vertical line at the present U-value.

2. Proceed horizontally right to the number of inches of insulation to be added.

3. Proceed vertically downward at this intersection to read the U-value after adding insulation.

4. This chart may also be used to determine the amount of insulation required by (1) drawing a line from the present U-value horizontally right; (2) drawing a line from the desired U-value at the top horizontal line vertically downward: (3) the intersection of these two lines will indicate the amount of insulation required to obtain the desired U-Value.

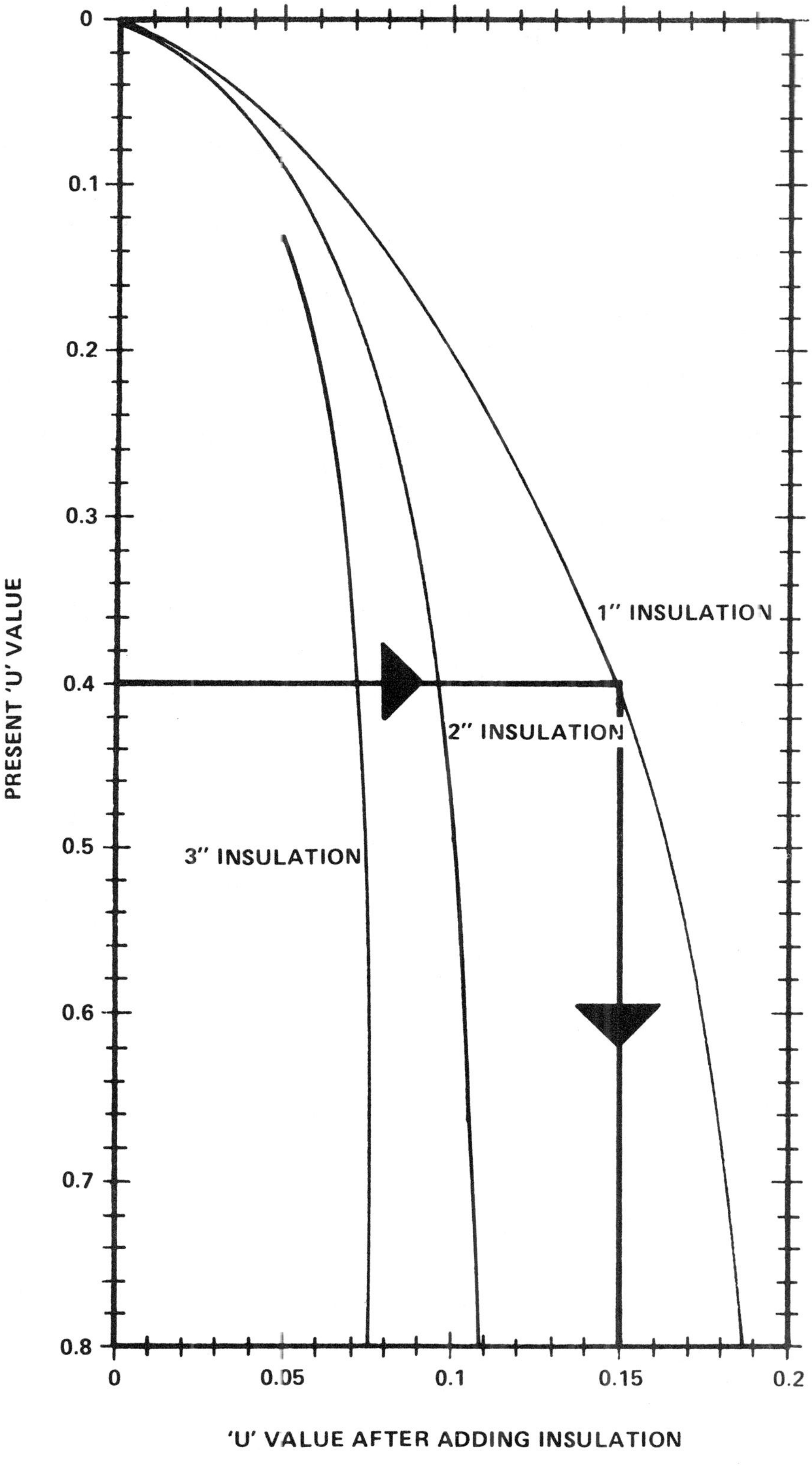

FIGURE 8-41. HEATING-EFFECT OF INSULATION ON "U" VALUE

Figure #8-42 Engineering Data

Source: Manufacturer's fan capacity tables for forward curve centrifugal fans.

Instructions for use of Nomograph

1. Enter the lower left hand vertical line at the fan volume.
2. Proceed horizontally right to the total static pressure line.
3. Proceed vertically upward at the intersection of this line to the number of hours of operation per year line.
4. Proceed horizontally left at this intersection and read the energy consumed in Btu x 10^6 per year.
5. Use Figure VII-42 for forward curve blades and Figure VII-43 for backward curved blades.

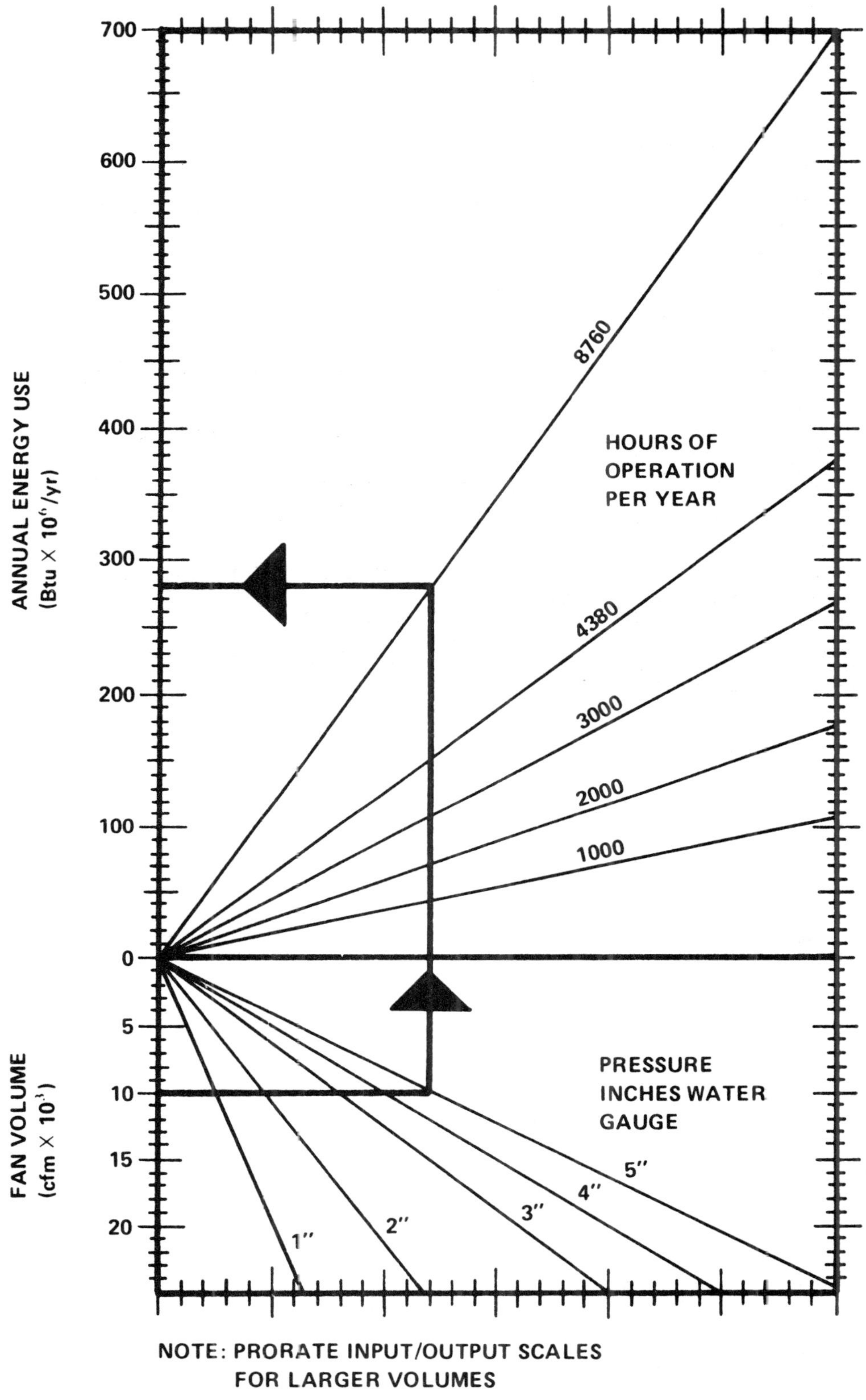

FIGURE 8-42. VENTILATION-ANNUAL ENERGY CONSUMED-CENTRIFUGAL FANS-FORWARD CURVE BLADES

Figure #8-43 Engineering Data

Source: Manufacturer's fan capacity tables for backward curve centrifugal fans.

Instructions for use of Nomograph

1. Use figure 8-42 for forward curve blades and figure 8-43 for backward curved blades.

2. Enter the lower left hand vertical line at the fan volume in cfm x 10 cubed.

3. Proceed horizontally right to the total static pressure line.

4. Proceed vertically upward at the intersection of this line to the number of hours of operation per year line.

5. Proceed horizontally left at this intersection and read the energy consumed in Btu x 10^6 per year.

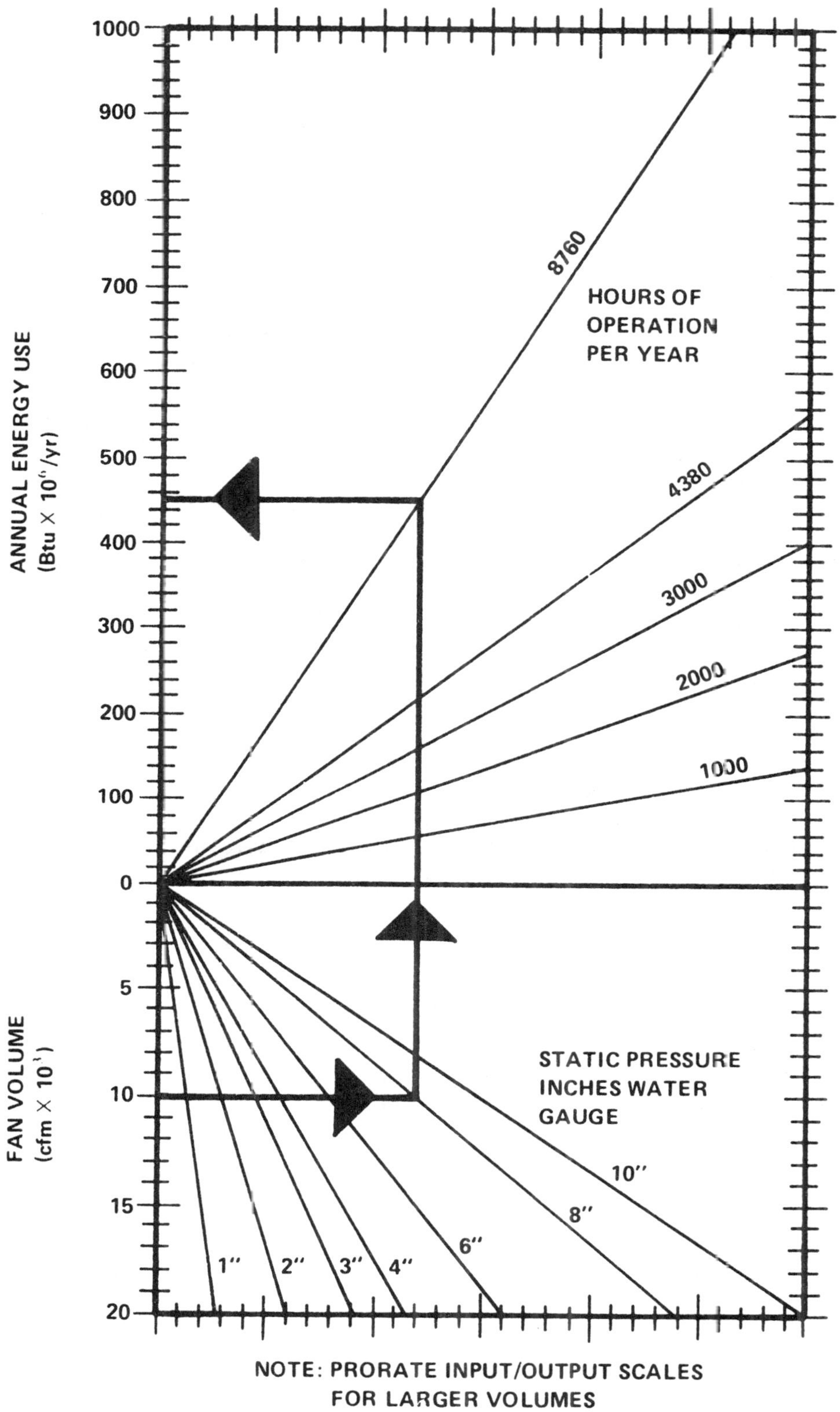

FIGURE 8-43. VENTILATION-ANNUAL ENERGY CONSUMED-CENTRIFUGAL FANS-BACKWARD CURVE BLADES

Figure #8-44 Engineering Data

This figure is based on the Standard Pump Formula:

$$\text{Brake Horsepower} = \frac{\text{lbs/min} \times \text{Feet of head}}{33{,}000 \text{ ft/lbs} \times \text{Pump efficiency}}$$

Pump efficiency was assumed ot be an average of 70%. Brake horsepower was converted to Btu's by the factor 2544.43 Btu/Horsepower Hour.

The upper half of the graph proportions the hours of pump operation per year from 100% (8760 hr.), down to 11% (1,000 hr).

Instructions for use of Nomograph

1. Use Figure 8-44 for gpm up to 500 and Figure 8-45 for gpm up to 2000.

2. Enter the nomograph on the lower left vertical line at the pump capacity in gpm.

3. Proceed horizontally right to the pumping head in feet.

4. Proceed vertically upward at this intersection to the number of operating hours per day.

5. Proceed horizontally left at this intersection to read the energy consumed in Btu x 10^6 per year.

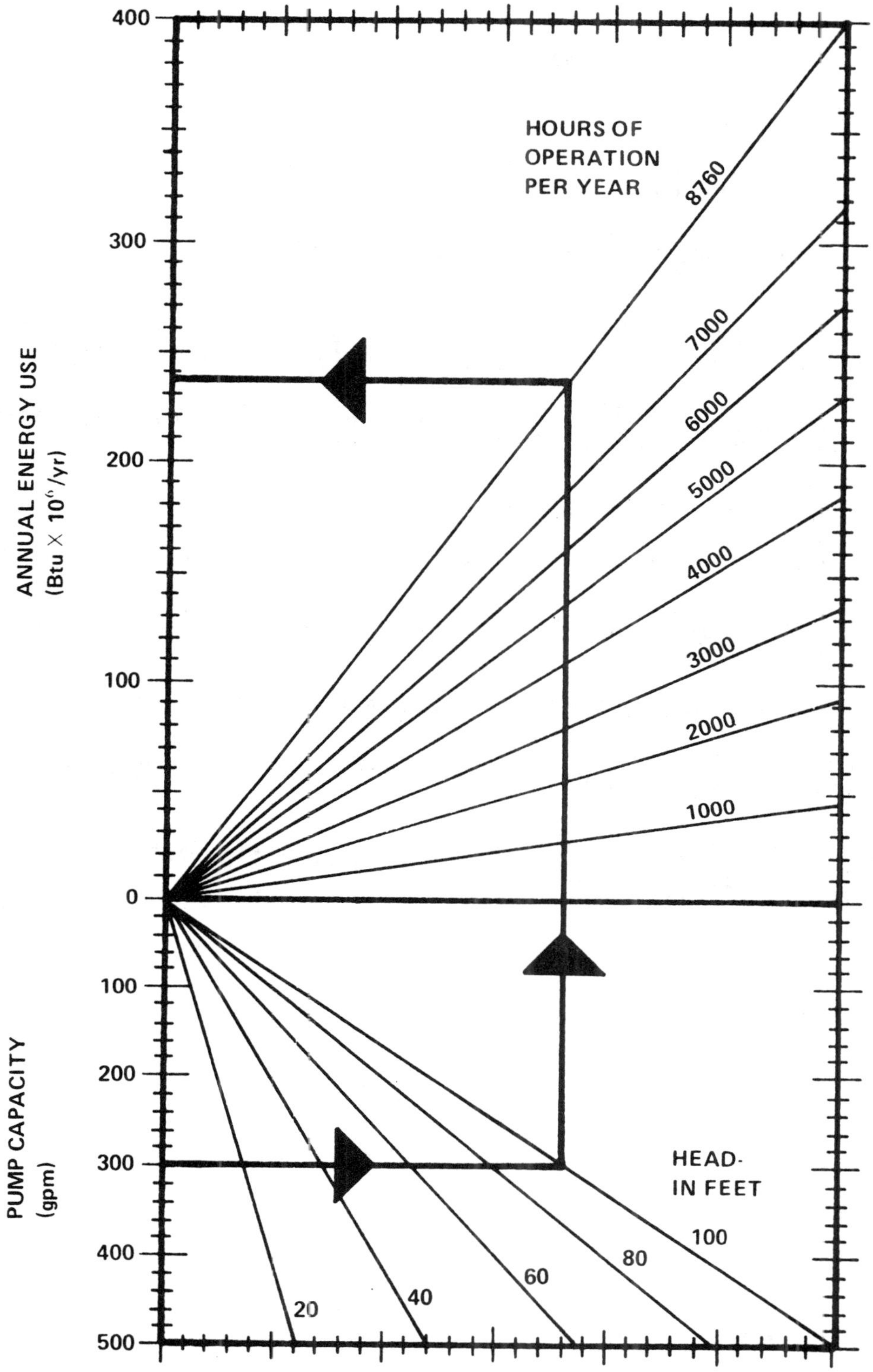

FIGURE 8-44. PUMPS-ANNUAL ENERGY CONSUMED-CENTRIFUGAL PUMPS UP TO 500 gallons per minute

Figure #8-45 Engineering Data

This figure is based on the Standard Pump Formula:

$$\text{Brake Horsepower} = \frac{\text{lbs/min} \quad \text{x Feet of head}}{\text{33,000 ft/lbs x Pump efficiency}}$$

Pump efficiency was assumed ot be an average of 70%. Brake horsepower was converted to Btu's by the factor 3544.43 Btu/Horsepower Hour.

The upper half of the graph proportions the hours of pump operation per year from 100% (8760 hr.), down to 11% (1,000 hr).

Instructions for use of Nomograph

1. Use Figure 8-44 for gpm up to 500 and Figure 8-45 for gpm up to 2000.

2. Enter the nomograph on the lower left vertical line at the pump capacity in gpm.

3. Proceed horizontally right to the pumping head in feet.

4. Proceed vertically upward at this intersection to the number of operating hours per day.

5. Proceed horizontally left at this intersection to read the energy consumed in Btu x 10^6 per year.

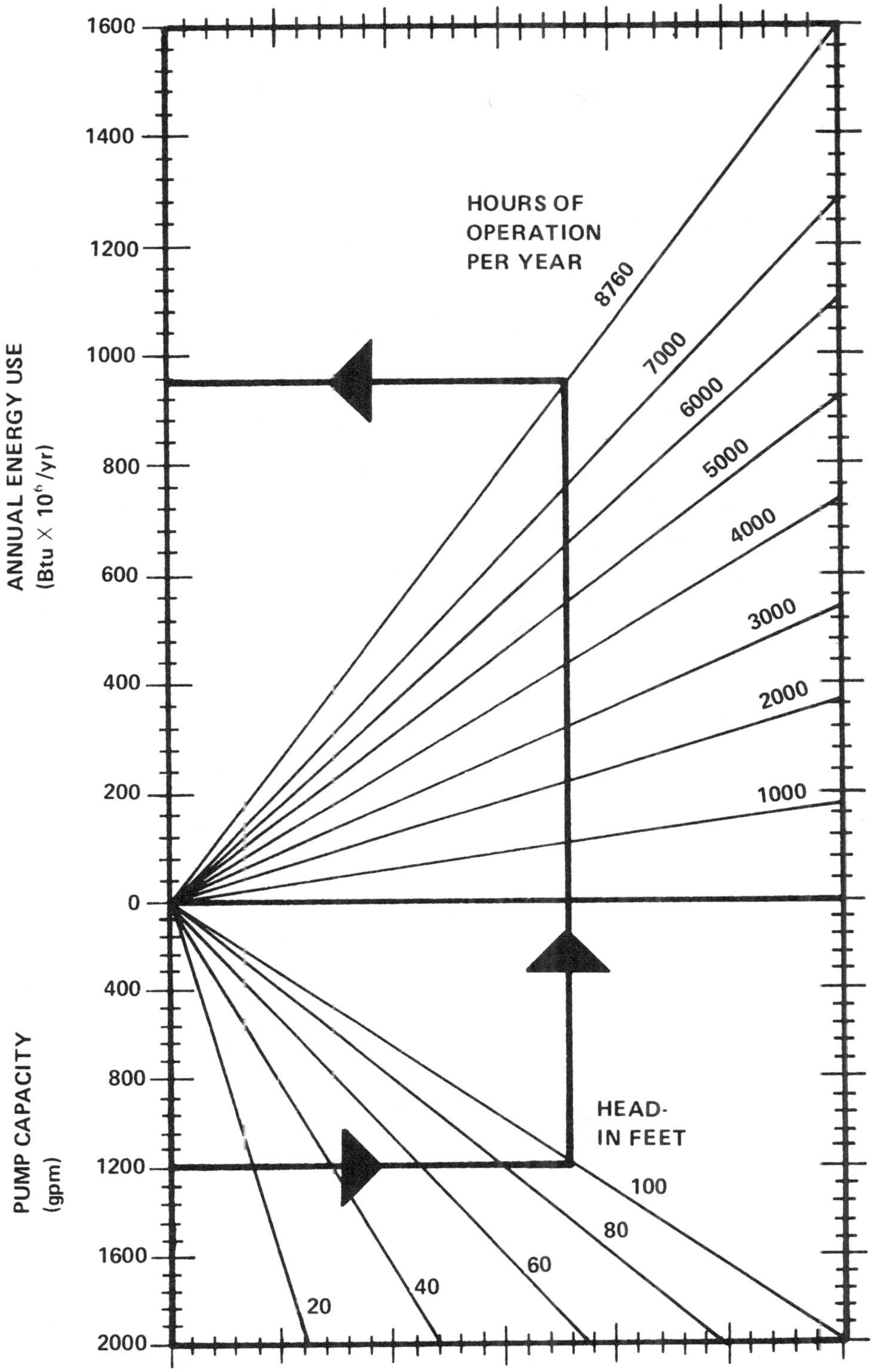

FIGURE 8-45. PUMPS-ANNUAL ENERGY CONSUMED-CENTRIFUGAL PUMPS UP TO 2000 gallons per minute

Figure #8-46 Engineering Data

Source: ASHRAE Handbook of Fundamentals
Assumes Rigid Insulation with K value of 0.27 at 75°F.

Instructions for use of Nomograph

1. Enter the chart on the lower horizontal line at the existing thickness of duct insulation.

2. Proceed vertically upward to the intersection of the temperature difference between the air inside the duct and air outside the duct.

3. Proceed horizontally left at this intersection to read the heat loss or gain in Btu per hour per square feet.

4. Multiply this by the number of square feet of duct and the operating hours per year to determine the annual heat loss or gain.

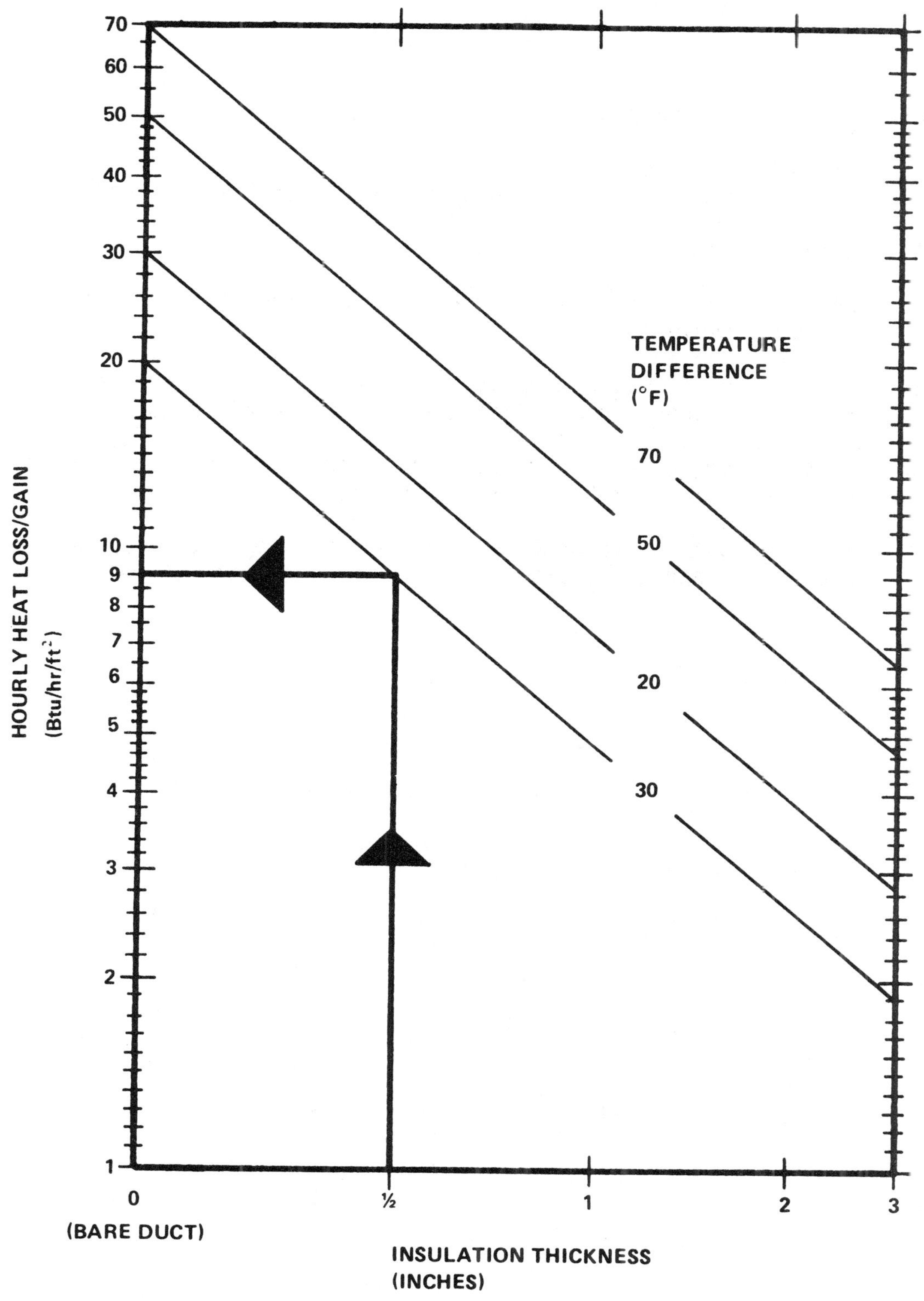

FIGURE 8-46. DUCT INSULATION-HEAT LOSS/GAIN FOR VARIOUS THICKNESS

Figure #8-47 Engineering Data

Figure is based on the addition of insulation with a thermal conductivity, k = 0.3 (Btu) (in.)/(sq.ft.) (hr) (degree F temperature difference) and an ambient air temperature of $68^{o}F$.

The following formula is the basis for determining the heat emitted (heat loss from the pipe:

$$q = \frac{T_s - T_a}{\frac{d_1}{2_k}\left(\log_e \frac{d_2}{d_1}\right) + \frac{d_1}{fd_2}}$$

Where:

q = heat emission from insulated pipe - Btu/(hr)x(sq ft of hot pipe surface)

Ts = temperature of pipe surface - degree F

Ta = ambient air temperature - degree F

d_1 = outside diameter of pipe - inches

d_2 = outside diameter of insulation - inches

k = thermal conductivity of insulating material - (Btu)X(in)/(sq ft)X (hr)X(degree F temperature difference)

f = surface coefficient - Btu/(sq ft)X(hr)X(degree F)

Instructions for use of Nomograph

1. Enter the nomograph on the lower vertical line at the pipe size in inches.

2. Proceed horizontally right to the intersection of the thickness of the insulation.

3. Proceed vertically upward at this intersection to the operating temperature line.

4. Proceed horizontally left to read the heat loss in Btu per hour per 10 foot of pipe length.

5. Multiply this by the pipe length and the number of operating hours per year to determine the annual heat loss.

6. Repeat this procedure for the added thickness of insulation and subtract to determine the annual savings per year by adding insulation.

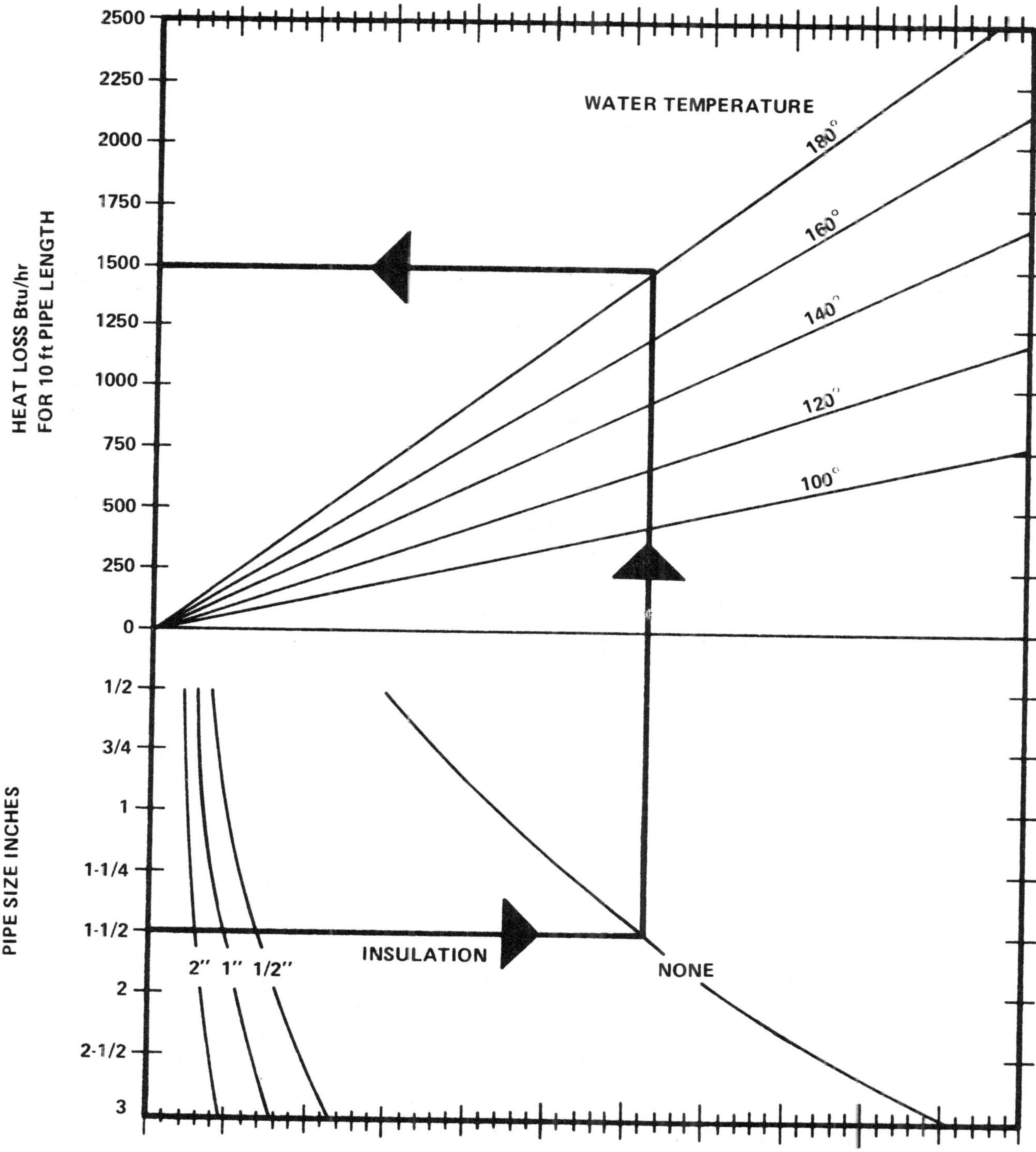

FIGURE 8-47. HEATING-HEAT LOSS FOR VARIOUS PIPE SIZES, INSULATION THICKNESS, AND WATER TEMPERATURES FROM 100°F TO 180°F

Figure #8-48 Engineering Data

Figure is based on the addition of insulation with a thermal conductivity, k = 0.3 X (Btu) X (in.)/(sq.ft.) X (hr) X (degree F temperature difference) and an ambient air temperature of 68°F.

The following formula is the basis for determining the heat emitted (heat loss from the pipe:

$$q = \frac{T_s - T_a}{\frac{d_1}{2k}\left(\log_e \frac{d_2}{d_1}\right) + \frac{d_1}{fd_2}}$$

Where:

q = heat emission from insulated pipe - Btu/(hr) (sq ft of hot pipe surface)

Ts = temperature of pipe surface - degree F

Ta = ambient air temperature - degree F

d_1 = outside diameter of pipe - inches

d_2 = outside diameter of insulation - inches

k = thermal conductivity of insulating material - (Btu) X (in)/(sq ft) X (hr) X (degree F temperature difference)

f = surface coefficient - Btu/(sq ft) X (hr) X (degree F)

Instructions for use of Nomograph

1. Enter the nomograph on the lower vertical line at the pipe size in inches.

2. Proceed horizontally right to the intersection of the thickness of the insulation.

3. Proceed vertically upward at this intersection to the operating temperature line.

4. Proceed horizontally left to read the heat loss in Btu per hour per 10 foot of pipe length.

5. Multiply this by the pipe length and the number of operating hours per year to determine the annual heat loss.

6. Repeat this procedure for the added thickness of insulation and subtract to determine the annual savings per year by adding insulation.

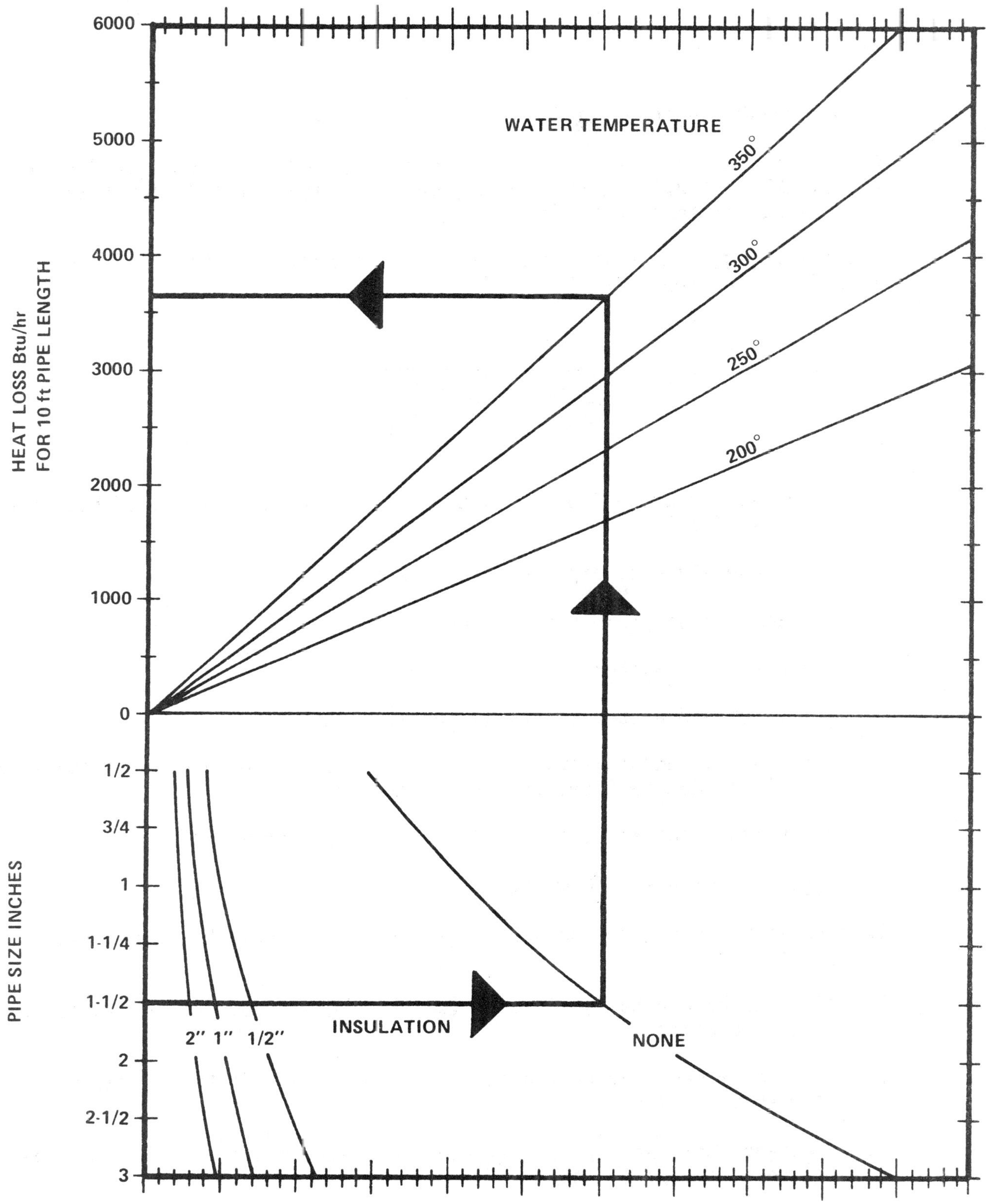

FIGURE 8-48. HEATING-HEAT LOSS FOR VARIOUS PIPE SIZES, INSULATION THICKNESS, AND WATER TEMPERATURES FROM 200°F TO 350°F

Figure #8-49 Engineering Data

Figure is based on the addition of insulation with a thermal conductivity, k = 0.3 X(Btu) X(in.)/(sq.ft.) X(hr) X(degree F temperature difference) and an ambient air temperature of 68°F.

The following formula is the basis for determining the heat emitted (heat loss from the pipe:

$$q = \frac{T_s - T_a}{\frac{d_1}{2_k}\left(\log_e \frac{d_2}{d_1}\right) + \frac{d_1}{fd_2}}$$

Where:

q = heat emission from insulated pipe - Btu/(hr) (sq ft of hot pipe surface)

Ts = temperature of pipe surface - degree F

Ta = ambient air temperature - degree F

d_1 = outside diameter of pipe - inches

d_2 = outside diameter of insulation - inches

k = thermal conductivity of insulating material - (Btu)X(in)/(sq ft)X(hr)X(degree F temperature difference)

f = surface coefficient - Btu/(sq ft)X(hr)X(degree F)

Instructions for use of Nomograph

1. Enter the nomograph on the lower vertical line at the pipe size in inches.
2. Proceed vertically upward to the thickness of the existing insulation.
3. Proceed horizontally left at this intersection and read the heat gain in Btu per hour per 10 feet of pipe length.
4. Multiply this by the length of pipe and the annual hours of operation to determine the annual heat gain.
5. Repeat the above procedure for the additional insulation to be applied.
6. The difference between the two represents the annual savings in Btu per year.

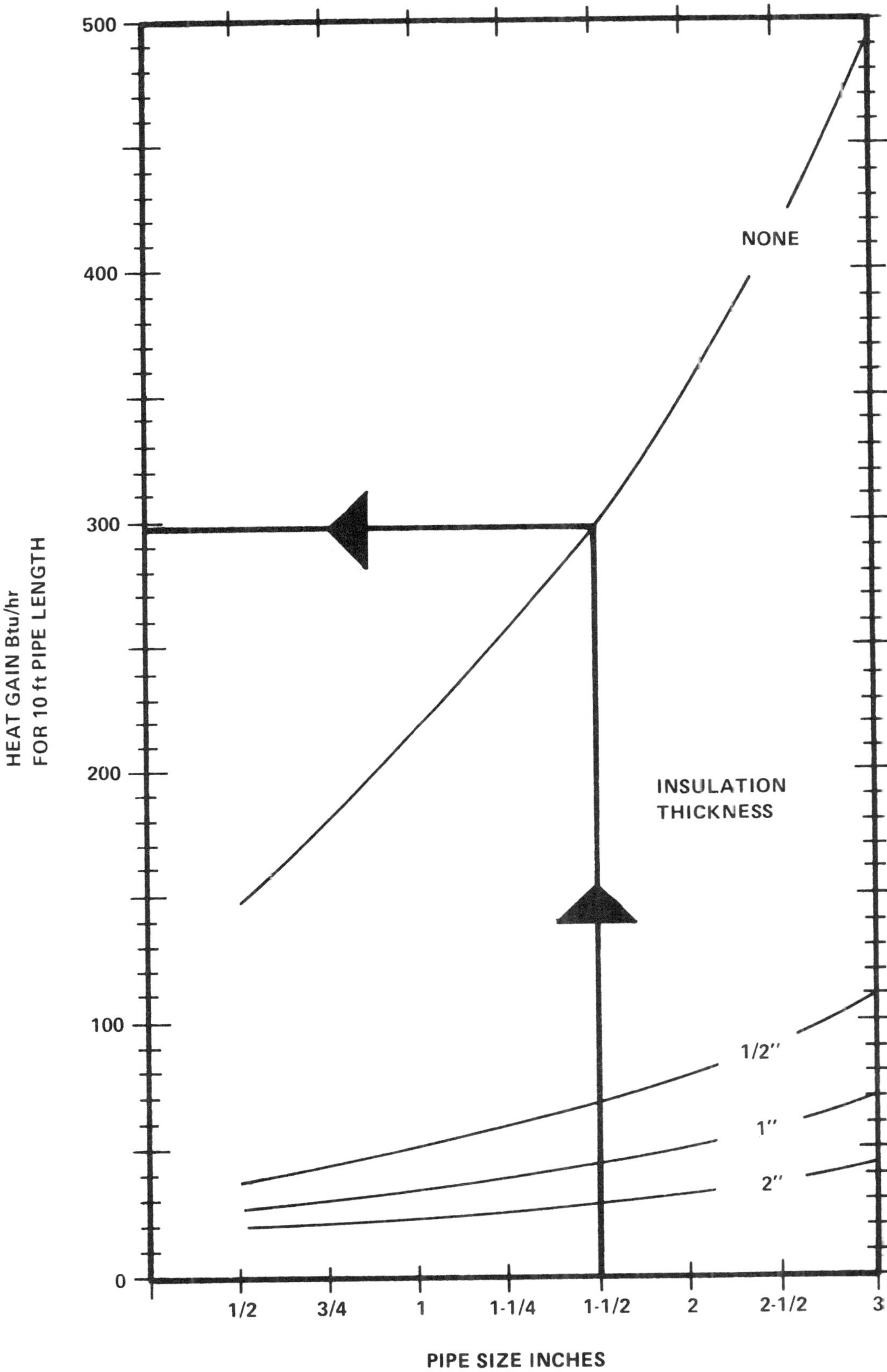

FIGURE 8-49. COOLING-HEAT GAIN FOR VARIOUS PIPE SIZES AND INSULATION THICKNESS 45°F WATER

Figure #8-50 Engineering Data

Source: Manufacturer's data

Instructions for use of Nomograph

1. Enter the chart on the lower horizontal line at the steam trap orifice size.

2. Proceed vertically upward to the intersection of the operating steam pressure.

3. Proceed horizontally left at this intersection to read the steam loss in pounds per hour.

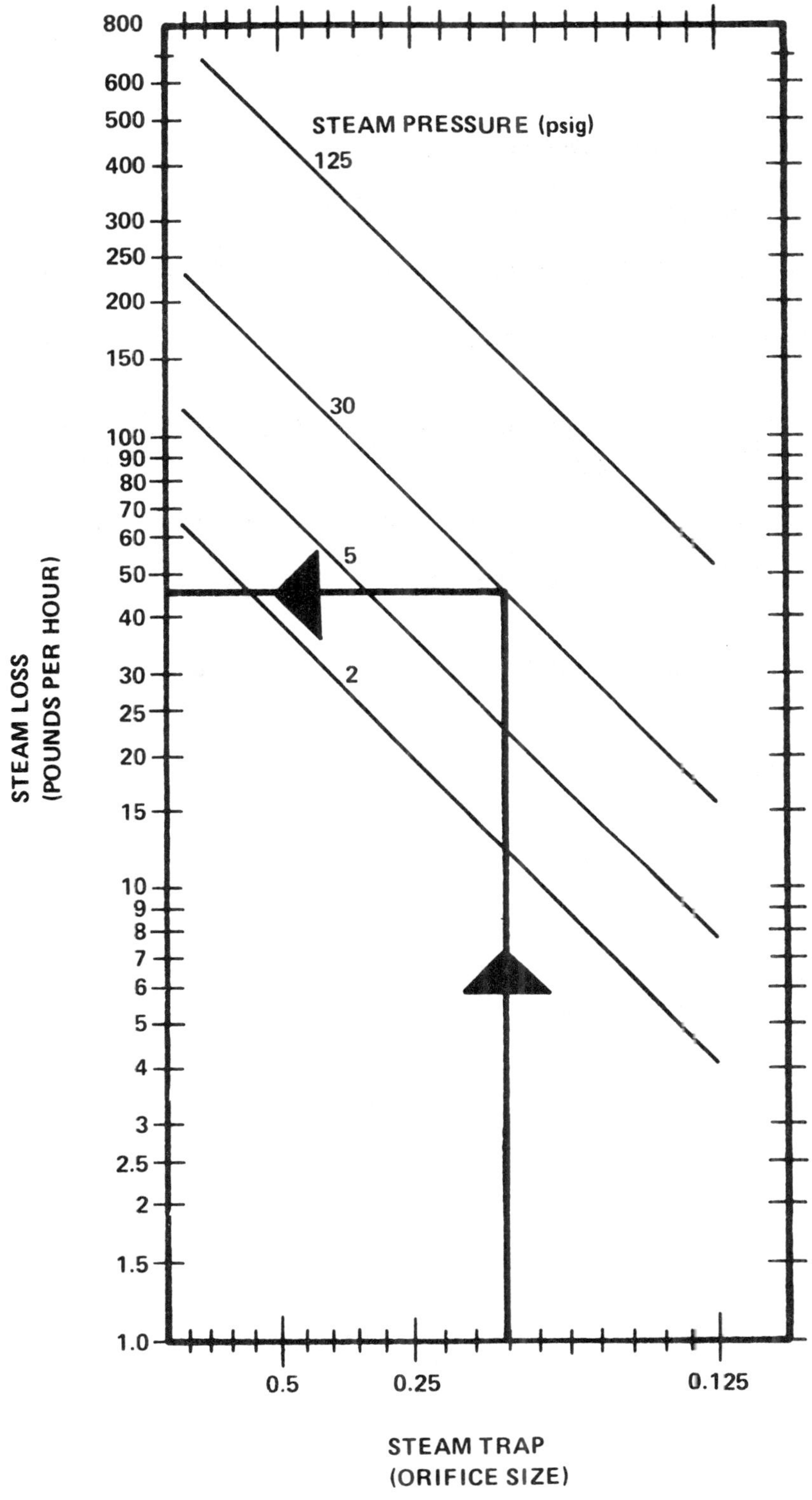

FIGURE 8-50. STEAM TRAPS-STEAM LOSS THROUGH LEAKING STEAM TRAPS

Figure #8-51 Engineering Data

Source: Manufacturer's data for different sizes and types of chillers

COP at various condensing temperatures was determined by the relationship:

$$COP = \frac{\text{Btu output}}{\text{Btu input}}$$

The change in COP was expressed in terms of the COP at nominal conditions and then plotted as a function of condensing temperature.

Instructions for use of Nomograph

1. Enter the chart on the lower horizontal line at the reduction in condensing temperature.

2. Proceed vertically upward to the intersection with the type of compressor line.

3. Proceed horizontally left at this intersection to read the percentage increase in coefficient of performance.

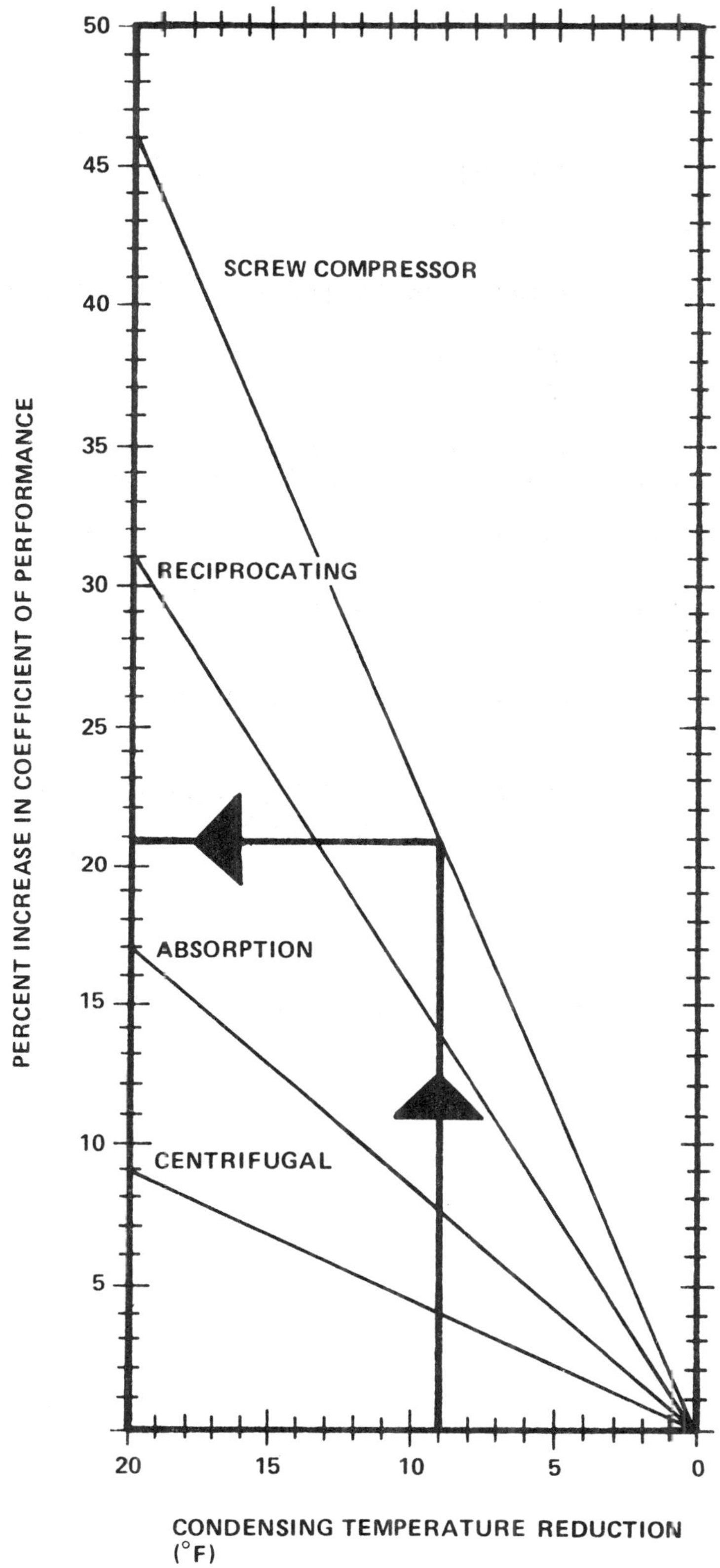

FIGURE 8-51. COOLING-EFFECT OF CONDENSER TEMPERATURE ON CHILLER COEFFICIENT OF PERFORMANCE - COP

Figure #8-52 Engineering Data

Source: Manufacturer's data for different sizes and types of chillers

COP at various condensing temperatures was determined by the relationship:

$$COP = \frac{\text{Btu output}}{\text{Btu input}}$$

The change in COP was expresed in terms of the COP at nominal conditions and the plotted as a function of condensing temperature.

Instructions for use of Nomograph

1. Enter the chart on the lower horizontal line at the present leaving chilled water temperature.

2. Proceed vertically upward to the intersection of the type of compressor line.

3. Proceed horizontally left at this intersection and read the percent increase in COP.

4. Repeat this for the maximum average chilled water temperature that can be used.

5. The difference between the two percentages represents the percent increase that may be achieved by increasing the chilled water temperature.

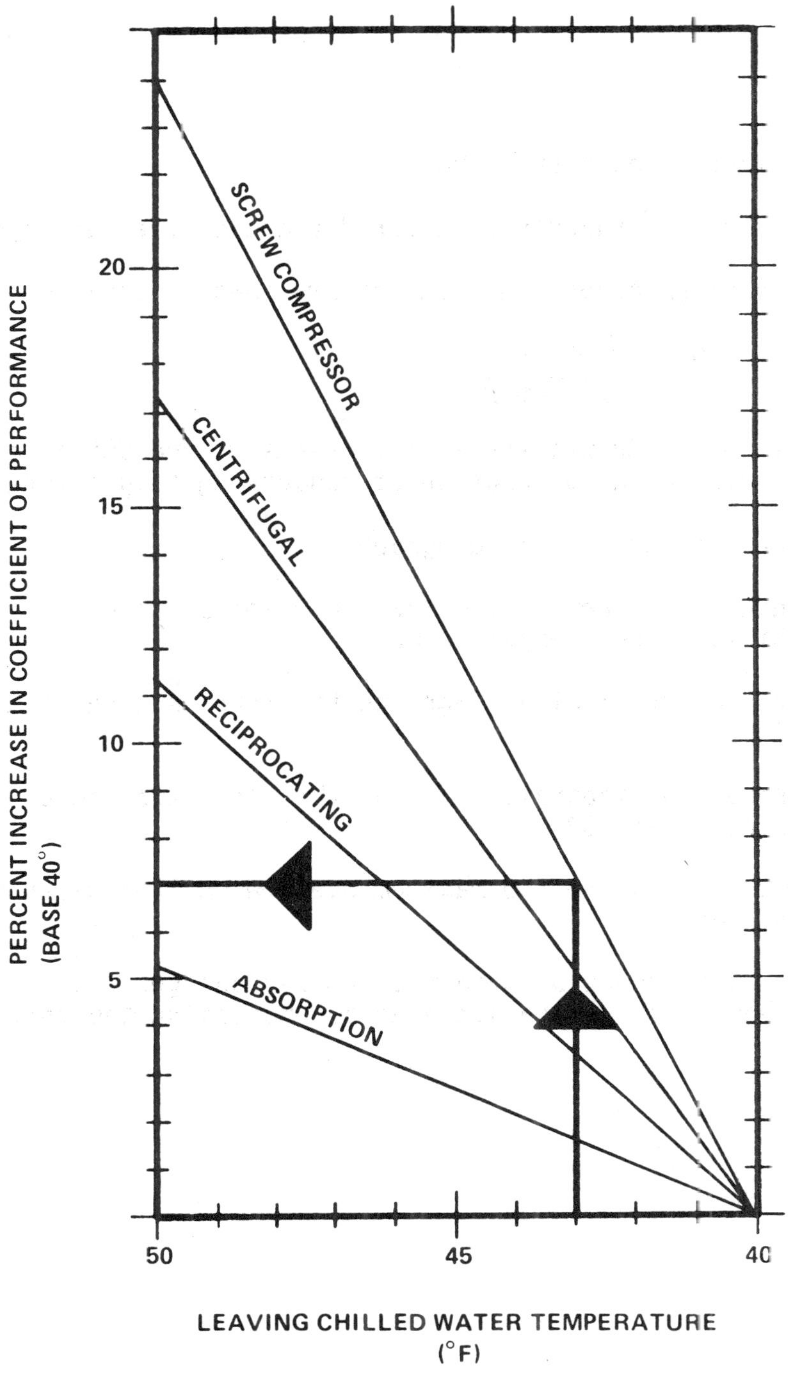

FIGURE 8-52. COOLING-EFFECT OF CHILLED WATER TEMPERATURE ON CHILLER COEFFICIENT OF PERFORMANCE - COP

Figure 8-53 Engineering Data

This graph is derived from manufacturers information and is typical for "dry heat of light" fixtures.

Instructions for use of Chart

1. Enter the chart on the bottom horizontal line at the cfm extracted through the fixture.

2. Proceed vertically upward to the proper block for the installed fixture.

3. At the intersection of this electric input line, proceed horizontally left to read the energy input to the fixture.

4. At the intersection of the light output line, proceed horizontally left to read the energy input required for the lighting output without the heat extraction.

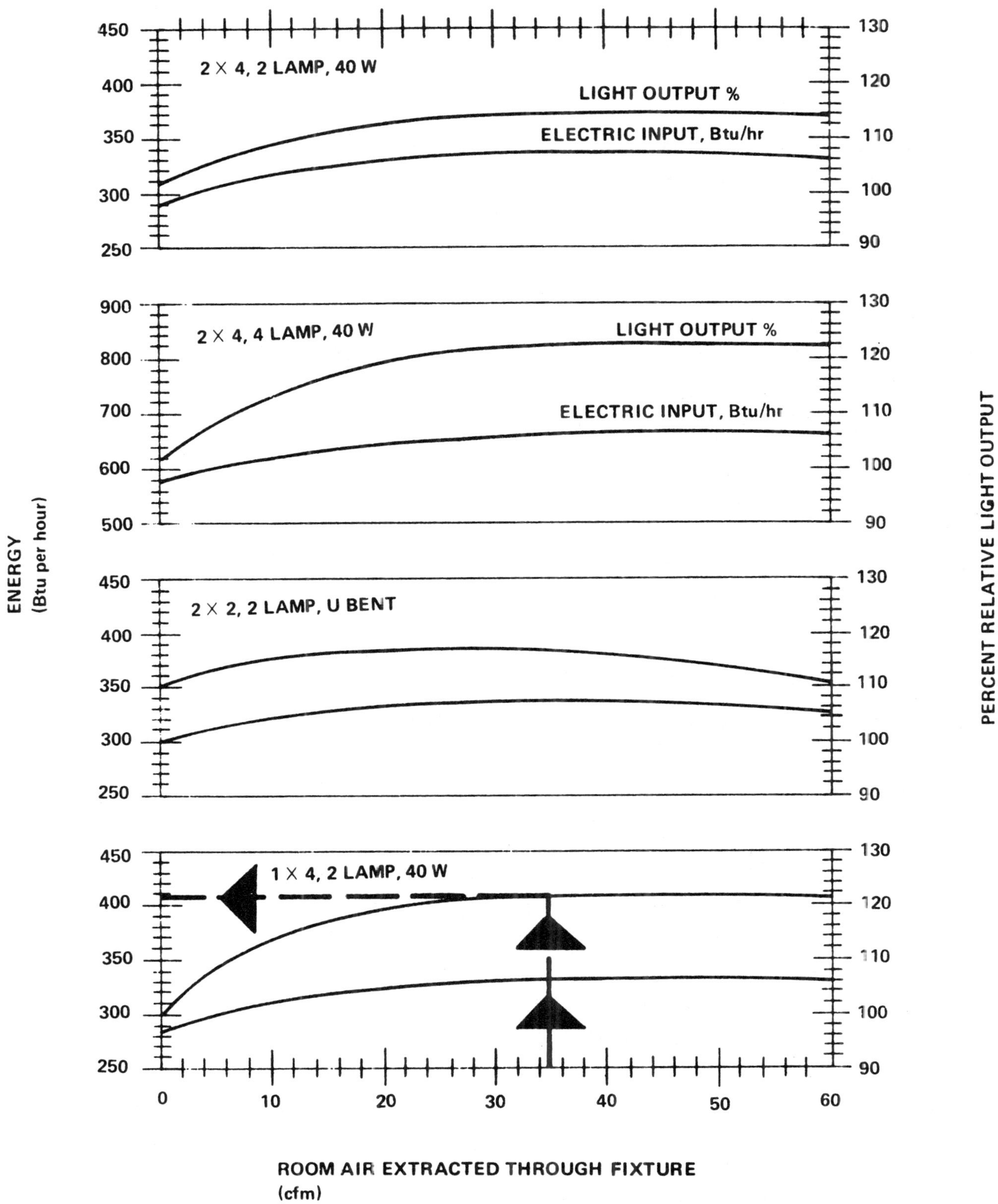

FIGURE 8-53. EFFECT OF HEAT TRANSFER ON LIGHT OUTPUT AND ELECTRICAL USAGE

Figure 8-54 Engineer Data

This graph is derived from manufacturers information and is typical for "wet heat of light" fixtures.

Instructions for use of Chart

1. Enter the chart on the bottom horizontal line at the inlet water temperature.
2. Proceed vertically upward to intersect the proper water flow line.
3. Proceed horizontally left to read the percentage of the power input that is picked up by the water.

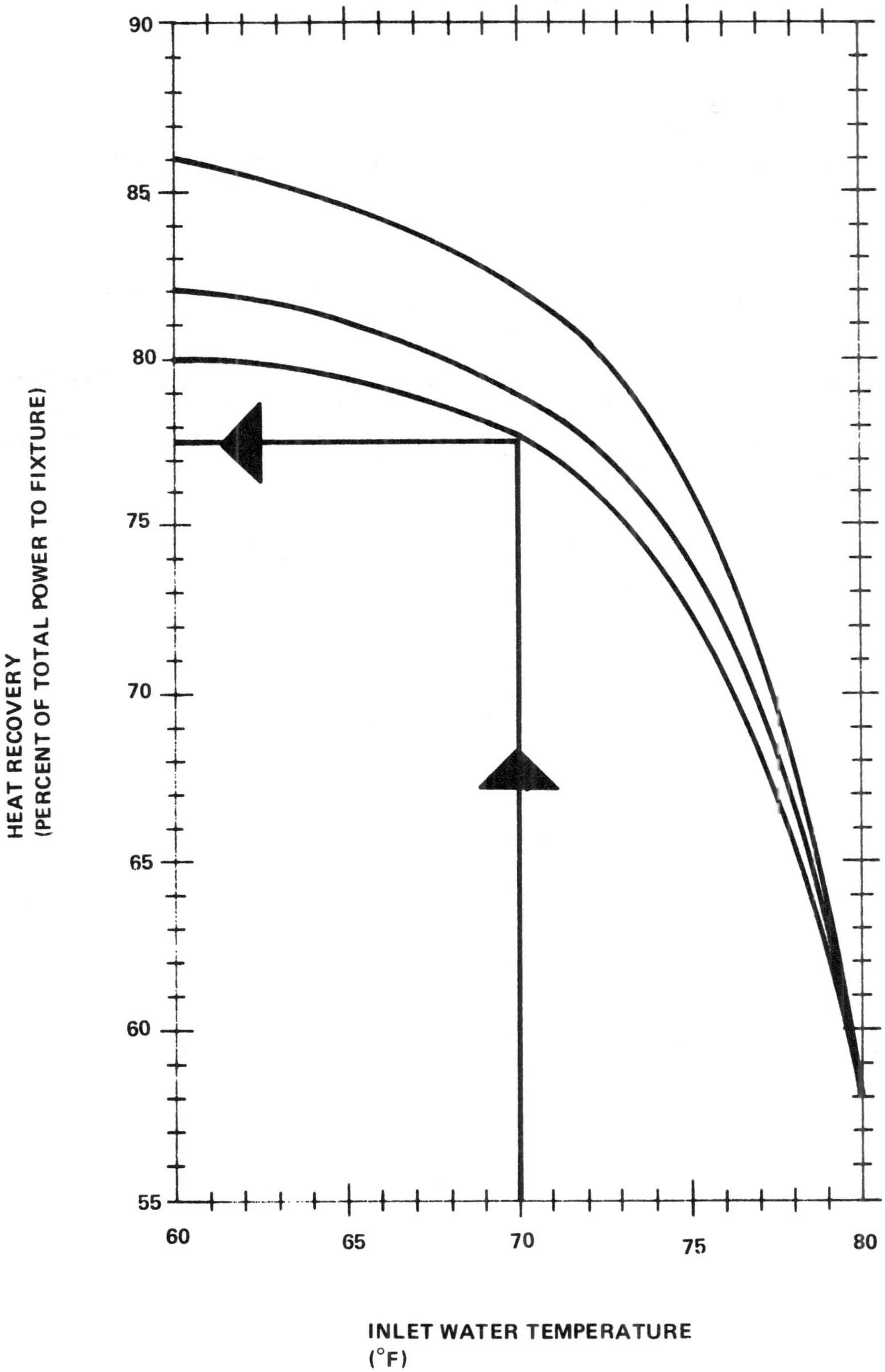

FIGURE 8-54. THERMAL PERFORMANCE OF "WET" HEAT OF LIGHT FIXTURE

Figure 8-55 Engineering Data

This graph is derived from manufacturers information and is typical of efficiency and power factor variations with motor loads.

Instructions for use of Chart

1. Enter the chart on the bottom horizontal line at percent of full load operation of the motor.

2. Proceed vertically upward to the intersection of the proper curve.

3. At the intersection of the efficiency, proceed horizontally left to read the operating motor efficiency.

4. At the intersection of the power factor curve, proceed horizontally left to read the operating power factor.

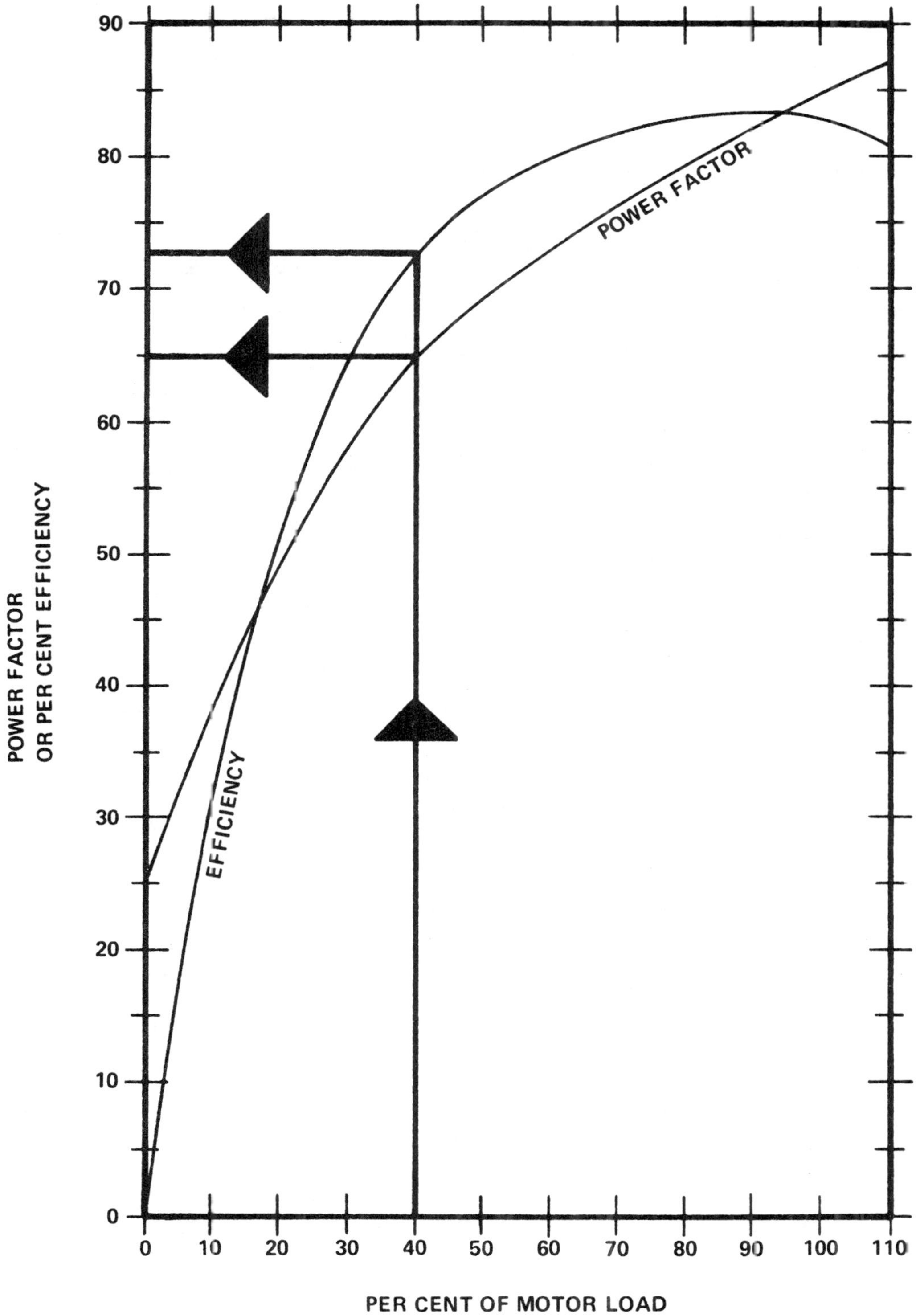

FIGURE 8-55. MOTORS-EFFICIENCY AND POWER FACTOR

YEARLY HEAT LOSS/SQUARE FOOT OF SINGLE GLAZING AND DOUBLE GLAZING

CITY	LATITUDE	SOLAR RADIATION LANGLEY'S	DEGREE DAYS	HEAT LOSS THROUGH WINDOW BTU/FT.^{2} YEAR					
				NORTH		EAST & WEST		SOUTH	
				SINGLE	DOUBLE	SINGLE	DOUBLE	SINGLE	DOUBLE
MINNEAPOLIS	45°N	325	8,382	187,362	94,419	161,707	84,936	140,428	74,865
CONCORD, N.H.	43°N	300	7,000	158,770	83,861	136,073	73,303	122,144	67,586
DENVER	40°N	425	6,283	136,452	70,449	117,487	62,437	109,365	59,481
CHICAGO	42°N	350	6,155	147,252	75,196	126,838	65,810	110,035	58,632
ST. LOUIS	39°N	375	4,900	109,915	56,054	94,205	49,355	84,399	45,398
NEW YORK	41°N	350	4,871	109,672	54,986	93,700	48,611	82,769	44,580
SAN FRANCISCO	38°N	410	3,015	49,600	25,649	43,866	23,704	41,691	23,239
ATLANTA	34°N	390	2,983	63,509	31,992	55,155	28,801	51,837	28,092
LOS ANGELES	34°N	470	2,061	21,059	11,532	19,487	10,954	19,485	10,989
PHOENIX	33°N	520	1,765	25,951	14,381	22,381	12,885	22,488	12,810
HOUSTON	30°N	430	1,600	33,599	17,939	30,744	17,053	30,200	16,861
MIAMI	26°N	451	141	1,404	742	1,345	742	1,345	742

YEARLY HEAT GAIN/SQUARE FOOT OF SINGLE GLAZING AND DOUBLE GLAZING

CITY	LATITUDE	SOLAR RADIATION LANGLEY'S	D.B. DEGREE HOURS ABOVE 78°F	HEAT GAIN THROUGH WINDOW BTU/FT.^{2} YEAR					
				NORTH		EAST & WEST		SOUTH	
				SINGLE	DOUBLE	SINGLE	DOUBLE	SINGLE	DOUBLE
MINNEAPOLIS	45°N	325	2,500	36,579	33,089	98,158	88,200	82,597	70,729
CONCORD, N.H.	43°N	300	1,750	33,481	30,080	91,684	82,263	88,609	76,517
DENVER	40°N	425	4,055	44,764	39,762	122,038	108,918	100,594	85,571
CHICAGO	42°N	350	3,100	35,595	31,303	93,692	83,199	87,017	74,457
ST. LOUIS	39°N	375	6,400	55,242	45,648	130,018	112,368	103,606	85,221
NEW YORK	41°N	350	3,000	40,883	35,645	109,750	97,253	118,454	102,435
SAN FRANCISCO	38°N	410	3,000	29,373	28,375	88,699	81,514	73,087	64,169
ATLANTA	34°N	390	9,400	59,559	50,580	147,654	129,391	106,163	87,991
LOS ANGELES	34°N	470	2,000	47,912	43,264	126,055	112,869	112,234	97,284
PHOENIX	33°N	520	24,448	137,771	97,565	242,586	191,040	211,603	131,558
HOUSTON	30°N	430	11,500	88,334	72,474	213,739	184,459	188,718	156,842
MIAMI	26°N	451	10,771	98,496	79,392	237,763	203,356	215,382	179,376

FIGURE 8-56. YEARLY HEAT LOSS/GAINS THRU SINGLE/DOUBLE GLAZING

IX. ENERGY CONSERVATION OPPORTUNITIES

SECTION IX
ENERGY CONSERVATION OPPORTUNITIES

IX. ENERGY CONSERVATION OPPORTUNITIES

A. Introduction

Presented in the following pages are 52 Energy Conservation Opportunities commonly found in existing American buildings which might be subject to the sort of energy management program described in the foregoing chapters of this manual. The intended function of these ECO's, as applied by users of the manual, is threeforld: (A) to provide the energy consultant with a systematic "checklist" of the types of ECO's likely to emerge during a building survey, and thus the types of ECO's to be looking out for; (B) to provide the energy consultant with specific procedural suggestion for implementing each ECO; (C) to provide the energy consultant with a uniform method for calculating the cost-benefit effectiveness of those ECO's (or PECO's) tentatively selected for implementation in the building under analysis.

"Theory," "Procedure," "Analysis"

Each ECO is accordingly divided into three separate sections. The "THEORY" section presents a concise statement of thermodynamic and engineering principles on which the ECO is based. The "PROCEDURE" section outlines specific suggestions for actually modifying a given building in such a manner as to produce the energy savings due to result from the ECO being discussed. The "ANALYSIS" section presents a method for analyzing the ECO in terms of future energy use and cost-benefit effectiveness as compared with the present energy consumption of the unmodified building.

It is in the latter section of each ECO primarily that the energy consultant will be directed to the various analytical "tools," including maps and nomographs, presented in Chapter VIII above. The reader will further note that the calculations of future energy use and cost-benefit associated with the analysis of each of the following ECO's presuppose that the energy consultant has (A) conducted a thorough survey of the present energy use of the building in question (as discussed in Chapter VI above) and (B) has thereby completed the various tabulations of present energy use inherent in such a survey. (In conducting the survey, of course, the energy consultant will already have become familiar with many of the forms, charts, and nomographs to which reference is made in the "ANALYSIS" sections of the following ECO's.)

Order of Application

In order to rationalize the application of these ECO's with actual building surveys and analyses, the seven different ECO groups have been arranged as nearly as possible in the order in which they should be investigated and implemented. For example, the ECO's have been arranged so that operational reductions will be calculated first; second, building load reductions will be calculated; then distribution load reductions, and finally, primary equipment load reduction. This is also the case with the ordering of individual ECO's within each group. In other words, the least expensive and logistically demanding ECO's-- corresponding with the "No Cost" and "Low Cost" categories discussed in Chapter VII --have been located first in the following pages, with the more expensive and problematic ECO's and groups of ECO's coming later.

Thus it is that the very first group of ECO's to be examined are those called "operational ECO's," the implementation of which generally requires nothing more complicated or expensive than "turning off," "turning down," or adjusting the controls of an "energy-user" presently consuming more energy than necessary. The second group of ECO's to be considered are "Lighting System ECO's," because they all address the reduction of the building while many of the ECO's within that group may be implemented at little cost. Conversely, the final group of ECO's to be considered are those dealing with distribution and primary equipment loads, and those most likely to incur higher costs.

Examples Of ECO'S In Actual Practice

In order to clarify the proper utilization of the various methods, formulae, procedures, and calculations described in the following pages, the reader will find in Chapter X a series of specific examples of some of these ECO's as they are applied to the analysis of a hypothetical office building in the American midwest. This hypothetical energy study, in addition to demonstrating the correct application of methods and procedures connected with the analysis of individual ECO's, demonstrates such other facets of a building study as the following: (A) tabulating present energy use, including the completion of survey and tabulation forms; (B) accounting for the "cumulative principle" in analyzing PECO's for future energy use and cost-benefit; (C) preparing the final statement of conclusions and recommendations for presentation to the building owner and/or energy management team.

B. Operational Energy Conservation Opportunities

B. Operational ECO's #1-7

Introduction

An enormous quantity of energy in existing buildings is wasted through improper and unnecessary operation of the building and its energy-consuming equipment. The following group of ECO's concerns this problem. It presents a body of opportunities for conserving energy which involve little more than "turning off," "turning down," and adjusting a building's "energy users" to improve efficiency without substantial modifications.

There are several reasons for implementing "operational" ECO's prior to any others. One is cost: the ECO's presented in this section will range in cost from nothing to very little. Another reason is immediacy of implementation: most of the ECO's in this section can be implemented prior to completion of the building study itself, thus providing an immediate "payback" in reduced utility bills. It is not unusual for a building owner to apply these early savings to other ECO's whose later implementation requires a greater investment.

(NOTE: The reader may observe that some of the following ECO's appear to belong in other ECO categories, e.g. "Lighting," "HVAC systems," etc. That is true: there are "operational" ECO's for virtually every major energy system in almost any building. It is for the reasons cited above that the following ECO's have been drawn from other categories and presented here for early analysis and implementation.)

ECO #1: REDUCE SYSTEM OPERATING HOURS

THEORY

Despite conventional wisdom of the past, there is little or no truth to the notion that it is cheaper and more efficient to leave a building system operating during slack or unoccupied hours than to shut the system down. This is a myth which arose in a period of relatively cheap energy resources, and it was rarely true even then. Nothing is cheaper nor more protective of energy resources than to avoid consuming those resources at every possible opportunity. This applies equally to all of the energy systems discussed below: HVAC, exhaust fans, lighting, and domestic water.

Most of these systems exist primarily for the comfort of occupants. This means that when a building or part of a building is unoccupied overnight, on weekends, and over holidays, the energy-using equipment serving that space should be turned off-- even if only for a relatively short time. There are exceptions to this rule, of course, as in locations where pipes may freeze or in the case of rare book archives and certain types of laboratories and medical facilities demanding constant temperature and humidity levels. Otherwise, during periods when a space is unoccupied, it should not be consuming energy.

PROCEDURE

HVAC Systems

There are many types of HVAC systems, each with its own peculiarities, and no general statement regarding the operation of those systems will apply to all of them. Most of those systems, however, are amenable to the following procedures for reducing operating hours:

(A) Turn off air-handling unit fans as soon as normal building hours are over. This may require turning off different fans at different times. It may also necessitate the continued operation of one or two fans to serve a small zone remaining occupied after the rest of the building has been vacated.

(B) Turn off zone pumps serving unoccupied areas.

(C) When all airhandling unit fans are turned off, turn off chilled and hot water pumps serving the building.

(D) Turn off chillers and boilers served by these pumps.

Of course, this equipment must be turned on again in plenty of time to have the space properly conditioned for renewed occupancy. This will require that chillers and/or boilers be turned on early enough to supply the needs of the earliest occupied zones. Zone pumps and air-handling units serving other zones, however, need not be turned on until later.

Exhaust Fans

Exhaust fans in office buildings are used primarily to ventilate toilet rooms, though certain other uses do exist, such as kitchen exhaust hoods, fume hoods, etc. Exhaust requirements in most cases are not continuous, and when the requirement does not exist, those fans should be cut off. With regard to toilet exhaust fans in a building normally occupied between 8 and 5, five days a week, those fans should be cut off at all other times - even in buildings partially occupied at odd hours. (The toilet exhaust function is generally not necessary under such circumstances.) Kitchen exhaust fans need only be operated while cooking is actually in process.

Lighting

Except in rare cases where a building is partially heated by its lighting system, the function of the lighting is to serve the needs of the building's occupants. Thus:

(A) When a space is unoccupied, even briefly, its lights should go off.

(B) Despite the widespread practice of leaving on lights after business hours for clean-up crews, it is much more efficient for the tenants to turn off lights as they leave and for security personnel to make sure that they are off on their first rounds. Clean-up crews can turn them back on as they are required for actual maintenance work.

(C) Lights in storerooms, vaults, lavatories, and other seldom-used areas should be turned off when not in use.

(NOTE: Lighting economy is very dependent on tenant cooperation, and it is often advisable to establish an on-going program of tenant energy-awareness with regard to turning off lights and electrical equipment when leaving a space in the building.)

ANALYSIS

In order to establish the potential energy savings and cost effectiveness (if investment is required) of the above procedures, carry out the following calculations and enter subsequent changes in figures on building survey and tabulation forms (as discussed in Chapter VII). Where a chart, table, or nomograph is required for computation, the reader will be referred to the appropriate document in Chapter VIII.

HVAC Systems

(A) Determine the new operating hours for the building's air-handling unit, fans, and pumps; if different from the present operating hours, substitute the new hours for the present hours on Distribution Load Survey Form (Figure 8-10) or Building Load Survey Form (Figure 8-9).

(B) Since a reduction in the operation of the air-handling unit will also reduce the outside air load for both heating and cooling, refer to Figure 8-30 and 8-34, using the new operating hours, determine new outside air factors. Enter these factors on Annual Heating and Cooling Tabulation Form (Figure 8-11).

(C) Likewise determine the new operating hours for exhaust fans; if different from present operating hours, enter the new hours on Distribution Load Survey Form (Figure 8-10) or Building Load Survey Form (Figure 8-9).

Domestic Water

(A) Determine the new operating hours for the domestic water recirculating pump; if different from the present hours, enter the new hours on Distribution Load Survey Form (Figure 8-10).

(B) While turning off the domestic water system during unoccupied hours will improve the overall efficiency of that system, the precise degree of improvement will depend on the particularities of the building under survey. A poorly insulated system would benefit more than a well insulated system. The energy consultant will have to determine the best means of calculating the energy savings which might be expected here.

Lighting

(A) Determine the new operating hours for the various lighting systems affected by this ECO.

(B) Substitute the new hours for the present hours on Building Load Survey Form (Figure 8-9)

Conclusion: All of the figures pertaining to future energy use with regard to this ECO should now be calculated and carried through the appropriated forms as explained in Chapter VII. This will determine whether the costs of implementing and maintaining this ECO are justified in terms of future savings for the building owner.

ECO #2: REDUCE SPACE LOADS DUE TO VENTILATION

Indoor Air Quality (IAQ) is an emerging issue of concern to building managers, operators and designers. Recent research has shown that indoor air is often less clean than outdoor air and Federal legislation has been proposed to establish programs to deal with this issue on a national level. This, like the asbestos issue, will have an impact on building design and operations. Americans today spend long hours inside buildings and building operators, managers and designers must be aware of potential IAQ problems and how they can be avoided.

IAQ problems, sometimes termed "Sick Building Syndrome," have become an acknowledged health and comfort problem. Buildings are characterized as "sick" when occupants complain of acute symptoms such as headache, eye, nose and throat irritation, dizziness, nausea, sensitivity to odors and difficulty in concentrating. The complaints may become more clinically defined so that an occupant may develop an actual "building-related illness" that is believed to be related to IAQ problems. Figure 9-1 illustrates[(1)] guidelines for some IAQ pollutants.

The most effective means to deal with an IAQ problem is to remove or minimize the pollutant source, when feasible. If not, dilution and filtration may be effective.

Dilution (increased ventilation) is to admit more outside air to the building. ASHRAE's 1981 standard recommended 5 CFM/person outside air in an office environment. The new ASHRAE ventilation standard, 62-1989, now requires 20 CFM/person for offices if the prescriptive approach is used.

Increased ventilation will have an impact on building energy consumption. However, this cost need not be severe. If an airside economizer cycle is employed and the HVAC system is controlled to respond to IAQ loads as well as thermal loads, 20 CFM/person need not be adhered to and the economizer hours will help attain air quality goals with energy savings at the same time. Incidentally, it was the energy cost of treating outside air that led to the 1981 standard. The superseded 1973 standard recommended 15-25 CFM/person.

Energy savings can also be realized by the use of improved filtration in lieu of the prescriptive 20 CFM/person approach. Improved filtration can occur at the air handler, in the supply and return ductwork, or in the spaces via self-contained units. Improved filtration can include enhancements such as ionization devices to neutralize airborne biological matter and to electrically charge fine particles, causing them to agglomerate and be more easily filtered.

Many buildings are ventilated at a rate far in excess of that necessary to maintain comfort, dilute odors, or meet code requirements. They are likewise ventilated 24 hours per day even when the building is unoccupied, resulting in a gross waste of energy. By way of reducing this waste, the building

---

[1]M. Black & W. Robertson, "Indoor Air Quality: Problems & Cures." Presented at the 13th World Energy Engineering Congress.

owner or operator can perform most of the procedures for shutting off ventilation at night or on weekends. But professional advice and help may be necessary to reduce the ventilation rate in all or portions of the building during occupied periods.

If ventilation is supplied to a building to provide makeup air for toilet, kitchens and other exhaust air quantities, analyze the exhaust systems and operate them only as needed. Savings in fan horsepower for exhaust fans, supply fans, and heating energy to temper the supply air will result. Cooling loads will be similarly reduced in the summer.

Figure 9-1

TABLE III

GUIDELINES FOR SOME IAQ POLLUTANTS

Pollutant	Concentration	Remarks
Asbestos	0.2 fibers/cm^3	OSHA standard set in July, 1986.
	0.1 fibers/cm^3	OSHA action level requiring monitoring programs; typical background levels in outdoor ambient air in urban areas are 0.00007 fibers/cm3.
Carbon Dioxide	1000 ppm	Japanese standard for buildings with floor space exceeding 3000 m^2 and HVAC system.
Carbon Monoxide	9 ppm	National Ambient Air Quality Standard average of 8 hours.
Formaldehyde	0.1 ppm	ASHRAE recommended limit based on comfort criteria which should protect all but hypersensitive individuals.
	0.4 ppm	HUD Standard for pressed wood products used in mobile homes, to prevent formaldehyde in indoor air from exceeding 0.4 ppm.
Nitrogen Dioxide	0.05 ppm	Annual National Ambient Air Quality Standard.
Ozone	0.08 ppm	Level of concern in World Health Organization criteria documents.
	0.12 ppm	National Ambient Air Quality Standard averaged over 1 hour.
Particulate	50 ug/m^3	National Ambient Air Quality Standard annual geometric mean.
	150 ug/m^3	National Ambient Air Quality Standard 24 hour average mean.

Pollutant	Concentration	Remarks
Radon	4pCi/L	U.S. Environmental Protection Agency technologically achievable target level.
	8 pCi/L (0.04 WL)	Remedial action level recommended by the National Council on Radiation Protection and Measurements.
Termiticides	1.0 ug/m^3 (Aldrin) 5.0 ug/m^3 (Chlordane) 2.0 ug/m^3 (Heptachlor) 10 ug/m^3 (Chlorpyrifos)	Recommended by the National Academy of Sciences Committee on Toxicology.
Volatile Organic Compounds	1-5 mg/m^3	Lars Molhave study levels suspected of causing sick building syndrome symptoms in some individuals. U.S. EPA Guideline.

ECO #3: CONTROL SPACE TEMPERATURE AND HUMIDITY AT ALL TIMES

THEORY

Because transmission heat gain and loss are directly proportional to the difference in temperature between outdoors and indoors, energy loss and gain can be minimized by maintaining the the indoor temperature at as nearly a constant value as possible. This constant value should be adjustable for occupied and unoccupied periods and for the heating and cooling seasons.

Heating Season

Moderate reductions in the indoor temperature for occupied periods during the heating season will conserve energy, although savings of course will not be as great as for unoccupied periods. Sharper reductions in the temperature of less critical areas, such as corridors and lobbies, will achieve more significant economies, and proportionate savings will be greatest in buildings which normally have shorter periods of occupancy (schools as opposed to stores), and in buildings or sections of buildings with the least internal heat gain (people and machines) and solar radiation to help with the heating load.

Cooling Season

The air conditioning systems in many buildings are designed to maintain 72°F dbt and 50% rh during peak loads in the cooling season, and usually this equipment is operated to achieve still lower temperatures during the base-load conditions which occur most of the time. Increasing the indoor temperature and humidity levels from 74°F dbt and 50% rh 78°F dbt and 55% rh will save approximately 13 percent of the energy required for cooling. The exact savings will depend on the volume of ventilation air and infiltration into a building, conduction losses, solar heat gain, internal loads of the building and the type of air conditioning system being used.

Humidification

During the heating season, humidification systems vaporize water into the dry ventilating air to increase its moisture and achieve the desired humidity within the building. This process requires a heat input of approximately 1000 Btu to vaporize each pound of water. The volume of moisture required to maintain a desired level of relative humidity is proportional to the amount of outdoor air entering the building, its dryness, and the natural moisture contributed by occupants.

Humidification systems are often designed not only to maintain the comfort and health of occupants, but to preserve materials and prevent the drying and cracking of wood, furniture, and other contents of the building. Unless

FIGURE 9-2 SUGGESTED INDOOR TEMPERATURES AND HUMIDITY LEVELS IN THE HEATING AND COOLING SEASON

	A Dry Bulb °F occupied hours maximum heating	B Dry Bulb °F unoccupied hours (set-back) heating	C Dry Bulb Temperature occupied*	D Minimum Relative Humidity occupied
1. OFFICE BUILDINGS, RESIDENCIES, SCHOOLS				
Offices, school rooms, residential spaces	68°	55°	78°	55°
Corridors	62°	52°	Uncontrolled	Uncontrolled
Dead storage closets	50°	50°	Uncontrolled	Uncontrolled
Cafeterias	68°	50°	75°	55°
Mechanical equipment rooms	55°	50°	Uncontrolled	Uncontrolled
Occupied storage areas, gymnasiums	55°	50°	78°	55°
Auditoriums	68°	50°	78°	50°
Computer rooms	65°	As required	75°	55°
Lobbies	65°	50°	82°	60°
Doctor offices	68°	58°	78°	55°
Toliet Rooms	65°	55°	80°	Uncontrolled
Garages	Do not heat	Do not heat	Uncontrolled	Uncontrolled
2. RETAIL STORES				
Department stores	65°	55°	80°	55°
Supermarkets	60°	50°	78°	55°
Drug stores	65°	55°	80°	55°
Meat markets	60°	50°	78°	55°

*Except where terminal reheat systems are used.

such uses of humidity are indispensable, the relative humidity should be strictly maintained at levels required by occupants alone.

PROCEDURE

Maintain dry-bulb temperature and relative humidity for various spaces at the levels suggested in Figure 9-2 ("Suggested Indoor Temperature and Humidity Levels") to reduce the heating and cooling load, as well as energy use and operating costs. Even greater savings may be realized, without serious discomfort, by maintaining higher levels in areas which are occupied only for short periods of time. Implementation of this ECO must be coordinated with administration and management personnel so that appropriate dress can be promulgated to avoid adverse user reactions.

Controls

Check the locations of all space thermostats and relocate thermostats affected by adjacent heat-producing sources such as lights, coffee pots, office equipment, etc. (or remove such heat-producing sources). Calibrate all thermostats and replace those with greater than a $\pm$ 2°F differential, so that accurate temperature control can be maintained.

Heating Season

In areas where it is not necessary to maintain high temperatures during occupied periods, i.e. corridors and lobbies, maintain even lower temperatures than for the other spaces. (See Figure 9-2 for suggested winter night setback temperatures.) Implement setback by resetting thermostats manually (if automatic setback control has not been installed), or by adjusting controls to suggested temperatures (if clock, day-night, or other automatic reset controls are available). Climate, type of system, and building construction will determine the length of the start-up period required to achieve daytime temperature levels. Experiment to decide upon the optimal setback temperature and start-up time for any particular building. If, in extremely cold weather, experience indicates that the heating system does not raise the temperature sufficiently by the time the building opens for the day, set temperatures back to a higher level for those periods of time only.

Cooling Season

The entire air conditioning system should be shut down at night, on weekends, and during other extended unoccupied periods. Local areas such as auditoriums, conference rooms, cafeterias, and other areas with short-term occupancy rates should be maintained at higher-than-design temperatures and humidities when unoccupied. It may be advisable to shut off the cooling units in these spaces altogether, allowing the temperature to rise freely until a few hours before the space is to be occupied again.

If controls are not available to effect the above measures automatically, they will have to be effected manually in approximately the following sequence. Shut down cooling or condenser fans, chilled water and condenser pumps, and supply and exhaust fans as well as chillers or compressors. If the cooling tower fans and pumps are interlocked with the compressors, they will shut off automatically when the compressors are shut off. Shut down boilers which supply steam or hot water to absorption units; do not maintain boiler water temperature or steam pressure when absorption units are inoperative. If, after shut-down, the air conditioning system is not capable of achieving and maintaining temperature and humidity conditions in hot spells, install a control to activate the equipment a few hours before occupancy instead of operating the system all night or throughout the weekend.

In certain climates such as parts of Arizona or California with a large diurnal swing (large temperature difference within 24 hours), night operation of the refrigeration system for periods of the summer may be economical. If conditions at night permit lower condensing temperatures, perform an analysis to indicate whether energy can be saved by operating the refrigeration system at night to precool the building or to store chilled water in the piping system.

An engineering and economic analysis, comparing refrigeration operation with economizer cooling, may be necessary to determine the most efficient mode of operation.

Humidification

Do not humidify during unoccupied periods. Winter humidity in cold climates drops 5 to 10 percent in buildings without humidifiers. In the absence of sufficient evidence to support the contention that higher rh levels are more comfortable or promote health, it is suggested that a 20 percent rh be maintained in all spaces occupied more than four hours per day. Shut off the humidifiers completely in all areas at night and during other unoccupied periods. If complaints of dryness and discomfort result, raise the humidity levels in 5 percent increments until the appropriate level for each area of the building is determined.

If higher relative humidities are necessary to reduce static electricity and eliminate shocks (30% rh will usually achieve this), raise the humidity levels only for those periods when shocks are a serious problem.

ANALYSIS

Energy Savings From Reduction of Winter Occupied Temperatures

To calculate the energy required after reducing the winter occupied temperatures, first reduce the present indoor thermostat setting and calculate savings by using the following formula:

$$HEC_F = 100 - \frac{(ti - to) - td}{(ti - to)} \times HEC_P$$

ti = present indoor thermostat setting from the Building Load Survey Form (Figure 8-9)

to = winter outdoor design temperature from the Climate Zone Form (Figure 8-7)

td = difference between present and reduced thermostat setting

HEC_P = present heating energy consumption from the Annual Heating and Cooling Tabulation Form (Figure 8-11)

HEC_F = future heating energy consumption, substitute for HEC_P on the Annual Heating and Cooling Tabulation Form (Figure 8-11).

Energy Savings From Reduction of Winter Unoccupied Temperatures

To estimate the energy savings of a night reduction of space temperature, perform the following calculations:

(A) Refer to Figure 9-20 to determine the recommended night temperature.

(B) Determine present Btu for heating from the Annual Heating and Cooling Tabulation Form (Figure 8-11).

(C) Where the building as presently operated does not provide a complete shut-down of outside air during the unoccupied periods, take the total heating Btu/yr and subtract from it the Btu/yr that are attributable to ventilation.

(D) Refer to Figure 8-20, "Energy Saved by Night-Setback", and using the heating consumption calculated in "C", determine the savings in Btu/yr at the order of magnitude required for entering the graph with the present Btu/yr.

(E) Subtract the savings from the present Btu for heating and substitute for the heating consumption on Annual Heating and Cooling Tabulation Form (Figure 8-11).

Energy Savings From Increasing Summer Occupied Temperatures

Use the multipliers in the following table to determine the energy saved in Btuh/1000 cfm of outdoor air (ventilation plus infiltration) by raising the dry-bulb temperatures at constant relative humidities. (All other factors are considered constant, and since the effect of raising the dry-bulb temperature on conductive heat gain is actually negligible, it has been disregarded here.)

FIGURE 9-3 ENERGY SAVINGS FROM INCREASED SUMMER SPACE TEMPERATURE*

Change in Dry Bulb Temp. °F From To	Btu Savings/1000 cfm Relative Humidity 50%	60%	70%
72 - 73	2,700	2,433	3,000
73 - 74	2,657	2,400	3,257
74 - 75	3,000	2,572	3,000
75 - 76	3,000	2,572	3,000
76 - 77	3,000	2,572	3,429

(A) From the above table, select the relative humidity column desired and determine the savings over the total temperature change by adding the Btuh/1000 cfm saved for each degree of increased temperature.

(B) Subtract this figure from the present figure for outside air cooling and substitute the new figure on the Annual Heating and Cooling Tabulation Form (Figure 8-11).

Energy Savings from Reduction of Winter Relative Humidity in Heated Spaces

If humidity reduction is a possibility, observe the following steps:

(A) Determine if humidity control can be lowered or eliminated by examining building requirements.

(B) Refer to Figure 8-28 to determine the energy used at the new lower humidity.

(C) Substitute this factor for the present humidity factor on the Annual Heating and Cooling Tabulation Form (Figure 8-11).

Energy Savings from Increasing Summer Relative Humidity in Conditioned Spaces

If humidity increase is feasible, observe the following steps:

(A) Refer to Figure 8-30 to determine if humidity can be raised from the present level.

(B) Using Figure 8-30, determine the energy used at the new higher humidity.

*With terminal reheat systems in operation, the indoor space conditions should be maintained at lower levels to reduce the amount of reheat and save energy. If cooling energy is not required to reduce temperature, maintain 74°F dbt instead of 78°F dbt.

(C) Substitute this factor for the present outside air cooling factor on Annual Heating and Cooling Tabulation Form (Figure 8-11).

Conclusion: All of the figures pertaining to future energy use with regard to this ECO should now be calculated and carried through the appropriate forms as explained in Chapter VII. This will determine whether the costs of implementing and maintaining this ECO are justified in terms of future savings for the building owner.

ECO #4: REDUCE FLOW AND TEMPERATURE OF HOT WATER

THEORY

Flow reduction: A primary benefit of reducing the quantity of hot water consumed in a building is that energy consumption will be reduced in direct proportion to the reduction in flow rate. A secondary benefit is the reduction in raw source energy required, due to the fact that less water will have to be treated in the water supply and sewage treatment plants, whether on-site or off-site.

For municipal facilities, the diminished energy requirements will result in lower operating costs than would otherwise be possible, which in turn will mean that fewer dollars will be needed to support the facility. In areas where there is a charge based on total water consumption flowing into the sewer, the reduction in consumption of water will result in direct savings to the building owner as well. (These latter arguments apply also to the use of cold water for domestic purposes, which should also be reduced in flow.)

Temperature reduction: Lowering the temperature of domestic hot water reduces storage and distribution losses through tanks and piping.

PROCEDURE

Flow reduction: Reduce hot water consumption wherever possible by using water restrictors at all faucets, showers, etc. Such consumption can be lowered to 1 1/4 or 1 1/2 gallons per person per day in office buildings today without inconvenience to the occupants. (Cold water flow rates should be similarly restricted.)

Temperature reduction: Determine the lowest temperature at which hot water will meet occupants' needs and reduce the temperature to this setting. Temperatures of 100°F will generally be more than adequate. Insufficient storage and/or heating capacity may require higher storage temperatures in order to maintain sufficient temperature.

ANALYSIS

Flow reduction: Using the desired flow, substitute this for the present flow on the Building Load Survey Form.

Temperature reduction: Temperature reduction increases the efficiency of the system by reducing parasitic losses. The system efficiency is the heater efficiency times the distribution efficiency.

On a seasonal basis, the following are average heater efficiencies:

(A) Oil-fired heating boilers used year-round, but with domestic hot water as the only summer load = .45.

(B) Oil-fired heating boilers used year-round with absorption cooling in the summer in addition to domestic hot water = .7.

(C) Gas-fired heating boilers used year-round, but with domestic hot water as the only summer load = 0.5.

(D) Gas-fired heating boilers used year-round with absorption cooling in the summer in addition to domestic hot water = .75.

(E) Separate oil-fired hot water heaters = .70.

(F) Separate gas-fired hot water heaters = .75.

(G) Separate electric water heaters = .95.

(H) Separate coal-fired water heaters = .45.

The distribution efficiency may be determined by measuring the temperature difference between the outlet of the tank and the outlet of the hot water tap. This temperature difference times the yearly water consumption in pounds obtained for the Building Load Survey Form (Figure 8-9),gives the distribution loss in Btu/year.

The distribution efficiency is then determined by:

$$\frac{\text{Building Load - Distribution Loss}}{\text{Building Load}}$$

where Building Load is the energy required to heat the water to the desired temperature (from Building Load Survey Form) Figure 8-9.

The system efficiency is determined by A thru H above times the distribution efficiency. This efficiency is used to determine the fuel requirements on the Building Load Survey Form Figure 8-9.

Lowering the water temperature 1°F reduces the distribution losses by approximately 1%. Calculate the new distribution efficiency resulting from the reduced temperature, multiply by the heater efficiency and substitute the new system efficiency on the Building Load Survey Form (Figure 8-9).

Conclusion: All of the figures pertaining to future energy use with regard to this ECO should now be calculated and carried through the appropriate forms as explained in Chapter VII. This will determine whether the costs of implementing and maintaining this ECO are justified in terms of future savings for the building owner.

ECO #5: ISOLATE OFF-LINE BOILERS

THEORY

Light heating-load demands on a multiple boiler installation are often met by one boiler on line with the remaining boilers idling on standby. Idling boilers consume energy to meet standby losses which can be further aggravated by a continuous induced flow of air through the idle boilers into the stack and up the chimney.

PROCEDURE

Unless a boiler is scheduled for imminent use to meet an expected increase in load, it should be secured and isolated from the operating boilers, as well as from the stack and chimney, by closing valves and dampers. Large boilers should be fitted with by-pass valves and regulating orifices to allow a minimum flow of air through the boiler, thus conserving fuel while the boiler is kept warm to avoid thermal stress when it is brought on-line again.

FIGURE 9-4 ISOLATING OFF-LINE BOILERS

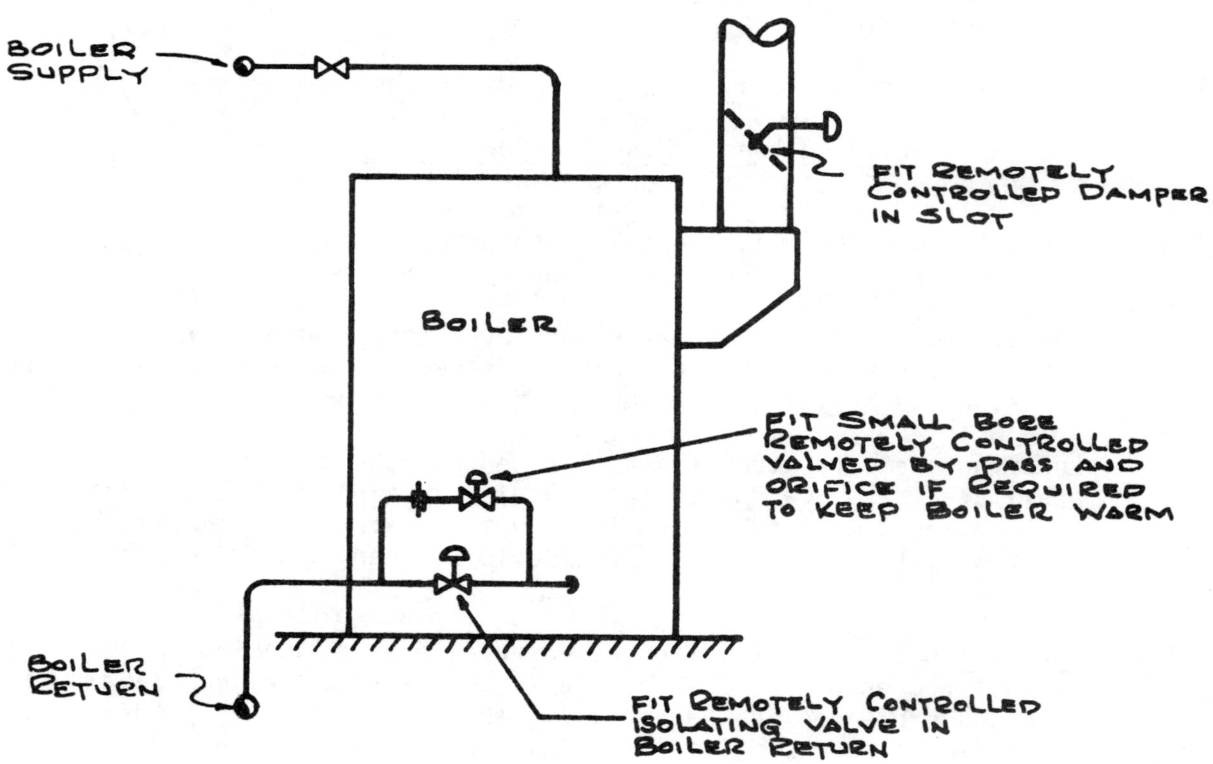

Note: (If a boiler containing water is isolated, it is particularly important that air flow through the stack be prevented, since it is possible for a back-flow of cold air to freeze the boiler.)

ANALYSIS

The total volume of wasted energy per year from idling boilers can be calculated on the basis of data for off-line boilers in the ASHRAE Guide:

(A) Determine from ASHRAE Guide the increased boiler efficiency to be expected from isolating off-line boilers.

(B) Substitute the new efficiency for the present efficiency on the Annual Heating and Cooling Tabulation Form (Figure 8-11).

Conclusion: All of the figures pertaining to future energy use with regard to this ECO should now be calculated and carried through the appropriate forms as explained in Chapter VII. This will determine whether the costs of implementing and maintaining this ECO are justified in terms of future savings for the building owner.

ECO #6: USE LOW TEMPERATURE CONDENSER WATER

THEORY

The efficiency ratio (COP) of chillers and compressors tends to increase as condensing temperature is reduced. COP is related directly to evaporating and condensing temperatures which, for a water chiller with cooling tower, are typically 40°F and 100°F respectively. If the evaporating temperature can be raised (by using chilled water at a higher temperature of, for instance, 50°F instead of 40°F) and/or if condensing temperatures can be reduced, the COP will increase and a greater cooling effect will result from the same power output.

In general, raising the chilled water temperature 10° and reducing the condensing water temperature 10° will result in an increased efficiency of 20 to 25 percent. Viewed in terms of seasonal and/or part-load conditions, the increased efficiency will be even greater. (NOTE: Each type of cooling unit has practical limits with regard to lowest acceptable condensing temperature, and this limit should not be exceeded, particularly with some of the older type absorption units.)

PROCEDURE

Where existing cooling installations operate at constant condenser water temperature and cooling tower fans are cycled on and off, modify the controls to operate cooling tower fans continuously whenever there is a machine on-line. Allow the condenser water temperature to drop as low as the refrigeration equipment will permit, at which point the cooling tower fans can be cycled on and off.

ANALYSIS

To calculate potential energy savings due to reduced condensing temperature, refer to Figure 8-51.

(A) Then establish the percentage increase in COP and calculate the revised COP.

(B) Substitute the improved COP for the present COP on the Annual Heating and Cooling Tabulation Form Figure 8-11.

Conclusion: All of the figures pertaining to future energy use with regard to this ECO should now be calculated and carried through the appropriate forms as explained in Chapter VII. This will determine whether the costs of implementing and maintaining this ECO are justified in terms of future savings for the building owner.

ECO #7: REDUCE OPERATING TIME OF ELEVATORS AND ESCALTORS

THEORY

Elevators and escalators account for one to four percent of the electrical energy of office buildings and large department stores. The amount of power required annually to operate an elevator is a function of the height of the building, the number of stops, passenger capacity and load factors, and efficiency of the hoisting mechanism. For example, a 2,500 lb. capacity "local" elevator making 150 stops per car-mile consumes about 5 kwh per car-mile, while an "express" elevator making 75 stops per car-mile consumes about 4 kwh. A 4,500 lb. capacity elevator in a 12-story department store stopping at every floor will use 13 kwh per car-mile.

Consumption will vary between elevators of the same capacity depending on the type of hoisting motor and control; whether the elevator is hydraulic, geared, or gearless; and the kind of service and percentage of load offset by the counterweight.

PROCEDURE

(A) Select speeds that are as slow as possible, while keeping maximum waiting time to no more than two minutes. Seek the assistance of the elevator manufacturer in studying traffic patterns and rescheduling elevator service. Where multiple elevators are installed, reduce the number in service during light traffic periods.

(B) Select the automatic control program for the lowest speed and heaviest loading, reducing the number of elevator units traveled per year to conserve energy.

(C) Where motor generator (MG) sets are installed in existing buildings, thus drawing power whether or not the elevator is running, de-energize the MG sets with 7-day timers for periods when some of the elevators are not required. Consider changing from MG sets to more efficient ICR controllers. (The same applies to escalators.)

(D) When choosing new elevators as replacements or additions, select the most efficient type for the application. The three major types of elevators are hydraulic, geared and gearless. The limitations of each must be considered in evaluating its efficiency. Hydraulic elevators are used for speeds of about 200 ft/min, geared elevators for about 300 ft/min and gearless elevators for higher speeds and high-rise buildings. Electric elevators with counter-weights use less energy than hydraulic elevators.

ANALYSIS

If an elevator or escalator can be turned off except during rush hours, calculate the savings by multiplying the off-hours times hourly consumption. Enter the new yearly consumption on the Distribution Load Survey Form (Figure 8-10).

The cost of implementing the above suggestions, (if investment is required), along with the circulation requirements of the particular building, must be carefully evaluated and compared with anticipated savings.

Example

Assuming 35% equivalent full-load operation, escalator energy consumption may vary from 1.3 kw per hour for a 32-inch wide model operating at 90 feet per minute with a 14-foot vertical rise, to 3.0 kw per hour for a 48-inch wide model operating at 90 feet per minute with a 25-foot vertical rise. Escalators consume energy whether they are carrying passengers or not.

Conclusion: All of the figures pertaining to future energy use with regard to this ECO should now be calculated and carried through the appropriate forms as explained in Chapter VII. This will determine whether the costs of implementing and maintaining this ECO are justified in terms of future savings for the building owner.

C. Lighting System Energy Conservation Opportunities

Page

C. ECO's #8-15: Lighting System

Introduction

Except in buildings that are electrically heated, lighting is a major consumer of electrical energy and accounts for the most significant portion of total energy usage in offices and retail stores. When lights are left on in areas which are unoccupied or unused for lengthy periods of the week-- in religious buildings and outdoor parking lots, for example --the inadvertent waste of energy often approaches or exceeds the energy used by other buildings during the same period. The cost of this waste for one year may equal the initial cost of installing automatic controls to eliminate it.

Electric lighting is also the major contributor to internal heat gain in stores and core areas of office buildings. Because it increases the cooling load, electric lighting causes an additional expenditure of energy to operate the refrigeration system, including air-handling units, chilled water, and cooling tower or condenser fans.

Possibilities for conserving energy through modifications in lighting will be discussed more thoroughly in the succeding ECO's. Some of the overall guidelines to be followed are:

(A) Reduce the illumination levels. This is a low or no cost ECO that normally only requires eliminating lamps or using lower wattage lamps.

(B) Under "Operational ECO's", lighting was turned off if switching existed. Further "off" time can result from new switching and or use of natural light.

(C) Improve the efficiency of lenses, lamps fixtures and ballasts. In many cases, the improved efficiency will permit reduced lighing levels.

Without major modifications to both the pattern of lighting fixtures and the fixtures themselves, it may be possible to attain a two watt per square foot conservation goal in existing offices and stores while maintaining proper lighting. Lighting levels on tasks have traditionally been measured by footcandles. A room with a high level of footcandles, however, does not necessarily have a high level of effective illumination. Certain impediments to visual acuity, such as veiling, reflection and disability glare, are not accounted for in footcandle measurements. The higher footcandle levels and added wattage may contribute nothing more than ineffective illumination.

Lighting Terminology

The Illuminating Engineering Society (IES), which in the past has set lighting standards based on footcandles, in 1972 adopted a measure of effective light called Equivalent Sphere Illumination (ESI). ESI is a measure of lighting for tasks that accounts for both quality and quantity in the lighting system. Due to the complexity and time involved in ESI calcuations, consider using computer simulation programs (now available through some fixture manufacturers) to approximate ESI footcandles. This method should be employed for areas with critical tasks. When high ESI lighting values are provided, they will yield very effective lighting at low energy input.

It is a general rule that all light sources except low pressure sodium, decline in their output of light (measured in lumens) as the remaining useful life of the lamp decreases. This decrease is called the Lamp Lumen Depreciation (LLD).

Dirt accumulation on lamps, fixtures, and lenses further reduces the quantity of light falling on the task. This reduction in available light is called the Luminaire Dirt Depreciation (LDD). Combined with the Lamp Lumen Depriciation (LLD), this comprises the maintenance factor (MF). The maintenance factor is used in lighting calculations to determine how many fixtures and lamps must be installed initially to give the designed illumination when lamp and dirt depreciation have set in. Such an installation will obviously provide more lumens on the task initially than required, and if light fall-off can be reduced by improved maintenance, some of the fittings and lamps can be removed without reducing illumination below the levels required.

These and other opportunities for reducing energy consumed and wasted by lighting are discussed in the following pages.

ECO #8: REDUCE ILLUMINATION LEVELS

THEORY

There are in most buildings numerous opportunities to conserve energy for lighting by reducing illumination levels where they need not be high and removing or disconnecting superfluous lighting altogether. A uniform modular lighting pattern of general illumination, throwing light equally on all areas regardless of task, may waste up to 50 percent of the energy used for lighting in a building. (The use of interior space often changes with time, and original patterns of lighting may be wasteful for current use.)

Quality of light is generally as important as quantity of light: a lighting system characterized by high quality illumination can reduce energy consumption for illumination by as much a 25 to 33 percent while maintaining the same visual environment as that provided by a system of lower quality. The efficacy of a lighting system is reduced by two factors: (A) discomfort glare caused by extremely uneven luminance in the field of view and (B) veiling reflections. Visual comfort probability (VCP) is a measure of discomfort glare. For satisfactory VCP brightness, ratios in the field of view should be restricted to 3:1 for immediate background and 10:1 for more remote background. To reduce veiling reflections, lighting fixtures should be kept out of the zone from which reflections originate-- the area in front of the task and in the viewing angle of incidence.

PROCEDURE

(A) Remove as many lamps as possible in situations where those remaining provide the desired level of illumination. When removing fluorescent or high intensity discharge lamps, also remove the ballast (or disconnect it in place). If it is left connected to the power source it will continue to consume energy even in the absence of lamps. When two-lamp fluorescent fixtures are mounted in a row, remove lamps in alternate fixtures within each row (rather than removing an entire row) to derive higher quality lighting. In order to maintain the recommended lighting level it may be necessary, after removing lamps, to use increased output lamps in place of some of the remaining lamps. Even so, there should be a net savings of energy. (Refer to Figure 8-35 to determine appropriate footcandle output with regard to common lamp replacements.)

(B) Consult Figure 9-5 ("Recommended Lighting Levels") for suggested levels in specific areas of the building. If several tasks requiring different levels of illumination occur within the same space, first consider their visual severity and then modify maintenance procedures, redecorate the area, and implement changes in the lighting system while reducing illumination levels to the appropriate level for each task.

(C) If one task with a critical lighting requirement is confined to a specific work area - i.e. drafting table, typewriter, desk top - in the midst of a larger work area with less critical requirements, provide a lower general illumination level for the overall area and a portable light at each critical task to raise the level of illumination locally. Use fluorescent portable lamps in favor of incandescent.

(D) Lower the lighting fixtures where increased efficiency will result. It is generally less costly to rehang fixtures than to lower the entire ceiling. The choice of options depends upon the comparative costs and benefits which should be analyzed for the proper alternative. The energy saved by dropping the mounting height from 14' to 9' in a 30' x 40' space, for example, while maintaining the same illumination level, would be 10 percent of the annual energy presently consumed for lighting the same area.

Also consider lowering stem or chain mounted fixtures in existing stores, supermarkets, warehouses, and high-ceilinged lobbies. In supermarkets, fixtures are often mounted on the ceiling, 14 to 18 feet above the floor, with the spaces lighted at 100 to 150 footcandles-- more light than necessary to read the labels of packages on the lower shelves. In cases like that, lowering fixture mounting heights and reducing the wattage of lamps may reduce annual lighting power and costs by 50 percent or more.

(E) Convert uniform lighting systems to non-uniform (or selective lighting), so that each distinct functional area within the building and the discrete tasks which occur within the same room are lighted only to the lighting levels and quality required for each task, and only for the period when the tasks occur. Refer to Figure 9-5 ("Recommended Lighting Levels").

(F) In order to select the most appropirate measures, make a careful analysis of task requirements, expected duration, possible future relocation, the quality of illumination required for specific tasks, and the frequency with which they may change.

(G) Measure existing light levels in the spaces being considered for improvement in efficiency. Light levels in standard footcandles can be determined with portable illumination meters such as a photovoltaic cell connected to a meter calibrated in footcandles. The light meter should be accurate to about ± 15 percent over a range of 30 to 500 footcandles and ± 20 percent for 15 to 30 footcandles. The meter should be color-corrected (according to the CIE Spectral Luminous Efficiency Curve) and cosine corrected. Generally, measurements refer to average maintained horizontal footcandles at the task or in a horizontal plane 30 inches above the floor.

Measurements should be made at any representative point between and under fixtures. An average of several readings may be necessary. Daylight should be excluded during illumination-level readings for a true determination of illumination level without light contribution from daylight.

FIGURE 9-5 RECOMMENDED LIGHTING LEVELS*

With proper attention to quality, the following levels should generally be adequate for the tasks cited:

(A) Circulation areas between work stations: 20 footcandles.

(B) Background beyond tasks at circulation area: 10 footcandles.

(C) Waiting rooms and lounge areas: 10-15 footcandles.

(D) Conference Tables: 30 ESI footcandles with background lighting 10 footcandles.

(E) Secretarial Desks: 50 ESI footcandles with auxiliary localized (lamp) task lighting directed at patper holder (for typing) as needed. In secretarial pools, 60 ESI footcandles.

(H) Kitchens: Non-uniform lighting with an average of 50 footcandles.

(I) Cafeterias: 20 footcandles.

(J) Snack Bar: 20 footcandles.

(K) Testing Labs: As required by the task, but background not to exceed 3 to 1 ratio in footcandles.

(L) Computer Rooms: As required by the task, (consider 2 levels, 1/2 and full). In computer areas, reduce general overall lighting levels to 30 footcandles and increase task lighting for critical areas for input. Too high a level of general lighting makes it difficult to read the self-illuminated indicators.

(M) Drafting: Full time 80 ESI footcandles at work station, part time 60 footcandles at work stations.

(N) Accounting Offices: 80 ESI footcandles at work stations

*All levels are average ESI footcandles unless otherwise noted. Where applicable, refer to health and safety codes and federal standards (OSHA) for minimum lighting specification.

The goal of the above standards is to reduce store and office lighting energy usage to less than 2 watts/sq. ft. gross floor area, or (2.5 watts/sq. ft. net area) and 1.5 watts/sq/ ft. for religious buildings. To determine net area, subtract from gross building floor area the corridors, storage rooms, lobbies, mechanical equipment rooms, stairwells, toilet rooms and other unoccupied, or seldom occupied areas.

The suggested illumination levels for office buildings listed above agree closely with new standards recommended by the General Services Administration for public office buildings. Keep in mind, however, that even lighting at lower intensities is very wasteful if lamps are burning when not needed.

ANALYSIS

In order to calculate the energy which might be saved by implementing one or more of the above suggestions:

(A) Determine the type(s) of lighting change(s) to be investigated.

(B) Calculate the reduced wattage to be consumed by the modified lighting system(s).

(C) Substitute this new wattage figure for the present figure on the Building Load Survey Form (Figure 8-10).

Conclusion: All of the figures pertaining to future energy use with regard to this ECO should now be calculated and carried through the appropriate forms as explained in Chapter VII. This will determine whether the costs of implementing and maintaining this ECO are justified in terms of future savings for the building owner.

ECO #9: USE ONLY NECESSARY ILLUMINATION

THEORY

Contrary to common belief, letting a light burn rather than turning it off NEVER saves electrical energy. Thus, when electric lighting is not required it should be turned off. (While frequent switching may in some cases shorten lamp life, the saving in electrical power will more than compensate.)

PROCEDURE

In order to reduce unnecessary illumination, take the following steps:

(A) When surveying the building, list all room areas and the number of occupants (and their activities) in each space and the length of time that each space is occupied and the location of switches. A careful analysis will reveal many opportunities for reducing the hours of operations of lightsor shutting off lights completely.

(B) Encourage users to turn off unused lights when not needed. If there are insufficient switches to turn off unwanted lights, install additional switches. Individual light switches should be considered particularly for the following areas: (1) spaces with multiple fixtures which are not all required all the time; (2) areas intermittently occupied; (3) areas where daylight is available for illumination (i.e., perimeter zones); (4) halls of worship and other large areas where occupancy may take up only some of the space at any one time.

(a) For adding switching in existing buildings, line-voltage or low-voltage surface-mounted type conductors are available which adhere to walls or ceilings. Ceiling switches (pull chain type) can be added where wall installations are not feasible.

(b) For large areas, add remote control lighting contactors or lighting relays (to operate multiple circuits) above the ceiling or at the lighting distribution panel. Remote switches can be either line-or low-voltage.

(c) Control site lighting and parking lot lighting with time switches and/or with photo-electric cells; the latter are also effective in controlling lighting fixtures at the interior perimeters of the building where daylight is sufficient for illumination.

(d) A frequency controlled relay, now available, can be added at individual fixtures or at the circuit-breakers which control a number of fixtures. The relay is controlled by an activator which superimposes a special command frequency over the existing wiring system. At present, two command frequencies are available. Each activator will handle 300 forty-watt rapid-start ballasts located within 500 feet of the activator which is mounted at the panel. The activator can be controlled by remote control or local switch, time clock or photo-electric switch. One activator can control the lights on any number of separate circuits.

(C) Add pilot lights outside all rooms that are infrequently used where there is no other external indication that lights have been left on. Pilot lights should also be added to indicate when loads are energized at remote locations such as site lighting, sign lighting, snow melting, ovens, blueprint and reproduction equipment, mechanical spaces, penthouses, etc.

Pilot lights can be surface mounted or recessed and can usually be added in parallel with the circuit or the load to be monitored. To conserve energy, use neon type pilot lights rather than incandescent lights. In buildings with conduit runs from the switch boxes to the lighting fixtures, it may be possible to pull a neutral wire down to an external switch (if not already available there) and then to install a combination switch and pilot light in the existing switch box.

(D) Multi-level ballasts with two or three levels of lamp control are available for 430 milliamp fluorescent lighting fixtures. A percentage reduction in power actually exceeds the percentage reduction in lumen output at the lower levels of multi-level ballasts.

(a) Two-level ballasts can replace single-level ballasts in exist-fluorescent installations without a change in the fixture and with only slight modifications, if any, to mounting and wiring. A three-level ballast can also be used with existing fixtures, but it requires mounting and wiring changes. (At present, level switching is available only at the ballast, and wall switches with relays must be used if remote switching is desired. Other equipment for remote control of multi-level ballasts without relays is currently in development.)

(b) The lumen output of 400 watt mercury vapor lamps can be reduced by 50 percent through modifications to the ballast system that permit switching from a 33-microfarad capacitor to a 22-microfarad capacitor. Lumens are reduced by 50 percent and wattage about 40 percent. The reduced power factor militates against a one-to-one reduction. If the Power Factor of the entire building is high, the lower Power Factor of the lamps need not be a serious problem.

(E) When frequent light-level changes are required in areas of transient occupancy, dimmers are a better alternative than multi-level switching, multi-level ballasts, or lamp replacement.

(F) To ensure the success of any switching program, instruct all occupants in the use of available switching. Ensure that any changes to switching are in compliance with electrical codes. Identify all ganged switches to correspond with the lights controlled. Color-code switches and institute a program of use (i.e. Blue 7 a.m. - 6 p.m., Red 9 a.m. - 1 p.m., etc.).

ANALYSIS

To calculate the savings which might be expected from the implementation of one or more of the above procedures:

(A) Select and analyze those procedures to be implemented.

(B) Calculate the new (reduced) total operating hours for each of the lighting systems to be modified.

(C) Substitute the new operating hours for the present operating hours on Building Load Survey Form (Figure 8-10).

Conclusion: All of the figures pertaining to future energy use with regard to this ECO should now be calculated and carried through the appropriate forms as explained Chapter VII. This will determine whether the costs of implementing and maintaining this ECO are justified in terms of future savings for the building owner.

Examples:

(A) The cost of adding switching depends upon the type of switching system selected and the existing fixture circuiting. Time switches cost from $30.00 to $150.00. Photo-electric cells cost about $75.00. Frequency controlled relays cost about $20.00 per relay, and for the associated controller approximately $300.00 per channel. A wall switch costs about $25.00. Low voltage relays cost as little as $6.00, and low voltage transformers cost approximately $25.00 each. A surface mounted neon pilot light can be installed over a door with wiring in surface raceway to a ceiling space and b.x. cable (12') to the nearest fixture in a room for as little as $25.00.

(B) Multi-level fluorescent ballasts cost about twice as much as standard ballasts, but if the low level is used most of the time, the cost may be repaid within a few years. The cost of installing a second capacitor in mercury vapor ballasts for multi-level switching varies widely when it can be done at all.

(C) Where new electrical fixtures are contemplated, consider using a 3 level lighting fixture in place of a 3 level ballast. These are offered as 3 lamp 3'x4' fixtures with standard 40 watt lamps. The center lamps are wired in tandem. A separate switching circuit is required for each light level. One manufacturer offers the following data:

3 lamps: 124 fc at 3.7 watts/sq ft
2 lamps: 83 fc at 2.5 watts/sq ft
1 lamp: 37 fc at 1.2 watts/sq ft
Reflections: Ceiling 0.80 - Walls 0.40 - Floor 0.6

(D) The least expensive application is dimmers for incandescent lights - materials cost about $30.00 for a 600-watt dimmer and about $125.00 for a 2000-watt dimmer. Fluorescent lamp dimming is more expensive as ballasts must be replaced and a wall control unit installed. Ballasts cost about $15.00 each, and $70.00 per dimmer control module which will handle up to twenty 40-watt lamps. Dimmers for 400- and 1000-watt mercury lamps, which also require ballast replacement and special control modules, are more costly than fluorescent dimmers. (Costs are highly variable, but may be approximately $75.00 for an additional ballast and from $400 to $2,500 for the control unit, depending on the number of lamps dimmed.

ECO #10: UTILIZE NATURAL LIGHT

THEORY

Natural light is particularly effective as a primary source of illumination in perimeter spaces. The amount of available daylight in a building is a function of operating hours, latitude, weather, time of year, air quality, window size and location, shading and glazing details, reflectivity of interior surfaces, and furnishings. Control of natural light, for effective use and integration with artificial light, is important, because the amount of natural light available varies. Control may be manual or automatic. Often blinds or drapes, which should be closed only for short periods of excessive sunshine or glare, remain closed all day with the electric lights turned on, when daylight could be used.

Emphasis should be placed on proper control of daylight; otherwise the heat gain imposed on the cooling system may outweigh savings in energy for lighting. Indirect daylighting, however, generates less heat per lumen of light than electrical lighting. When equivalent lumens from daylight replace electrical lighting, savings in cooling load will occur.

The luminous efficacy (lumens per watt) of various light sources as compared to daylight are shown below (no allowance for ballast of luminaires):

Low Pressure Sodium	183 lumens/watt
Natural light	120 lumens/watt (varies)
High pressure (HD)	105-120 lumens/watt
Metal halide	85-100 lumens/watt
Fluorescent	67-91 lumens/watt
Mercury vapor	56-63 lumens/watt
Incandescent	17-22 lumens/watt

PROCEDURE

Natural illumination may be improved through better use of the following:

<u>Skylights</u>: Modify or reactivate skylights to provide a daylight supplement to electric lighting without glare or excessive heat. The amount of illumination available from skylights varies with the following factors:

(A) Duration, volume, and intensity of available sunlight; adjacent shading; ground reflectance.

(B) Area and location of skylight.

(C) Distance from floor to skylight.

(D) Transmission properties of skylight.

(E) Design and structure of skylight with regard to orientation of transparent or translucent surfaces.

(F) Reflectance of interior surface of skylight well. (Refer to IES Handbook, Reference 40, for methods of measuring solar illumination through skylights.

In order to reduce heat loss, heat gain, and glare, consider the following options:

(A) Paint skylights with white or translucent paint.

(B) Add prismatic lighting lenses below the skylights to reduce glare.

(C) Provide an external louver above the skylight to reduce solar heat gain, heat loss, and glare.

(D) Add tinted or reflective film to admit sunlight while reducing glare.

(E) Install a movable shutter above the skylight to reduce heat gain and glare in summer, while allowing solar radiation to be used for offsetting heat loss in winter.

(F) Paint interior of light well with a light color to increase illumination. (Caution: Where skylights are used as an automatic smoke hatch, modifications must not change this function.)

Windows: When major building renovations or additions are planned, consider adding windows with multiple glazing and thermal barriers in cold climates. The number, size, and type of windows to be used and the shielding or screening which best accompanies them can be determined from the recommendations for window treatment and selection of types in ECO #17.

Reflectors: Install reflectors at windows to increase the amount of daylighting. Horizontal or vertical reflectors located on the exterior of the building can often be added without significantly changing the appearance of the building. The effectiveness of windows for daylighting can be increased by 25 percent or more with proper reflector devices. (The light contributed by the reflecting surface should be directed onto the ceiling. Diffuse daylighting provides more lumens per watt than the artificial sources it replaces and the heat gain and load on the air conditioning system will be less than for an equal quantity and quality of electric lighting.)

Paint light wells and skylighted interior courts, lighter colors to increase reflectance onto windows and adjacent surfaces. At windows close to grade, a border of white stone or concrete serves well as a reflector.

Other Factors To Consider in Daylighting:

(A) If electric lighting is turned off when daylight from skylights is available, as much as 40 watts of electric power can be saved per sq ft of skylight area. The amount of energy saved for lighting using skylights must be analyzed in comparison to the increased amount of energy required for heating and the increase or reduction in heat gain and cooling load in order to determine the net yearly increase or reduction in energy due to skylights. Especially consider installing skylights in existing buildings when reroofing--if the energy conservation balance is favorable --as the cost of installation is lowest at that time.

(B) Compared to opaque surfaces, windows or skylights reduce the energy required for lighting, reduce the energy required for cooling by reducing heat gain from lights, reduce heating load in the winter by permitting more solar radiation to enter, but may increase cooling load in summer due to solar radiation and conduction losses (well-shaded windows reduce this liability).

(C) When using daylighting, therefore, light levels must be controllable; if they are significantly higher than required, the excess heat gain (which increases the cooling load) could offset the energy saved in electric lighting. Provide shades, drapes, or blinds at the windows to allow control of the intensity of light.

ANALYSIS

To calculate the savings which might be expected form the implementation of one or more of the above suggestions:

(A) Select and analyze the procedure(s) to be implemented.

(B) Calculate the new (reduced) operating time of the lighting system based on increased use of natural light.

(C) Substitute the new operating time for the present operating time on the Building Load Survey Form (Figure 8-9).

Conclusion: All of the figures pertaining to future energy use with regard to this ECO should now be calculated and carried through the appropriate forms as explained in Chapter VII. This will determine whether the costs of implementing and maintaining this ECO are justified in terms of future savings for the building owner.

ECO #11: IMPROVE THE EFFECTIVENESS OF EXISTING LIGHTING FIXTURES

THEORY

By maintaining a higher light output from existing fixtures, it may be possible to reduce the wattage of lamps in each fixture and, in some cases, to reduce the number of fixtures in service without a reduction in illumintion. A number of variables affect system maintenance. The most important are lamp lumen depreciation (LLD) which is the drop in lumen output over lamp life due to lamp characteristics; and luminaire (lighting fixture) dirt depreciation (LDD), which is the drop in lumen output due to the accumulation of dirt on the interior reflective surfaces of the fixture and lens (see Figure 8-36).

The combination of LLD and LDD results in an average reduction in light output of about 25 percent from fluorescent fixtures (varying with room atmosphere and lamp types) over time with the maintenance procedures in general practice today. In some cases, louvers or lenses may not be required, and their removal will permit increased fixture output. Improving the reflectance of wall and ceiling surfaces through the use of lighter colors, regular cleaning, etc. will also improve the effectiveness of existing lighting.

PROCEDURE

Improve lighting maintenance as follows:

(A) Replace existing lamps with new ones having a lower LLD over the rated life of the lamp (initial light output is often increased also). Fluorescent lamps with better LLD factors may cost somewhat more, but the investment can usually be recovered by lower expenditures for energy and maintenance. Refer to Figure 8-37 for lamp lumen depreciation, and to Figure 8-39 for fluorescent group relamping. Group relamping will give a higher level of maintained footcandles throughout the lives of the clustered lamps.

(B) Improve the maintenance factor by setting up a maintenance schedule for regularly cleaning fixtures and lamps and by instituting a relamping program. Clean a fixture at least every time it is relamped. When relamping and cleaning are combined, the additional labor is very small. In extremely dirty atmospheres, clean the fixtures between lamp replacements as well. Clean the lenses and interior housings. Use an anti-static compound to reduce electrostatic dirt accumulation. (Refer to Figures 8-36 and 8-38 for the effects of lamp and fixture cleaning and improved maintenance factors.)

(C) When no critical reading or writing tasks are involved, and where some glare from exposed lamps is acceptable (check with codes where applicable), remove louvers or lenses from fixtures. Consider this option, for example, in corridors, storage areas, stores with high ceilings, equipment rooms, snack bars, etc.

(D) Increase the reflectance of ceilings, walls, and floors by cleaning or by painting with colors of higher reflectance: more illumination on tasks with the same lumen output will result. Increasing the number of lamps and luminaires without adverse effect on commended lighting levels. Greater reflectances enhance the performance of natural daylight as well as artificial lighting. For example, a 25x25 ft room with a 9-ft ceiling and a ceiling reflectance of 80%, wall reflectance of 70% and floor reflectance of 30% will provide 35% more footcandles (with no increase in power consumption) than the same room with a ceiling reflectance of 50%, wall reflectance of 30% and floor reflectance of 10%. Caution: high gloss finishes may reduce visual comfort by increasing glare. Discretion should be used when redecorating.

Using the improved lighting depreciation and reflectances, reduce the wattage in fixtures to provide the recommended lighting levels (Figure 9-4).

ANALYSIS

In order to calculate the savings which might be expected from the implementation of one or more of the above suggestions:

(A) Select the procedure(s) to be implemented.

(B) Fefer to the Figures 8-36, 8-37, 8-38, and 8-40 for assistance in analyzing the effects of the selected changes. Minimum lighting levels must be maintained (see Figure 8-38) under the lowest lumen output of the fixtures. If, for example, the maintenance factor (Figure 8-38) can be improved from .60 to .70, the decreased output of the fixture changes from 40% to 30%. Assuming the lighting level is correct under present conditions, an opportunity exists to reduce the wattage by 14%. This procedure for determining wattage reduction is similar for all of the procedures.

(C) Calculate the new (reduced) total wattage demand of the modified lighting system(s).

(D) Substitute the new wattage figure for the present figure on the Building Load Survey Form (Figure 8-9).

<u>Conclusion:</u> All of the figures pertaining to future energy use with regard to this ECO should now be calculated and carried through the appropriate forms as explained in Chapter VII. This will determine whether the costs of implementing and maintaining the ECO are justified in terms of future savings for the building owner.

ECO #12: USE MORE EFFICIENT LENSES

THEORY

Lenses are provided to reduce glare (i.e. increase the visual comfort probability). The use of lenses reduces the amount of light from the lamps, and that reduction for a particular fixture is reflected in the coefficient of utilization. The visual comfort probability, the coefficient of utilization, the candlepower distribution, and a number of other characteristics depend partly on the type of lens that is used. Various type of lenses are available to achieve desired candlepower distributors and visual comfort probabilities. In some cases, efficiency may vary among various lens types otherwise providing similar characteristics. A more efficient lens can increase the effective light output of each fixture, thus allowing a reduction in the lamp wattage and/or the number of fixtures required.

PROCEDURE

Remove lenses where they are not required for glare control (corridors, toilet rooms, non-critical task areas), improving the fixture's coefficient of utilization (CU) and permitting a reduction in the wattage required (with lower wattage lamps) without reduction in the number of footcandles on the task or background.

Replace inefficient lenses and consider both efficiency and quality when choosing lenses for new fixtures. To evaluate the efficiency of lenses in existing fixtures, use the fixture manufacturer's CU for a standard fixture. Compare the CU of the present lens to the CU of lenses being considered as repleacments. The improvement in light levels will be directly proportional to improvement in CU. In choosing lenses, the following should be considered:

(A) Evaluate the needs of the task being lighted, as well as the characteristics of the fixture and lamp involved.

(B) Evaluate comparitive lens efficiencies by comparing CU's. In making such comparisons, use the same room reflectance and room cavity ratio with the CU's (from manufacturer's tables) for the various lenses available.

(C) Where visual performance is critical, it may be better to use a less efficient lens with better glare control. The quality of lighting will be improved, which may permit a reduction in footcandles (and wattage) for the same visual acuity.

(D) When higher efficiency light sources such as high pressure sodium are used, less efficient lenses providing more effective control may be required.

The relative efficiency and general characteristics of the most common lenses, compared to prismatic plastic, are as follow:

Lenses for Fluorescent Fixtures

(A) Prismatic plastic and glass: Generally the most efficient of its type for the degree of glare control. Best suited for all finished office space applications.

(B) Plastic 1/2" x1/2" or 3/8" x 3/8" louver: Very inefficient. Should be used only where extreme dirt build-up is a problem. About 33 percent less efficient than prismatic plastic.

(C) Parabolic wedge louver: Very inefficient in transmitting light but provides excellent visual comfort and glare control. This lens is practical for use where the task requires a high level of visual acuity with low glare but may increase veiling reflections where the task is specular. About 45 percent less efficient than prismatic plastic.

LENSES FOR INCANDESCENT AND HID

(A) Fresnel Type Lens: The most efficient of its type, best suited for all recessed fixture applications.

(B) Opal or White Diffusers: Much less efficient in delivering light to a work surface. Generally 30 percent less efficient than Fresnel type lens. Good light diffusion.

ANALYSIS

To calculate the savings which might be achieved through the installation of more efficient lenses: (A) determine the new (reduced) annual wattage demand based on the proposed new lenses; (B) substitute the new wattage for the present wattage on the Building Load Survey Form (Figure 8-9).

Conclusion: All of the figures pertaining to future energy use with regard to this ECO should now be calculated and carried through the appropriate forms as explained in Chapter VII. This will determine whether the costs of implementing and maintaining this ECO are justified in terms of future savings for the building owner.

ECO #13: INSTALL MORE EFFICIENT LAMPS

THEORY

The efficiency of lamps is measured in lumens or usable light produced per watt input. Selecting lamps with higher lumens per watt often permits the removal of some lamps, provided the lumens produce the required footcandles. More efficient lamps also impose reduced heat loads on the air conditioning system. In supermarkets and produce markets, reduced lighting reduces loads on commercial refrigeration systems as well. In winter, any heat lost by the reduction in wattage can generally be supplied more efficiently by the heating system itself.

Such energy savings can often be amplified by modifying an existing fixture to accept a different type and higher efficiency lamp. (Figures 8-40 and 9-6 indicate the lumens per watt for various lamp types.) More efficient lamps save energy by permitting use of lower wattage lamps or a reduction in the number of fixtures used. Lamp efficiencies vary between lamps consuming the same number of watts and even between lamps of the same type. Different colors, shapes, gaseous fills, cathode construction and internal coatings cause those variations. For example, a fluorescent 40-watt, T-12 lamp, rated at 430 millamperes, can have a rating from around 53 lumens to around 92 lumens per watt. A natural white fluorescent lamps at 2100 lumens (53 lumens/watt) provides 1/3 less foot candles than cool white fluorescent lamps at 3200 lumen (80 lumens/watt) Accordingly, it may be desirable to relamp with lower wattage lamps as an alternative to removal of lamps.

Esthetics, size, efficiency, color, initial cost, operation, and maintenance cost all must be considered in selecting lamps. Energy conservation in lighting is not synonymous with sacrificing quality. Quality is not a measure of quantity alone: it is obtained by the intelligent application of many interrelated factors. Correctly applied, energy conservation can increase the quality of lighting while reducing operating costs.

PROCEDURE

Convert incandescent fixtures or lower efficiency High Intensity Discharge fixtures to accommodate more efficient lamps where feasible. Conversion from incandescent to mercury vapor or some other high-intensity discharge source requires the addition of a ballast (when self-ballasted lamps are not used). Ballast can be added at each fixture or mounted remotely (provided the wire from fixture to ballast is sized properly). Fixtures which use mercury vapor and metal halide lamps can often be converted to high pressure sodium where color rendition is not critical. Conversion from any other High Intensity Discharge source to high pressure sodium will require installation of a new ballast.

The variation in efficiency of lamps (measured in lumens per watt) between incandescent lamps and high pressure sodium vapor lamps is about 1 to 7 (an extreme example). The intended function of a given lamp, of course, will sometimes preclude the application of a higher-efficiency lamp. For example, using a very bright concentrated light source, i.e. a high pressure sodium vapor lamp, in an office with a low ceiling would not provide comfortable lighting. Instead, fluorescent lamps with outputs of up to about 93 lumens per watt will be both effective and energy conserving. A 20-watt incandescent lamp in a fixture directly attached to a paper holder at a typist's desk could provide 60 footcandles on a manuscript with low ambient background lighting. The same illumination level would require 11 watts of fluorescent lighting from a ceiling fixture, even though the fluorescent lamp produces more lumens per watt.

Consider the use of higher-efficiency lamps for site lighting, parking lots, canopies, advertising signs, display counters and other special applications, as well as for interior lighting.

ANALYSIS

To calculate the savings which might be expected from the installation of higher-efficiency lamps, refer to Figures 8-40 and 9-6 Then:

(A) Determine the new (reduced) annual wattage demand of the lighting system based on the proposed modifications.

(B) Substitute the new wattage for the present wattage on the Building Load Survey Form (Figure 8-9).

Conclusion: All of the figures pertaining to future energy use with regard to this ECO should now be calculated and carried through the appropriate forms as explained in Chapters VII. This will determine whether the costs of implementing and maintaining this ECO are justified in terms of future savings for the building owner.

FIGURE 9-6 RELAMPING CHANGES TO INCREASE LIGHTING EFFICIENCY WITHOUT MODIFICATIONS TO LUMINAIRES OR BALLASTS

(All lamps operating at 120 volts)

Incandescent Lamps

Present Light Source						Replacement Light Source				
Watts	Lamp Description	Lamp Volts	Lumens @ Socket Volts	Watts @ Socket Volts	Hrs. Life @ Socket Volts	Watts	Description	Lumens	Hours Life	Reduction in Watts
40	40 A/99	130	323	35	7,000	25	25 A	235	2,500	10
60	60 A/99	130	597	53	7,000	40	40 A	455	1,500	13
						40	40 A/99	420	2,500	13
75	75 A/99	130	770	66	7,000	54	54 A	775	3,500	12
						55	55A	670	2,500	11
						60	60 A	870	1,000	6
						60	60 A/99	775	2,500	6
100	100 A/99	130	1,147	88	7,000	75	75A/99	1,190	750	13
100	100 A21/99	130	1,109	88	7,000	75	75 A/99	1,000	2,500	13
150	150 A23/99	130	1,779	132	7,000	90	90 A	1,290	3,500	42
150	150/99	130	1,771	132	7,000	92	92 A	1,490	2,500	40
						100	100 A	1,750	750	32
						100	100 A/99	2,500	1,490	32
150	150 R/FL	120	1,870	150	2,000	75	75PAR/FL	765	2,000	75
150	150PAR/FL	120	1,740	150	2,000	250	Q 250 Par38	3,220	3,000	50
200	200 A/99	130	2,626	176	7,000	150	150 A	2,880	750	26
200	200/99 IF	130	2,510	176	7,000	150	150 A23/99	2,310	2,500	26
				135		135	135A	2,100	3,500	41
				41		138	138 A	2,300	2,500	38
						160	H3B160/SS/M	2,700	20,000	16

FIGURE 9-6 (con't)

Incandescent Lamps

Present Light Source						Replacement Light Source				
Watts	Description	Lamp Volts	Lumen @ Socket Volts	Watts @ Socket Volts	Hrs.Life @ Socket Volts	Watts	Description	Lumens	Hrs. Life	Reduction in Watts
60	60/99IF Extended Service	120	740	60	2,500	54	54/99IF Extended Service	645	2,500	6
60	60A19/35 Industrial Service	120	670	60	3,500	54	54A19/35 Industrial Service	590	3,500	6
100	100/99IF Extended Service	120	1,480	100	2,500	90	90/99IF Extended Service	1230	2,500	10
100	100A21/35 Extended Service	120	1,280	100	3,500	90	90A21/35 Extended Service	1090	3,500	10
150	150A/99IF Extended Service	120	2,350	150	2,500	135	135A/99IF Extended Service	1990	2,500	15
150	150A25/35 Industrial Service	120	2,150	150	3,500	135	135A25/35 Industrial Service	1790	3,500	15
300	300	120	4,900	300	3,000	250	250PS-35 Self-Ballasted Mercury	4800	11,000	50
1000	1000	120	18,300	1000	3,000	750	750R-57 Self-Ballasted Mercury	17650	15,000	250
1500	1500	120	28,400	1500	3,000	1,250	1250 Bt-56 Self Ballasted Mercury	38000	15,000	250

FIGURE 9-6 (con't)

Incandescent Lamps

Present Light Source				Replacement Light Source				
Watts	Description	Beam Candle-power	Hrs. Life	Watts	Description	Beam Candle power	Hrs. Life	Reduction in Watts
100	Reflector Floodlight R-40	800	5,000	75 PAR-38	Projector Floodlight	1,430	5,000	25
150	Reflector Floodlight R-40	1,200	5,000	100PAR-38	Projector Floodlight	2,230	5,000	50
150	Reflector Floodlight R-40	1,200	5,000	75 PAR-38	Projector Floodlight	1,430	5,000	75
150	Reflector Floodlight R-40	1,200	5,000	100 BR-40	Floodlight	1,200	5,000	50
200	Reflector Floodlight R-40	1,600	5,000	150PAR-38	Projector Floodlight	3,450	5,000	50
300	Reflector Floodlight R-40	2,450	5,000	200PAR-38	Projector Floodlight	4,560	5,000	100
500	Reflector Floodlight R-40	3,600	5,000	250PAR-38	Projector Floodlight	5,850	5,000	250

FIGURE 9-6 (con't)

Incandescent Lamps

Present Light Source						Replacement Light Source				
Watts	Lamp Description	Lamp Volts	Lumens @ Socket Volts	Watts @ Socket Volts	Hrs. Life @ Socket Volts	Watts	Description	Lumens	Hours Life	Reduction in Watts
300	300 M/99 IF	130	3,996	264	7,000	200	200A	4,010	750	64
300	300/99 IF	130	3,996	264	7,000	200	200A/99	3,410	2,500	64
						300	HSB 300/SS/M	7,800	20,000	0
300	300	120	5,820	300	1,000	300	HSB 300/SS	7,800	20,000	0
500	500/99 IF	130	6,984	440	7,000	300	HSB 300/SS	7,800	20,000	140
500	500/99 IF	120	9,070	500	2,500	450	HSB 450/SS	9,500	16,000	50
750	750/99	130	10,934	660	7,000	500	500	10,850	1,000	160
						500	500/99	9,070	2,500	160
						450	HSB 450/SS	9,500	16,000	210
750	750 R 52	120	13,000	750	2,000	750	HSB 750R/120	14,000	16,000	0
1,000	1,000/99	130	15,246	880	7,000	750	750	17,040	1,000	130
						750	750/99	14,200	2,500	130

FIGURE 9-6 (con't)

Flourescent Lamps

Present Light Source				Replacement Light Source				
Watts	Description	Color	Lumens	Watts	Description	Color	Lumens	Reduction in Watts
40	F40T12-48"	CW	3150	34	F40/RS/EW	CW	2800	6
40	F40T12-48"	WW	3200	34	F40/RS/EW	WW	2900	6
75	F96T12Slim line	CW	6300	60	F96T12/EW	CW	5220	15
75	F96T12Slim line	WW	6400	60	F96T12/EW	WW	5340	15

FIGURE 9-6 (con't)

High Intensity Discharge Lamps

Present Light Source					Replacement Light Source					
Watts	Description	Color	Lumens	Hours Life	Watts	Description	Color	Lumens	Hours Life	Reduction in Watts
	Mercury Vapor					Mercury Vapor				
400	H33CD-400	Clear	21,000	24,000	300	H33CD-300	Clear	14,000	16,000	100
	H33G1-400/DX	Deluxe White	23,000	24,000	300	H33GL-300/DX	Deluxe White	15,700	16,000	100
400	Mercury Vapor		23,000	24,000	360	High pressure sodium		34,200	12,000	40
175	Mercury Vapor		8,500	24,000	150	High pressure sodium		12,000	12,000	25

ECO #14: INSTALL MORE EFFICIENT FIXTURES

THEORY

The lamp and fixture efficiency together determine the quantity of light transmitted into a space for each watt of power consumed. Fixture performance is indicated by its coefficient of utilization under specific room conditions, and includes the effect of lenses or reflectors. The total fixture performance in light output per watts input depends upon both lamp lumens per watt and the fixture coefficient of utilization (CU). (Consult manufacturers for the CU of particular fixtures under consideration.)

PROCEDURE

When a more efficient lamp source would suit the application but conversion of the fixture to utilize this source is not possible, consider replacing the fixture itself. Select the most efficient light source for the application, then choose a fixture with high-efficiency performance.

ANALYSIS

To calculate the savings which might be achieved through the installation of more efficient light fixtures: (A) determine the new (reduced) annual wattage demand of the lighting system based on proposed modifications; (B) substitute the new wattage for the present wattage on the Building Load Survey Form (Figure 8-9).

Conclusion: All of the figures pertaining to future energy use with regard to this ECO should now be calculated and carried through the appropriate froms as explained in Chapter VII. This will determine whether the costs of implementing and maintaining this ECO are justified in terms of future savings for the building owner.

Examples: Here is a series of comparative energy savings to be obtained from certain typical fixture replacements.

	EXISTING FIXTURES	REPLACEMENT FIXTURES
STORE	66-500 watt incandescent down lights	20-400 watt metal halide surface modules
	Total KWH/yr. = 122,100	Total Kwh/yr = 34,040
	Total savings = 88,060 kwh/yr	72% savings

	EXISTING FIXTURES	REPLACEMENT FIXTURES
	17-1000 watt incandescent pendant fixtures	14-400 watt metal halide pendant fixtures
	Total kwh/yr = 62,900	Total kwh/yr = 24,050
	Total savings = 38,850 kwh/yr	62% savings
PARKING AREA	12-1500 watt tungsten halogen lamp floodlights	4-1000 watt metal halide cluster
	Total kwh/yr = 27,000	Total kwh/yr = 6,480
	Total savings = 20,520 kwh/yr	76% savings
SIGN	2-1500 watt tungstem halogen floodlights	2-400 watt metal halide floodlights
	Total kwh/yr = 12,000	Total kwh/yr = 3,680
	Total savings = 8,320 kwh/yr	69% savings
SECURITY	6-500 watt tungsten halogen floodlights	6-150 watt high pressure sodium lights
	Total kwh/yr = 12,000	Total kwh/yr = 3,920
	Total savings = 8,080 kwh/yr	67% savings
SUPER MARKET	250-4 lamp 40 watt strip fluorescent	70-400 watt high pressure sodium down lights (where change in color rendition is acceptable)
	Total kwh/yr = 146,250	Total kwh/yr = 84,000
	Total savings = 62,250 kwh/yr	43% savings
OFFICE	Building Area	1,000,000 sq. ft.
	Office Area (fluorescent)	700,000 sq. ft.
	Support Areas (incandescent)	
	Mechanical Areas	150,000 sq. ft.
	Circulation and Storage	150,000 sq. ft.

	EXISTING	REPLACEMENT
Mechanical Area Lighting Energy Consumption	Primarily incandescent lamps used	90% fluorescent lamps used

	117,500 kwh/yr	70,500 kwh/yr
Circulation and Storage Area Lighting	85% incandescent 15% fluorescent	15% incandescent 85% fluorescent
Energy Consumption	518,400 kwh/yr	324,00 kwh/yr
Yearly Energy Consumption in these Portions of the Building	635,900 kwh/yr	394,500 kwh/yr
Total Energy Saved	635,900 kwy/yr - 394,500 kwh/yr 241,400 kwh/yr....38% savings	

ECO #15: INSTALL MORE EFFICIENT, HIGH POWER FACTOR BALLASTS

THEORY

If the power-factor of the ballast used in a luminaire (fluorescent or High Intensity Discharge) differs from that of the ballast used in the actual photometry of the luminair, the light output will differ by the same amount. The standard ballast consumes 12 to 16 watts; more efficient types consume approximately 10 watts.

The power factor of the ballasts is almost independent of the watts measured when ballasts are connected to the same load; the lower power-factor ballasts, however, contribute most to a reduction in the building's power-factor. Indeed, some utility companies have established penalty clauses in their rate schedules for lower power-factor installations. New ballasts should be of the high power-factor type. (See ECO #51 for overall power factor considerations.)

PROCEDURE

Initiate a program of replacing ballasts that need replacement with very efficient and high power-factor ballasts. Where fixtures have been delamped, disconnect from the line all ballasts that no longer supply any lamps.

ANALYSIS

When uncertified ballasts are used, there may be no reliable data available. The savings that may result from any of the following options should be carefully compared to the cost of replacing the ballasts under consideration: increasing the power factor of the system, i.e. reducing reactive power; reducing energy consumed by highly efficient ballasts; saving energy by disconnecting ballasts; increasing light output through exact ballast factors.

To calculate the savings which might be achieved through replacement of lighting ballasts: (A) determine the new (reduced) annual wattage demand based on the proposed modifications; (B) substitute the new wattage for the present wattage on the Building Load Survey Form (Figure 8-9).

<u>Conclusion</u>: All of the figures pertaining to future energy use with regard to this ECO should now be calculated and carried through the appropriate forms as explained in Chapter VII. This will determine whether the costs of implementing and maintaining this ECO are justified in terms of future savings for the building owner.

D. Building Envelope Energy Conservation Opportunities

D. Building Envelope ECO's: #16 - 20

Introduction

A major factor in the annual volume of energy consumed (and wasted) due to "building load," as discussed in Chapter III, is the energy efficiency of the building shell or envelope. The building envelope typically comprises such components as roof, exposed floors, exterior walls, windows, and doors. The relative efficiency of a building envelope is a function of (A) temperature difference between the microclimate (the behavior of weather and seasons in the immediate vicinity of the building and interior space conditions); (B) building and component orientation; (C) the ability of the envelope to resist undesirable or untimely losses and gains of indoor heat due to conduction, convection, infiltration and isolation (solar radiation).

The performance of a building envelope in both resisting and making use of heat transfer due to the above principles is a function of envelope design, construction, operation and maintenance, and the characteristics of materials used in the building's exterior components. It is toward these components that the energy consultant must look for whatever conservation opportunities may exist in relation to the building envelope.

The rate of heat transfer through a building envelope is expressed as a "U" value (Btuh/sq ft of surface per degree of difference between indoor and outdoor temperatures). It is the object of many of the ECO's discussed below to improve the "U" value of the building envelope to the maximum extent feasible given constraints of microclimate, condition of building, and investment budget. (Most of the ECO's studiously avoid envelope modifications calling for significant architectural alterations due to the cost of such changes.)

In view of the importance of microclimate in the relative efficiency of a building envelope, the following ECO's should always be studied in conjunction with the "Climatic Zone Maps" in Chapter IX, along with whatever additional and more localized climatic data are available to the energy consultant. Regional variations in this regard can be quite significant. In offsetting wintertime heating loads through increased solar radiation into a building, for example, it is useful to note that buildings in Minnesota receive approximately 25 percent more annual radiation on south exposures and 15 percent more on east exposures than on north exposures. In Florida, however, the volume of radiation striking all four exposures is virtually equal.

The successful implementation of the following ECO's should result in measurable reductions in the building load of the structure under analysis. This in turn will reduce both the distribution load and the primary equipment load of the structure, leading to annual savings in energy consumption and utility bills which should assuming a reasonable "payback period" exceed the cost of implementing the ECO's.

ECO #16: REDUCE TRANSMISSION OF HEAT THROUGH WALLS

THEORY

Heat transmission through a wall is a function of its resistance to heat flow, modified by the effects of solar radiation and wind velocity on the outside surfaces. The mass of the wall produces a thermal inertia which delays the impact of outdoor temperature changes on conditioned indoor spaces. Heat loss through walls is greatest of course in colder climates, but it varies everywhere according to wall orientation, with the highest losses occurring through north walls and the lowest through south walls, due to the beneficial effects of the sun. The difference is more pronounced in walls having a high absorption coefficient. (The tendency of a particular wall to lose heat is expressed as a "U" value, or heat transfer coefficient, which is further discussed below.)

PROCEDURE

The overall "U" value of a wall may be decreased by adding insulation material to the inside or the outside surface, or by filling the cavities within the wall structure. The addition of insulation to the outside of walls can be accomplished with relative ease in one or two story buildings, but becomes increasingly difficult for the higher stories because of the requirement for access staging. Insulation added to the outside of walls must be weatherproof and vapor-sealed to prevent deterioration of its insulating properties due to the ingress of moisture.

External insulation may be added in the form of prefabricated insulating panels finished to give the most advantageous absorption coefficient. A design problem likely to occur when adding insulation to the outside surface of walls is treatment of window openings and door openings; it must be resolved architecturally for each individual case. The appearance of the building will be changed and may be subject to municipal approval in meeting local fine arts standards or covenant requirements.

Addition of insulation to the inside surface of walls can be accomplished with the same ease for single and multi-story buildings (as access is from the floor of each story treated). It may, however, entail moving furniture and fittings away from the wall and could interrupt normal operation of the building. Internal insulation must be protected from moisture by a vapor barrier and from degradation by wear and tear either by an integral finished surface or by a protective finish such as wood paneling or gypsum plasterboard. Design problems likely to occur when adding insulation to the internal surfaces of walls are treatment of window and door openings, architraves and reveals, the junction between the insulation, floor and ceiling; and repositioning equipment such as heaters, electric receptacles, etc., which are recessed in the existing wall.

Insulation can be added easily to a wall having an unobstructed internal void by filling the space either with granular insulation or with plastic foam. Neither process will alter the appearance of the building and either can be accomplished with minimum inconvenience to the occupants. Each process requires holes cut or core-drilled through the wall to give access to the internal void and to assure that all voids are completely filled. Granular insulation should be blown in under pressure to assure a compressed fill as it tends to pack down over time. Foams must be carefully mixed to minimize shrinkage.

There are many specialist firms that provide these complete insulation services. Investigation should be made to ensure that the wall materials are compatible with the type of insulation considered and that interaction will not occur that could damage either the wall structure or the insulation. Some wall cavities in older buildings were designed to prevent driving rain from penetrating to the inside skin. In such cases, use an insulating material that will not interfere with this function, or coat the outside wall surface with epoxy resin or similar waterproofing.

ANALYSIS

To determine whether additional wall insulation is worthwhile, it is first necessary to know the "U" value of the untreated wall. By examination, determine the type and thickness of the individual components of the wall, e.g. the brick, sheetrock, plaster, etc. Using this information and the thermal characteristics listed in ASHRAE *Fundamentals*, compute the overall "U" value of the wall. This calculated value should then be checked by the following procedure:

(A) Measure indoor and outdoor air temperatures with an accurately-calibrated thermometer at night or on a sunless day.

(B) Measure at the same time the inside surface temperature of the wall with an accurately-calibrated contact thermometer.

(C) Calculate the rate of heat flow through the wall from the inside air temperature, inside surface temperature, and inside film coefficient (assume R of 0.6), using this formula:

$$\text{Heat Flow} = \frac{\text{Inside air temperature minus inside surface temperature}}{\text{Film Coefficient}}$$

Using the calculated rate of heat flow, now calculate the overall "U" value using this formula:

$$U = \frac{\text{Heat Flow}}{\text{Inside air temperature minus outside air temperature}}$$

For example, if the indoor temperature is 68°F, the outdoor temperature is 20°F, and the inside wall surface temperature is 59.4°, then:

$$\text{Rate of heat flow} = \frac{68-59.4}{0.6} = 14.3 \text{ Btuh}$$

$$\text{Overall "U" value} = \frac{14.3}{68-20} = 0.298, \text{ or about } 0.3$$

This method allows for normal variation of the external film coefficient, but for best results, measurements should be made when the wind speed is at an average rate. Since the calculated and the measured "U" values of the wall generally will differ somewhat, an average of the two figures should be used.

Before potential savings can be calculated and the implementation of this ECO be justified, it is necessary to know the "U" values of the walls both before and after adding insulation. Furthermore, having established the "U" value of existing walls, a decision must now be made as to how much insulation should be added and where it should be applied, considering the investment and the anticipated savings from the reduction of energy used. This decision will be governed not only by economic analysis but by the practicality of coordinating the insulation with the existing structure, which may require an architectural solution based on the particular circumstances of the building.

To determine the future "U" value, use Figure 8-41. If the resulting "U" value is insufficient to produce the desired savings, Figure 8-41 can then be used to determine the thickness of insulation needed to improve the "U" value to the desired figure.

Future heating economies: In order to calculate the cost-benefit effectiveness of this ECO in terms of future heating ecoomies (having determined the desired new "U" value of the wall in question), refer to Figure 8-25 or 8-26 and (A) establish the new "Heating Factor" based on the improved "U" value; (B) substitute the new factor for the present factor on the Annual Heating and Cooling Tabulation Form (Figure 8-11).

Future cooling economies: In addition to savings achieved by reducing winter heat losses, wall insulation will also reduce summer heat gains. This extra savings will be small, but nevertheless should be recognized and added to the heating savings when assessing the economic feasibility of this ECO, as it may make the difference between whether the ECO is or is not worthwhile. To determine the cooling savings due to insulation, refer to Figures 8-22, 8-23, and 8-24. Then: (A) determine the new "Cooling Factor" based on the improved "U" value; (B) substitute the new factor for the present factor on the Annual Heating and Cooling Tabulation Form (Figure 8-11).

Conclusion: All of the figures pertaining to future energy use with regard to this ECO should now be calculated and carried through the appropriate forms as explained in Chapter VII. This will determine whether the costs of implementing and maintaining this ECO are justified in terms of future savings for the building owner.

ECO #17: REDUCE TRANSMISSION OF HEAT THROUGH WINDOWS AND SKYLIGHTS

THEORY

Heat transfer through glass is affected by many environmental factors, of which the most significant are: solar radiation; outdoor/indoor temperature difference (thermal conduction); velocity and direction of air flow (across interior and exterior glass surfaces).

Heat loss through windows is highest in areas of high degree days, is modified by orientation, and is greater through windows of northern exposure than of southern exposure, due to the absence of direct north sunlight. Direct and diffuse solar radiation through windows is the largest single component in heat gains through glass. Heat gain due to conduction, by comparison, is relatively insignificant, and it is rarely if ever worthwhile to convert single glazing to double glazing for the sole purpose of reducing conductive heat gain. It is often worthwhile, however, to convert single to double glazing for the benefits obtained in the heating season, and this in turn will reduce conduction gains in the summer. (These benefits, though small, should be included in any economic analysis of double glazing on a yearly basis.)

PROCEDURE

There are many methods by which the heat losses and gains of glass areas can be reduced. Each method and/or combination of methods should be evaluated for its effectiveness, impact on building appearance and function, and economic justification, i.e. in repaying the cost of installation by the amount of energy conserved within a reasonable period of time. The following methods are among the more common.

Storm windows: May be applied either to the outside or inside of existing windows. If they are mounted outside, then multi-story buildings three stories and higher with fixed glazing will have to be scaffolded, but if they can be fitted inside, the use of special staging will be unnecessary. On the other hand, lower buildings can easily be fitted with external storm windows without scaffolding and at less cost; this would not entail moving furniture and disrupting normal work. If existing windows and frames are of poor construction and allow a high rate of infiltration, the storm windows should be fitted on the outside where possible. This will reduce infiltration without incurring the extra cost of caulking the window. If storm windows are added to the existing windows, consider the use of reflective or tinted glass to reduce solar gain and glare in summer. (NOTE: The space between the storm window and the existing window should be vented and provided with weepholes to prevent moisture buildup.)

Double glazed windows: Existing single glazed windows may be converted permanently to double glazed windows by the addition of a new glazing frame to accept the additional pane of glass. Where the existing window frame is in good condition and the glazing system permits, a single sheet of glass may be replaced by a sealed double glazing unit. Where the existing window frame is in poor condition and is scheduled for renovation or replacement, a double glazed unit should be considered. (NOTE: The space between the two panes should be vented or otherwise protected against moisture build-up.)

Thermal barriers: Most commercial, office, and religious buildings are unoccupied for more hours per week than they are occupied, and a major portion of the unoccupied hours occur at night when heat loss is highest. During these unoccupied hours, daylight is either not available or not required and windows can thus be covered with thermally insulated barriers or shutters to reduce heat transmission losses. Thermally insulated barriers may be applied to the inside of the building, so arranged as to slide in a track or fold back into the window reveal or onto the face of the wall. When closed, the barriers should provide a reasonable seal over the window but need not be air tight. Barriers may be purchased ready-made or may be cus' n-made to fit the windows. The bulk of the cost of thermal barriers is represented by labor for construction and fitting rather than by cost of materials. Thermal barriers should therefore be selected or designed to have the best insulating value attainable for a given thickness.

The following table shows the composite "U" values obtained by the barrier/window assembly for various thicknesses of insulation. It is assumed that the insulation has a "k" of 0.25 which is typical for glass wool, beaded polystyrene and polyurethane foam.

Insulation Thickness	COMPOSITE "U" VALUE Single Glazing	Double Glazing
1/2"	0.28	0.23
1"	0.18	0.16
1-1/2"	0.13	0.12
2"	0.11	0.10

Solar control devices: A variety of these are available for installation outside the window, inside the window or on the window surface itself to reduce solar gain in proportion to the window's shading coefficient (refer to Figure 9-7 and to ASHRAE Guide for information shading coefficients). External sun screens which prevent direct sunlight from falling upon the glass surface are the most effective method of controlling solar heat gain. (Internal shading devices are least effective, as sunlight has already entered the building and only a proportion can be reflected back through the window.) External shading may be achieved by fitting eyebrows over the tops of windows or fins at the sides of windows. The size and disposition of these must be determined for each installation based on the sun's altitude and azimuth for the building in question. Eyebrows are most effective over windows with a southern orientation and can be designed to provide total shading from high altitude sun during the middle of the day. They are not as effective, as vertical side fins, however, for east and west orientations.

FIGURE 9-7 SHADING COEFFICIENTS

GLASS	
1/8" Clear Double Strength	1.00
1/4" Clear Plate	0.93 - 0.95
1/4" Heat Absorbing Plate	0.65 - 0.70
1/4" Reflective Plate	0.23 - 0.56
1/4" Laminated Reflective	0.28 - 0.42
1" Clear Insulating Plate	0.80 - 0.83
1" Heat Absorbing Insulating Plate	0.43 - 0.45
1" Reflective Insulating Plate	0.13 - 0.31

SHADING DEVICE	WITH 1/4" CLEAR PLATE GLASS	WITH 1" CLEAR INSULATING GLASS
Venetian Blinds - Light Colored, Fully Closed	0.55	0.51
Roller Shade - Light Colored, Translucent, Fully Drawn	0.39	0.37
Drapes - Semi-Open Weave, Average Fabric Transmittance and Reflectance, Fully Closed	0.55	0.48
Reflective Polyester Film	0.24	0.20
Louvered Sun Screens - 23 Louvers/In.	0.15 - 0.35	0.10 - 0.29
- 17 Louvers/In.	0.18 - 0.51	0.12 - 0.45

External louvered sun screens may be fitted close to the outside surface of the windows and, where possible, within the window reveal itself. Operable windows may permit installation of external sun screens without the use of special staging. The louvered sun screens may be fixed permanently in position or can be arranged to slide in channels or made removable so as to admit solar radiation in winter.

Tinted or reflective glass may be added to existing glazing to convert it from single to double or may be used to replace the existing glazing. Reflective polyester films may also be applied directly to the inside surface of the glass. These films are self-adhesive and require careful application to avoid bubbles or trapped air and the subsequent formation of Newton Rings. The film is easily abraded, which reduces performance and appearance. Film on the inside of a building subjected to normal wear and tear will have a reduced useful life. If the window is double-glazed, apply the film to the inside surface of the outer layer of glass.

Internal shades of various types, i.e. venetian blinds, roller shades, drapes, etc., may be fitted to the inside of the windows. If drapes are used, they should be fire resistant and preferably of woven fiberglass which has a higher reflectance. Although internal shades are not as effective as other methods of solar control, they are relatively inexpensive and easily adjustable, permitting full visibility through the window and maximum use of natural light when solar radiation has ebbed.

ANALYSIS

Upon determining which of the above methods would most effectively reduce solar heat gains and/or losses through glazing for the building in question, refer to Figures 8-19 and 8-20. Then: (A) establish the new heating and/or cooling factor based on improved window treatment; (B) substitute this factor for the present factor on the Annual Heating and Cooling Tabulation Form (Figure 8-11). (NOTE: to determine the energy saved by thermal barriers, refer to Figure 8-21 and execute the above procedure.)

Future Cooling Economies: In addition to energy savings through reduced heat loss, double-glazing and other window treatments also reduce heat gains in summer, resulting in a slight reduction in cooling loads which should be considered in a cost-benefit analysis of this ECO. In most cases, too, double-glazing or storm windows will reduce infiltration, and subsequent savings in both heating and cooling energy shoud likewise be taken into consideration before making a decision on window treatment. (For infiltration, see ECO #20.)

Thermal barriers will also reduce the cooling load, because during unoccupied hours, when the cooling system is completely shut down, the barriers will help contain the previously cooled air inside the conditioned space, thus reducing energy consumed by cooling-system start-up in the morning.

To determine the cooling energy which might be saved by double-glazing and/or shading devices, refer to Figures 8-16, 8-17, and 8-18. Then: (A) establish the new cooling factor based on improved window treatment;

(B) substitute the new factor for the present factor on the Annual Heating and Cooling Tabulation Form (Figure 8-11). (NOTE: The loss and gain of radiant heat through windows and skylights often has a significant effect on the occupants of a building and therefore on the air temperatures required to keep the occupants comfortable. Room temperatures of 70° or even 76° can feel uncomfortable when the occupant is located near a cold window surface. The converse is true where direct solar radiation on occupants neutralizes the effect of the cooling system. In situations where these conditions have led to excessive room temperatures-- either higher or lower than otherwise necessary --to compensate for heat loss or gain through glass, the savings to be expected from normal thermostat settings should be included in the cost-benefit analysis of this ECO. The energy consultant should likewise bear in mind that any change in the volume or quality of solar radiation through glazing will affect the natural lighting level of a given space. This effect should also be considered in an analysis of this ECO, as it may require an increase in artificial lighting levels.)

<u>Conclusion</u>: All of the figures pertaining to future energy use with regard to this ECO should now be calculated and carried through the appropriate forms as explained in Chapter VII. This will determine whether the costs of implementing and maintaining this ECO are justified in terms of future savings for the building owner.

ECO #18: REDUCE TRANSMISSION OF HEAT THROUGH ROOF

THEORY

Heat transfer through a roof/ceiling unit is a function of its resistance to heat flow, modified by the effects of solar radiation (reduces heat loss) and wind (increases heat loss) on the outside surface. The effect of solar radiation is influenced by the absorption coefficient of the outside surfaces; dark colors (high absorption coefficient) will reduce the heat loss more than light colors (low absorption coefficient).

The mass of the roof with its thermal inertia will mitigate heat loss by delaying the impact of outdoor temperature changes on conditioned spaces indoors. Conduction and solar gains through roofs can form a considerable part of the cooling load. The most effective means of reducing heat transfer through a roof is to lower the "U" value of the roof by adding insulation. For buildings located in high degree hour zones, with large roof areas in proportion to floor area (such as supermarkets and other single-story buildings), consideration should be given both to insulating the roof and to reducing its absorption coefficient.

PROCEDURE

Where the existing roof is sound and directly accessible from an attic space or ceiling void, polyurethane foam or mineral fiber may be sprayed on the under-surface, with rigid batt or other applicable insulation for the inside surface. Any of the above could be laid on the top of a ceiling in an attic space, but attention should be paid to uninsulated ducts or pipes or attics, where reduced temperatures due to ceiling insulation may result in excessive heat losses and/or freezeups. It is suggested that these ducts or piping also be insulated.

The resistance or "R" value of the insulation should be a minimum of 6, though often a layer of insulation with a higher "R" value, while not adding considerably to the first cost, provides impressive savings.

Where a roof has deteriorated and is scheduled for extensive repair, deteriorated insulation may be replaced and certain rigid foam insulations may be applied on top of the roof, not only reducing the "U" factor but also the thermal stress on the repaired membrane. Where roofing must be replaced, new rigid insulation may be added to reduce the "U" factor. Such "rigid" insulation may be fiberglass, fiberboard, cellular glass, perlite boards, or other materials embeded in a solid mopping of hot asphalt and protected with the new roof memebrane.

Any paint or reflective finish applied to reduce absorption must be compatible with the existing roof and capable of withstanding abrasion. The absorption coefficient of roofs may also be reduced by adding a surface layer of white pebbles or gravel. Care must be taken, however, that the weight of the additional layer does not exceed the load-bearing capacity of the roof and that gravel-stops are fitted around water drains and the perimeter of the roof itself. If insulation is added to the surface of the roof for the purpose of reducing heat loss and/or gain, select the most desirable absorption coefficient.

Soffits of roof overhangs and roof facias, often left uninsulated, can result in excessive heat loss, and they should be insulated at the same time as the main roof. Where the soffit is accessible, sprayed or rigid insulation may be used, and where it is inaccessible, a hole can be cut in the structure for blowing in mineral fiber.

The use of roof sprays is advisable when solar heat on the roof is excessive, generally under the following conditions:

(A) On buildings with a large roof-to-floor-area ratio, i.e., one or two story office buildings with large roof areas or stores where heat gains through the roof constitute a major portion of the cooling load.

(B) Where there is more than 70% possible sunshine during the cooling season.

(C) High dry-bulb conditions accompanying the solar conditions.

(D) Low wet-bulb conditions coincident with high dry-bulb conditions and sunshine.

(E) Where economic analysis proves such use practicable. (See data provided in the <u>ASHRAE Handbook of Fundamentals</u>.)

ANALYSIS

In evaluating the cost-effectiveness of adding insulation to a roof, both heating and cooling benefits should be studied. In either case it is first necessary to know the "U" value of the existing structure. The procedure for calculating and measuring the "U" value of walls, outlined earlier in this section, may also be used for roofs. Having established the "U" value of the existing roof, a decision must be made as to whether additional insulation is required, how much should be added, what type of insulation should be used, and where it should be applied. These decisions, which are influenced by the characteristics of each particular building, should involve a minimum of architectural disruption.

The cost of adding insulation to roofs varies with each particular solution, and to make an accurate assessment for a particular case, quotations should be obtained from local contractors or pricing manuals.

<u>Future heating economies</u>: Having determined the existing "U" value of the roof, refer to Figure 8-41 to determine the new "U" value. To

determine the heating energy saved by the insulated roof, refer to Figure 8-28. Then: (A) determine the new heating factor based on the improved "U" value; (B) substitute this factor for the present factor on the Annual Heating and Cooling Tabulation Form (Figure 8-11).

Future cooling economies: In addition to savings achieved by reducing winter heat losses, roof insulation will also reduce summer heat gains. To determine the cooling energy which might be saved by insulating a roof and/or changing its absorption coefficient, refer to Figure 8-27. Then: (A) determine the new cooling factor based on the improved "U" value; (B) substitute the factor for the present factor on the Annual Heating and Cooling Tabulation Form (Figure 8-11).

Conclusion: All of the figures pertaining to future energy use with regard to this ECO should now be calculated and carried through the appropriate forms as explained in Chapter VII. This will determine whether the costs of implementing and maintaining this ECO are justified in terms of future savings for the building owner.

ECO #19: REDUCE TRANSMISSION HEAT OF THROUGH FLOORS

THEORY

Heat transfer through slab-on-grade floors occurs mainly around the perimeter. Very little heat is transferred through the center of the floor, due to the insulating effect of the earth, which is strongest at the center and weakest at the perimeter areas. Heat loss and gain through suspended floors over unheated spaces, however, occurs evenly across the entire floor area roughly in proportion to the difference between the inside and outside temperatures. Transmission loss through floors can be especially significant when they are located over unheated crawlspaces or where they project out from the building below.

Since existing buildings, it is generally impracticable to add insulation to the tops of floors, consideration of this ECO will be restricted to insulation added below suspended floors and at the edges of slab-on-grade floors.

PROCEDURE

Uninsulated slab-on-grade floors should be insulated around the perimeter. Insulation material should be placed vertically along the outside edge of the floor and should extend down at least two feet below the floor surface. Insulation materials may be rigid board or foam as described for walls. Insulation below ground should be applied in a bedding of hot asphalt, and existing flashing at ground-floor level may have to be extended to cover the top edge of the insulation. A suspended floor above an unheated space (garage, crawl-way, etc.) may be insulated on the underside by applying spray foam, or rigid insulation as described for roofs. (NOTE: The insulation of a floor above an unheated crawl-space may increase the danger of freezing or heat loss form pipes or ducts in the unheated spacel It is often advisable, therefore, to insulate exposed pipes or ducts at the same time the floor is insulated.)

The energy savings likely to result from edge-insulation of slab-on-grade floors is difficult to compute. In general, however, when condensation occurs at a floor's perimeter in cold weather, or when a floor-surface temperature close to the outside wall is greater than 10° lower than the indoor temperature, insulation will usually be cost-effective.

ANALYSIS

To evaluate the benefits of adding insulation to suspended floors, it is first necessary to know the present "U" value of the existing construction, which may be determined by using the same formula discussed previously for walls (See ECO #16).

Having established the "U" value of the existing floor, a decision must be made as to whether additional insulation is required, how much, and what type. As with roofs and walls, the cost of adding insulation to floors will vary in each particular case depending upon the circumstances and architectural considerations.

Future heating economies: To determine the new "U" value of the insulated floor, refer to Figure 8-41. Enter the graph at the present "U" value, intersect with the appropriate line for new insulation thickness, and read out the new "U" value. To determine the heating energy saved by insulating suspended slab floors, refer to Figure 8-29. Then: (A) determine the new heating factor based on the improved "U" value; (B) substitute this factor for the present factor on the Annual Heating and Cooling Tabulation Form (Figure 8-11). (NOTE: The negative effect of cold floors on occupants is often greater than for walls or windows due to proximity, relative surface area, and actual occupant contact with the floors. This factor and its impact on occupant's demands for higher room temperatures should be accounted for where it occurs and included in a cost-benefit analysis of this ECO.)

Future cooling economies: The insulation of floors to reduce a building's heating load will also conserve energy during the cooling season, a factor to be considered in calculating the feasibility of this ECO. To determine the cooling energy likely to be saved by this ECO, refer to Figure 8-24. Then: (A) determine the new cooling factor based on the improved "U" value of the floor; (B) substitute this factor for the present factor on the Annual Heating and Cooling Tabulation Form (Figure 8-11).

Conclusion: All of the figures pertaining to future energy use with regard to this ECO should now be calculated and carried through the appropriate forms as explained in Chapter VII. This will determine whether the costs of implementing and maintaining this ECO are justified in terms of future savings for the building owner.

ECO #20: REDUCE SPACE LOAD DUE TO INFILTRATION

THEORY

Infiltration of cold outside air into a building through cracks, openings, gaps around windows and doors, etc. increases the building load to such an extent that it is often responsible for as much as 25 percent of a buildings annual energy consumption. While infiltration tends to increase with wind velocity, penetrating the building at its windward exposure, negative pressure on the roof and leeward exposures can also force cold air inside through other exposures, open doors, and passageways. Tall buildings are subject to a stack effect which results from the difference between indoor and outdoor temperatures. The stack effect originates in open vertical spaces such as stairwells, elevator shafts, and service shafts. The potential for a stack effect in tall buildings is always present and can only be minimized through the procedures discussed below.

To maintain equilibrium, the quantity of air entering a building either by uncontrolled infiltration or by ventilation must equal the quantity of air leaving the building either by exfiltration or by ventilation exhaust. Buildings frequently have unbalanced ventilation systems which cause the rate of exhaust to exceed the rate at which air is introduced into the ventilation system, causing infiltration to increase (to make up the balance) beyond the level which would otherwise occur.

PROCEDURE

The fresh air ventilation and exhaust system should be examined and a balance sheet drawn up. Where exhaust exceeds the outdoor air requirements, the rate of exhaust should be reduced, which in turn will reduce infiltration. Ideally, the quantity of outdoor air for ventilation should exceed the total exhaust from the building by 10 percent to promote exfiltration rather than infiltration, except in very "tight" buildings where this difference will be reduced to avoid excess pressurization that can cause doors to remain open, etc. Infiltration, unlike ventilation systems, cannot be turned off at night or during weekends (although it may be retarded when exhaust fans are shut down). It can, however, be reduced at all times.

Caulking, weatherstripping, and vestibules: Particularly effective measures include caulking cracks around window and door frames and weatherstripping windows and doors. In buildings that experience heavy and continuous traffic through external doors, infiltration may be reduced by building vestibules for each external door to form an "air-lock." The vestibule should be sufficiently long that the external door closes before the internal door is opened. Depending on the particular characteristics of the building, the vestibule may be constructed either inside or outside the building. Vestibule doors may be manually operated and self-closing, or, if traffic is particularly dense, automatically opened and closed by pressure pads or photo-electric cells.

Revolving Doors: Another solution here is to replace single-swing doors with revolving doors. Although more effective when used in conjunction with vestibules, revolving doors may also be used alone. (When fitting revolving doors, it is always necessary to provide an auxiliary hinged door for use by handicapped persons, as well as to allow bulky objects, baby carriages, strollers, etc. to be taken in an out of the building.

Vertical shafts and stairs: To reduce the effect of potential stack action present in vertical shafts and stairs, these areas should be sealed from the rest of the building. Open stairwells that connect with circulation spaces at each floor level should be provided with walls and self-closing doors to isolate them. Access holes into vertical service shafts should be provided with gasketed ceiling covers. Holes at the shaft wall, to allow the passage of pipes and ducts at each floor level, should be sealed and sleeves around the pipes and ducts packed with asbestos rope or other suitable material.

Elevator shafts: Elevator shafts are usually provided with a vent at the top into the equipment room. This vent is necessary as a smoke release, but it can be responsible for large air movements through the elevator doors up the shaft into the equipment room and then outside through badly fitted doors and windows in the equipment room. Although the equipment room is designated as unoccupied space, the windows and doors should be weatherstripped to reduce the effect of stack losses through elevator shafts.

ANALYSIS

1. Vestibules and/or Revolving Doors: To calculate the energy which might be saved by installing vestibules or revolving doors, it is first necessary to determine the estimated number of passages through a given entrance to the building under survey. This figure can be determined by (A) counting the times the door is opened per hour on an average day; (B) reference to prior experience with similar buildings by the energy consultant or other members of the energy management team. Having established the estimated passages per hour, refer to Figure 8-15. Then: (A) determine the new cfm of infiltration based on improved doors; (B) substitute this figure for the present cfm on the Building Load Survey Form (Figure 8-11).

Future cooling economies: To determine the potential heat-energy savings of this ECO, transfer the above new figure for cfm to the appropriate space on the Annual Heating and Cooling Tabulation Form (Figure 8-11) (assuming infiltration exceeds outside air introduced through the HVAC system).

Future cooling economies: In addition to the savings achieved by reducing winter heat losses, reduction of infiltration will also reduce sensible and latent heat gains in summer. Depending on the severity of the cooling season, this extra saving can be substantial and should be included in a cost-benefit analysis of this ECO. As a guideline to this procedure, and assuming that summer wind velocity is only half that of winter: (A) divide the winter cfm rate by two, giving an approximate cooling season infiltration rate in cfm; (B) enter this figure in the appropriate space on the Annual Heating and Cooling Tabulation Form (Figure 8-11) (assuming infiltration exceeds the volume of outside air introduced through the HVAC system).

2. Weatherstripping of Windows and Doors: The effect of infiltration around windows and doors is identical to that resulting from door openings, and the method of calculating the potential energy savings due to weatherstripping is similar to that described above.

Future heating economies: Upon determing the type of weatherstripping to be used for the building in question, refer to Figure 8-14. Then: (A) determine the new rate of infiltration in cfm based on weatherstripping; (B) substitute this new figure for the present figure on the Building Load Survey Form (Figure 8-9)

Future cooling economies: Use the same procedure described in item 1 above (halving the winter infiltration in cfm), or calculate the cooling season infiltration rate by reference to Figure 8-14 (in conjunction with the wind velocity at the building in question). The rest of the procedure is identical with that described in 1 above.

Vertical shafts and stairs: Infiltration due to stack conditions can be estimated using the methods described in the ASHRAE Handbook of Fundamentals.

Conclusion: All of the figures pertaining to future energy use with regard to this ECO should now be calculated and carried through the appropriate forms as explained in Chapter VII. This will determine whether the costs of implementing and maintaining this ECO are justified in terms of future savings for the building owner.

E. Distrubution Systems Energy Conservation Opportunities

Page

E. Distribution System ECO's: #21 - 25

Introduction

The energy consumed and wasted through a building's distribution system, as discussed in Chapter III, is often called a "parasitic" energy load because it contributes nothing directly to the quality of the indoor environment. In fact, the energy required to distribute air and water through a building is counter-productive in relation to cooling, since the heat generated in distribution must be removed by the cooling system.

As the term "distribution" implies, the sole function of a distribution system is to transport heat energy (including chilled air and water) from the source of generation to the point of end-use or consumption. It is in the network of pipes, ducts, fans, pumps, and grilles involved in this "transport" that energy waste occurs, and it is in this network that improvements must be made if the "parasitic" energy load of a building is to be reduced.

Transport energy losses are inherent in the design of virtually any distribution system. These losses, and indeed the parasitic load itself, are a function primarily of three basic factors in the thermal performance of a building and its distribution system: (A) rate of flow through pipes and ducts; (B) resistance to flow; (C) losses due to conduction and infiltration through poor insulation, leaks, cracks, and other defects in piping and ductwork.

Rate of flow-- the volume and velocity of fluids in passage through channels is a function of building load, itself a function of the total demand of the energy-consuming sectors ("functional areas") of the building. A reduction in building load, therefore, will often reduce the rate of flow through a distribution system, and this in turn will reduce the resistance to flow. It is owing to this cause-effect relationship that most of the ECO's discussed below should be examined and implemented only after the building load has been reduced to the maximum extent feasible (as discussed in Chapter III under "Cumulative Load Reductions").

Among the key options to be considered in reducing parasitic energy losses are the following:

(A) Reduce hours of fan and pump operation.

(B) Improve performance of terminal devices to reduce their resistance to fluid flow and increase their heat transfer characteristics.

(C) Lower resistance to flow in duct and piping systems to reduce required horsepower for fans and pumps.

(D) Decrease fluid leaks and thermal losses from piping, air-handling equipment, and other vessels holding or transporting hot or cold water, air or steam.

These and other options will be examined at greater length in the following section of ECO's.

ECO #21: REDUCE ENERGY CONSUMPTION FOR FANS BY REDUCING AIR FLOW RATES AND RESISTANCE TO AIR FLOW

THEORY

The implementation of energy conservation measures which reduce heating and or cooling loads first will permit subsequent reductions in the volume of air to be handled by the HVAC system. Whenever the volume of air is reduced, the resistance to air flow in that section of the system is also reduced, since the resistance of any system depends upon the quantity of air passing through the system as well as the characteristics of the various portions of the system, including: (a) intake louvers and air outlet duct caps or elbows; (b) heating and cooling coils; (c) filters; (d) ductwork system, balancing dampers, and exhaust hoods; (e) registers, grilles, and diffusers; (f) fan scroll and housing.

If, as is often the case, the original HVAC system has been oversized for anticipated loads (due to a conservative "safety" or reserve factor in the design), or if the original HVAC loads have been reduced, the opportunities to reduce air flow rates, volume, and resistance will generally exceed the potential for further reduction in "building" loads.

Resistance to Air Flow

(A) Filters: The resistance to air flow through filters is a function of filter construction, type of media, area of media/unit volume (proportional to actual velocity through the media-- not to be confused with face velocity) and the dirt-load at any given time. In general, resistance to air flow increases with filter efficiency, although there are some high-efficiency filters which also have low resistance (see Figure 9-1).

(B) Coils: Resistance to air flow through coils depends largely upon the number of rows or depth of the coils required for adequate heat transfer. Cooling coils are usually many rows deep because of the small temperature differences between coil and air, resulting in a high resistance to air flow. Heating coils, which have a higher temperature difference, are usually fewer rows deep than cooling coils and impose less resistance to air flow. Both coils reside in the air stream and impose combined resistance year-round, even though only one coil may be used at any given time.

Air Flow Rates

Reducing the resistance to air flow in an HVAC system results in an increase of air delivery, which in turn permits a reduction in fan speed to bring air volume down to the original level. With multi-vane centrifugal fans (the type most commonly used in HVAC systems), the power input (brake horsepower or bhp) varies directly with the cube of the speed, and any reduction in speed, which is proportional to air volume, results in a substantial decrease in power input.

The basic operating laws for centrifugal fans are as follow: (a) the volume varies directly with the speed; (b) the pressure varies directly with the square of the speed; (c) the power input (bhp) varies directly with the cube of the speed; (d) the volume varies directly with the square of the pressure. If it is possible to reduce the volume of air passing through a system by 10 percent, the savings in power will be about 27 percent.

PROCEDURE

To conserve energy for fan horsepower, first reduce heating and cooling loads; second, reduce the resistance to air flow in each of the six areas listed above in THEORY (paragraph one); third, measure the increased volume which results and calculate the new air volume required to meet new loads; fourth, reduce fan speed accordingly; fifth, change fan motors if necessary.

(A) Resistance to Air Flow

Filters: A primary cause of filter resistance to air flow is simply dirty filters. In addition to implementing a rigorous program of filter cleaning and maintenance, install a manometer across each filter bank to indicate when the filters should be changed. If the building utilizes multiple air systems, consider the use of an alarm sytem to report filter conditions at a central point. (To determine the "dirty" resistance limit of the filters being utilized, consult the manfacturer.)

Many filters impose high resistance due to their inherent characteristics and/or because they exceed required filtration standards. Figure 9-10 tabulates 25 different filter types in common use, showing installation and operating costs (based on 1974 cost figures, hence to be used only for purposes of approximate comparison). Determine filtration requirements in terms of NBS atmospheric Dust Spot Efficiency, taking into account the ambient air conditions and space air quality requirements. (Except for special applications such as computer rooms and food service, a 50 percent NBS dust rating is generally adequate.) Then check present filter installation and determine whether an alternative type of filter will meet building needs at operating costs sufficiently reduced to provide an economic payback.

Coils:

Where the cooling medium is chilled water and the heating medium is hot water, it may be possible to remove the heating coil and repipe the cooling coil to provide both heating and cooling (though not simultaneously). Eliminating the heating coil will reduce the system resistance and allow savings in fan horsepower to be achieved.

A side benefit of using the cooling coil for both heating and cooling is that the extra heat transfer surface of the cooling coil provides opportunities to lower the hot-water temperatures and use low-grade waste heat.

FIGURE 9-8 AIR RESISTANCE THROUGH COILS

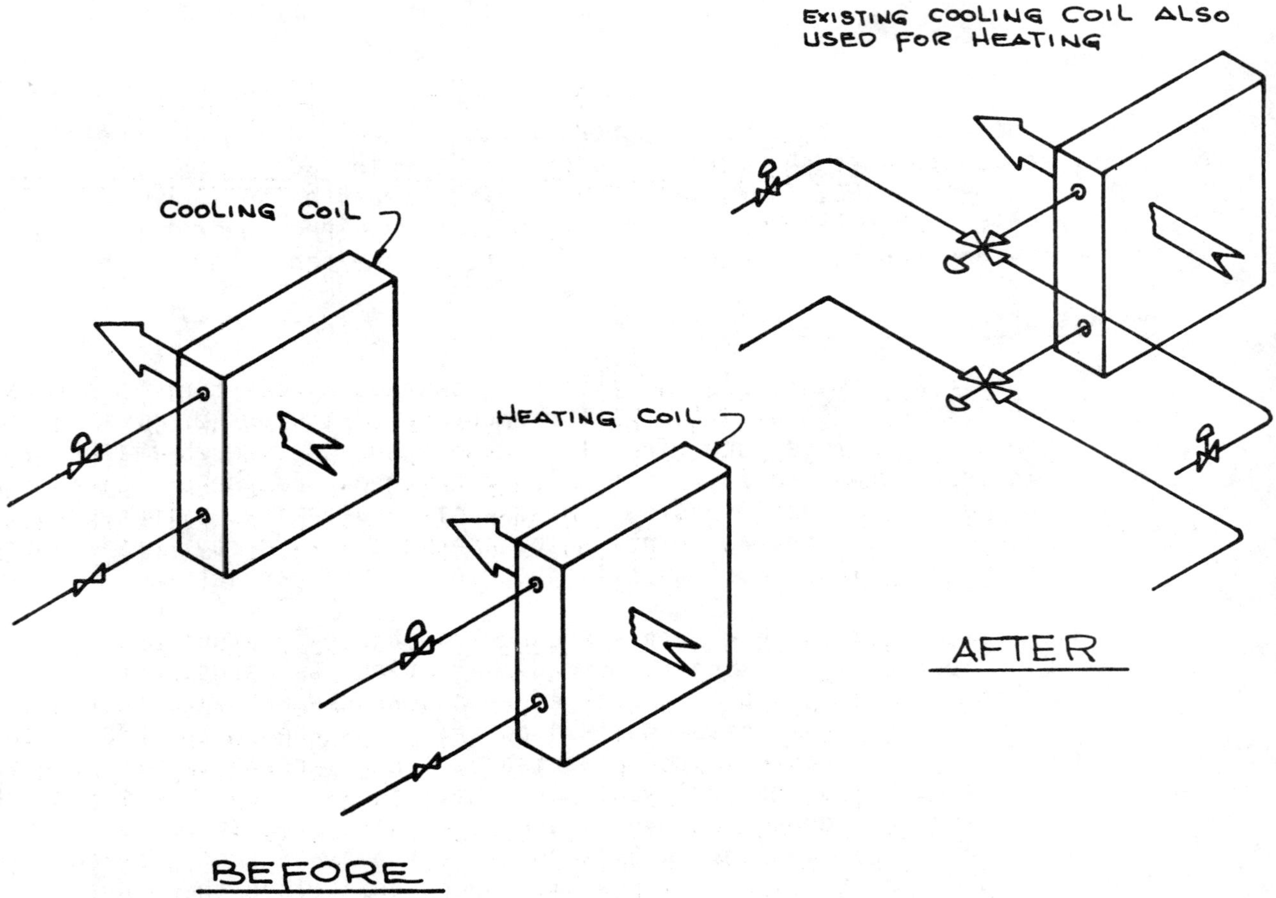

Ducts: Reduce high resistance to air-flow caused by duct fittings on the inlet and discharge side of fans by modifying the shape of the fittings. (See ASHRAE "Equipment Guide" for recommendations.)

Where space permits and velocities exceed 2000 ft/minute, modify abrupt changes in sections of duct work to long tapered fittings and install turning vanes in square bends.

Supply and exhaust fans must operate at a pressure sufficient to overcome the resistance to air flow through the ducts serving the outlet farthest from the fan (index outlet). Where the index outlet is remote from other parts of the system and is served by a long run of duct, it imposes a power consumption penalty on the whole of the system. It is often worthwhile to replace this section of duct with one of larger cross-section and lower resistance.

FIGURE 9-9 AIR RESISTANCE THROUGH BRANCH DUCTS

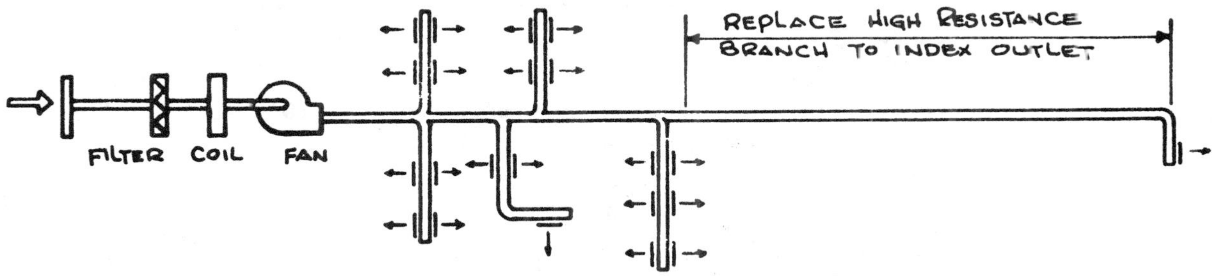

Grilles, registers, dampers: To achieve correct volumes at each grille or register, adjust dampers in low resistance branches until all branches are of a resistance equal to the index run. When systems are initially started, all dampers are often closed more than they need be, adding unnecessary resistance. It is also a common and wasteful practice to reduce fan volume by closing down dampers on the fan inlet or outlet rather than by reducing fan speed.

To alleviate such difficulties, if necessary, rebalance duct the system by first opening fully any dampers which are not used to redirect air flow between branches or outlets (main dampers close to fan). Identify the index outlet and fully open any dampers between this outlet and the fan. Measure the volume at the index outlet and adjust branch dampers successively (starting with the next longest run) until proportional volumes are achieved at each outlet. (This is a process of trial and error, as each damper adjustment will affect flow rates in branches already adjusted, but two or three successive adjustments of the whole system will give good balance.)

(B) Air Flow Rates

To reduce the volume of an air system, reduce the fan speed in proportion to the load change. To reduce the speed of belt-driven fans, change the size of the motor sheave. It may be necessary also to change the drive belts if adjustment will not take up the slack.

To reduce the speed of direct driven fans, exchange the drive motor for a lower speed type. This may be possible only if the desired fan speed falls within the motor speed ranges available or if a compromise speed is acceptable.

Where the air system both heats in winter and cools in summer, it is unlikely that one flow rate will be ideal for both seasons. The conflict between heating and cooling requirements can be resolved by changing

FIGURE 9-10 AIR FILTER TYPE COMPARISON

Filter Type	1. Cleanable 2" Thick (High Vel.)	2. Throw-away 2" Thick	3. Automatic Roll Type	4. Pleated Filter 2" Thick
Size of Filter Bank				
No. High x No. Wide	2 x 3	2 x 5	-	2 x 4
Height x Width	4' x 6'	4' x 10'	5'4" x 5'	4' x 8'
Average Efficiency				
NBS Atmospheric				
Dust Spot	8-10%	10-15%	20-25%	36%
Other Methods	-	-	-	-
Filter Life (Hours)				
Prefilter	600	480	3750	2000
Afterfilter	-	-	-	-
Pressure Drop (in. W.G.)				
Initial	0.08	0.10	0.40	0.14
Final	0.40	0.30	0.40	0.60
Average	0.24	0.20	0.40	0.37
Initial Installation[1,8]				
Equipment	$328	$107	$ 780	$124
Installation	78	91	91	91
Electric Wiring	-	-	215	-
	$406	$198	$1086	$215
Cost of One Set-Replacement Media [8]				
Prefilter[2]	-	$ 21	$ 87	$ 49
Afterfilter[2]	-	-	-	-
Replacement Time - Man Hours				
Prefilter	4	2	2	3
Afterfilter	-	-	-	-
Annual Operating Cost [8]				
Material	-	$381	$203	$216
Labor @ $7/hr.	$533	333	43	120
Electric Power[3,7]	126	104	210	194
Miscellaneous	-	-	-	-
Total[4]	$659	$818	$456	$530

FIGURE 9-10(con't)

Filter Type	5. Bag Type Cartridge	6. Bag Type Cartridge	7. Bag Type Cartridge	8. Bag Type Cartridge w/ Prefilter
Size of Filter Bank				
No. High x No. Wide	2 x 2	2 x 2	2 x 3	2 x 3
Height & Width	4' x 4'	4' x 4'	4' x 6'	4' x 6'
Average Efficiency				
NBS Atmospheric Dust Spot	38-40%	45%	50-55%	50-55%
Other Methods	-	-	-	-
Filter Life (Hours)				
Prefilter	6600	2920	5500	-
Afterfilter	-	-	-	9000
Pressure Drop (in.W.G.)				
Initial	0.30	0.25	0.30	0.30
Final	0.75	1.00	0.80	0.80
Average	0.525	0.625	0.55	0.55
Initial Installation[8]				
Equipment	$113	$182	$351	$351
Installation	78	78	78	78
Electric Wiring	-	-	-	-
	$191	$260	$429	$429
Cost of One Set-Replacement Media[8]				
Prefilter	-	-	$306	-
Afterfilter	-	$ 46	-	$306
Replacement Time - Man Hours				
Prefilter	4	2	3	-
Afterfilter	-	-	-	3
Annual Operating Cost[8]				
Material	$ 83	$137	$486	$298
Labor @ $7/hr.	25	55	44	26
Electric Power[3,7]	276	336	288	288
Miscellaneous	-	-	-	-
Total	$384	$528	$818	$612

FIGURE 9-10(con't)

Filter Type	9. Auto-Roll Bag Type Afterfilter	10. Rigid Cartridge	11. Rigid Cartridge w/Prefilter	12. Bag Type Cartridge
Size of Filter Bank				
No. High x No. Wide	-	2 x 3	2 x 3	2 x 3
Height & Width	4'8" x 6'8"	4' x 6'	4' x 6'	4' x 6'
Average Efficiency				
NBS Atmospheric Dust Spot	50-55%	55-60%	55-60%	80-85%
Other Methods	-	-	-	-
Filter Life (Hours)				
Prefilter	3,750	3,500	-	-
Afterfilter	10,000	-	4,500	4,000
Pressure Drop (in.W.G.)				
Initial	0.70	0.20	0.20	0.37
Final	1.25	1.00	1.00	1.00
Average	0.975	0.60	0.60	0.685
Initial Installation [8]				
Equipment	$1,175	$332	$332	$378
Installation	169	78	78	78
Electric Wiring	215	-	-	-
	$1,559	$410	$410	$456
Cost of One Set-Replacement Media [8]				
Prefilter	$ 72	$278	-	-
Afterfilter	306	-	$278	$337
Replacement Time-Man Hours				
Prefilter	2	3	-	-
Afterfilter	3	-	3	3
Annual Operating Cost [8]				
Material	$ 434	$ 697	$546	$ 737
Labor @ $7/hr.	68	69	53	60
Electric Power	512	316	316	360
Miscellaneous	-	-	-	-
	$1,014	$1,082	$915	$1,157

FIGURE 9-10 (con't)

Filter Type	13. Bag Type Cartridge w/Prefilter	14. Automatic Roll with Bag Type Afterfilter	15. Rigid Type Cartridge	16. Rigid Type Cartridge with Prefilter
Size of Filter Bank				
No. High x No. Wide	2 x 3	-	2 x 3	2 x 3
Height x Width	4' x 6'	4'8" x 6'8"	4' x 6'	4' x 6'
Average Efficiency				
NBS Atmospheric	80-85%	80-85%	85-90%	85-90%
Dust Spot	-	-	-	-
Other Methods				
Filter Life (Hours)				
Prefilter	-	3,750	-	-
Afterfilter	5,500	6,000	2,500	3,500
Pressure Drop (in. W.G.)				
Initial	0.37	0.77	0.30	0.30
Final	1.00	1.50	1.00	1.00
Average	0.685	1.135	0.65	0.65
Initial Installation[8]				
Equipment	$378	$1,203	$367	$367
Installation	78	169	78	78
Electric Wiring	-	215	-	-
	$456	$1,587	$445	$445
Cost of One Set - Replacement Media[8]				
Prefilter	-	$ 72	-	-
Afterfilter	$337	337	$332	$332
Replacement Time - Man Hours				
Prefilter	-	2	-	-
Afterfilter	3	3	3	3
Annual Operating Cost[8]				
Material	$536	$ 658	$1,161	$ 829
Labor @ $7/hr.	44	83	96	69
Electric Power	360	596	342	342
Miscellaneous	-	-	-	-
Total	$940	$1,337	$1,599	$1,240

FIGURE 9-10 (con't)

Filter Type	17. Electrostatic Auto-Roll Afterfilter	18. Bag Type Cartridge	19. Bag Type Cartridge w/Prefilter	20. Electrostatic Bag Type Afterfilter
Size of Filter Bank				
No. High x No. Wide	-	2 x 3	2 x 3	-
Height x Width	4'8" x 8'8"	4' x 6'	4' x 6'	4'8" x 6'8"
Average Efficiency				
NBS Atmospheric Dust Spot	90%	93-97%	93-97%	93-97%
Other Methods	-	-	-	-
Filter Life (Hours)				
Prefilter	2,500	-	-	2,500[6]
Afterfilter	4,000	2,500	3,500	12,000
Pressure Drop (in. W.G.)				
Initial	0.45	0.48	0.48	0.55
Final	0.45	1.00	1.00	1.00
Average	0.45	0.74	0.74	0.775
Initial Installation[8]				
Equipment	$2,990	$437	$437	$2,730
Installation	169	78	78	156
Electric Wiring	397	-	-	215
	$3,556	$515	$515	$3,101
Cost of One Set - Replacment Media[8]				
Prefilter	$2	-	-	$ 2
Afterfilter	2	$ 3	$3	3
Annual Operating Cost[8]				
Material	$157	$1,411	$1,008	$ 294
Labor @ $7/hr.	104	96	69	83
Electric Power	236	388	388	408
Miscellaneous	-	-	-	-
	$497	$1,895	$1,465	$ 785

FIGURE 9-10(con't)

Filter Type	21. HEPA without Prefilter	22. HEPA with 85% Prefilter	23. Activated Charcoal
Size of Filter Bank			
No. High x No. Wide	2 x 5	2 x 5	1 x 5
Height x Width	4'0" x 10"	4'0" x 10'0"	2'0" x 10'0"
Average Efficiency			
NBS Atmospheric			
Dust Spot	100%	100%	-
Other Methods	99.97%DOP	99.97%DOP	-
Filter Life (Hours)			
Prefilter	-	-	-
Afterfilter	6,000	24,000	8,760
Pressure Drop (in.W.G.)			
Initial	1.00	1.00	0.35
Final	2.00	2.00	0.35
Average	1.50	1.50	0.35
Initial Installation [8]			
Equipment	$1,300	$1,300	$2,600
Installation	390	390	975
Electric Wiring	-	-	-
	$1,690	$1,690	$3,575
Cost of One Set - Replacement Media [8]			
Prefilter	-	-	
Afterfilter	$ 975	$ 975	$ 598
Annual Operating Cost [8]			
Material	$1,424	$ 356	$ 598
Labor @ $7/hr.	79	20	109
Electric Power	788	788	184
Miscellaneous	-	-	-
	$2,291	$1,164	$ 891

FIGURE 9-10(con't)

Filter Type	24. 95% DOP Rated Filter w/Prefilter	25. 95% DOP Rated Filter w/85% Prefilter
Size of Filter Bank		
No. High x No. Wide	2 x 5	2 x 5
Height x Width	4'0" x 10'0"	4'0" x 10'0"
Average Efficiency		
NBS Atmospheric		
Dust Spot	99%	99%
Other Methods	95% DOP	95% DOP
Filter Life (Hours)		
Prefilter	-	-
Afterfilter	6,000	24,000
Pressure Drop (in. W.G.)		
Initial	0.50	0.50
Final	1.00	1.00
Average	0.75	0.75
Initial Installation 8		
Equipment	$1,300	$1,300
Installation	390	390
Electric Wiring	-	-
	$1,690	$1,690
Cost of One Set - Replacement Media 8		
Prefilter	-	-
Afterfilter	$ 975	$ 975
Replacement Time - Man Hours		
Prefilter	-	-
Afterfilter	6	6
Annual Operating Cost 8		
Material	$1,424	$ 356
Labor @ $7/hr.	79	20
Electric Power	394	394
Miscellaneous	-	-
	$1,897	$ 770

FOOTNOTES

1. Spare set of replacement cells was included in installed cost of cleanable filters.
2. Allowance for replacement filter price increase averaging 15% over 20 years was included.
3. Fan power was figured at $0.03 per Kwh and 59%* overall efficiency.
4. Operating costs based on 8760 hours of operation per year.
5. Prefilter costs are not included in columns 8, 11, 19, 21, 22, 24, 25. Add costs from columns 1, 2, 3, 4, or 5 to arrive at the total cost.
6. Costs are for 10,000 Cfm system. Cost per Cfm will be slightly lower for smaller systems and higher for larger systems.
7. Power absorbed by filter to overcome its resistance to air flow.
8. All costs are based on national costs for 1974. Current cost data should be substituted for the specific location involved.

* Fan efficiency typically varies from 50 - 70%

pulley sizes for each season, by installing a two-speed motor, or by adopting a compromise air volume for both seasons and adjusting the supply-air temperature to suit.

ANALYSIS

The cost-benefit analysis of this ECO must of course be conducted in terms of the two main groups of actions outlined above.

(A) Reducing Resistance to Air Flow

(1) Total all of the steps contemplated to reduce air flow resistance to determine the new system static pressure. This reduction may be calculated by using Figure 9-10, Manufacturer's Data and ASHRAE Guide.

(2) Determine the new energy use/yr from Figure 8-42 and 8-43 using the new system static pressure.

(3) Substitute this for the present energy use on Distribution Load Survey Form (Figure 8-10).

(B) Reducing Air Flow

(1) Calculate the new air flow in cfm based on measures to reduce the building load.

(2) Calculate the new kw required by fans to satisfy the new air flow requirements using this formula:

$$\text{New kw} = \text{Initial kw} \times \frac{\text{New cfm}^3}{\text{Initial cfm}}$$

(3) If reduced air flow will be accomplished by closing dampers or installing inlet vanes, calculate the new kw for fans with this formula:

$$\text{New kw} = \text{Initial kw} \times \frac{\text{New cfm}}{\text{Initial cfm}}$$

(4) Substitute this new kw for the present kw on the Distribution Load Survey Form (Figure 8-10).

Conclusion: All of the figures pertaining to future energy use with regard to this ECO should now be calculated and carried through the appropriate forms as explained in Chapter VII. This will determine whether the costs of implementing and maintaining this ECO are justified in terms of future savings or the building owner.

ECO #22: REDUCE PUMP ENERGY BY REDUCING RESISTANCE AND FLOW RATES

THEORY

Energy transport losses are inherent in the design of any water distribution system, but they can be minimized by reductions in flow rate through the system and system resistance to water flow.

Water system flow rates, as with air flow rates discussed in ECO #21, are a function of building loads. Minimizing building loads will allow flow rates to be reduced accordingly. (To avoid repetitive actions, every effort should be made first to reduce building cooling and heating loads.)

FIGURE 9-11 PUMP SYSTEM CURVE

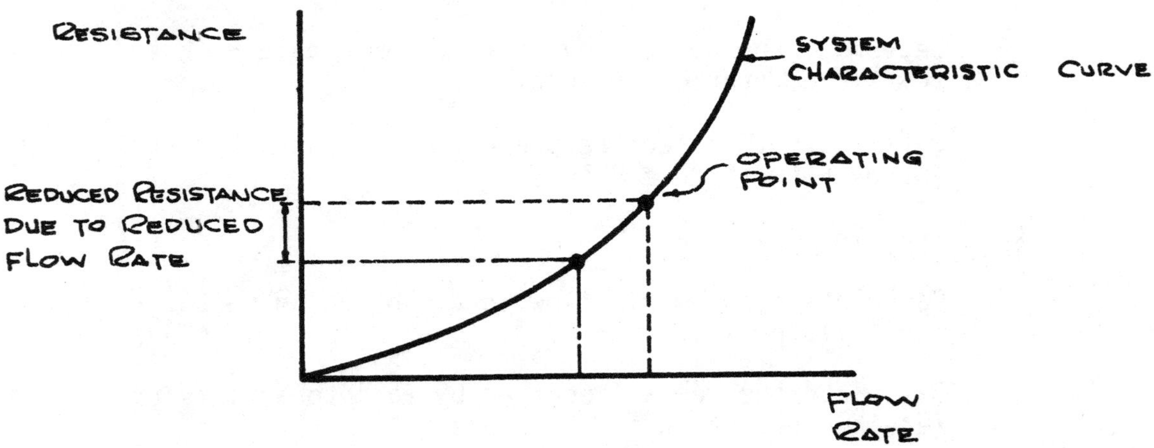

The characteristics of any given system are pre-determined by the lengths and sizes of pipes and the sizes and shapes of fittings (bends, tees). The resistance to flow is a function of velocity (or flow rate), and system characteristics. With any change in flow rate, the system resistance will change (according to the laws of fluid flow), and the operating point will move accordingly on the system characteristic curve.

For instance, a piping system with a flow rate of 2,000 gpm at 150 ft resistance, changed to 1,800 gpm, will have a new resistance of:

$$\frac{150 \times (1800)^2}{(2000)^2} = 121.5 \text{ ft}$$

If it is not possible to reduce the flow rate, the remaining option to reduce the power required for a given flow rate is to alter the system characteristic curve.

FIGURE 9-12 PUMP SYSTEM CURVE

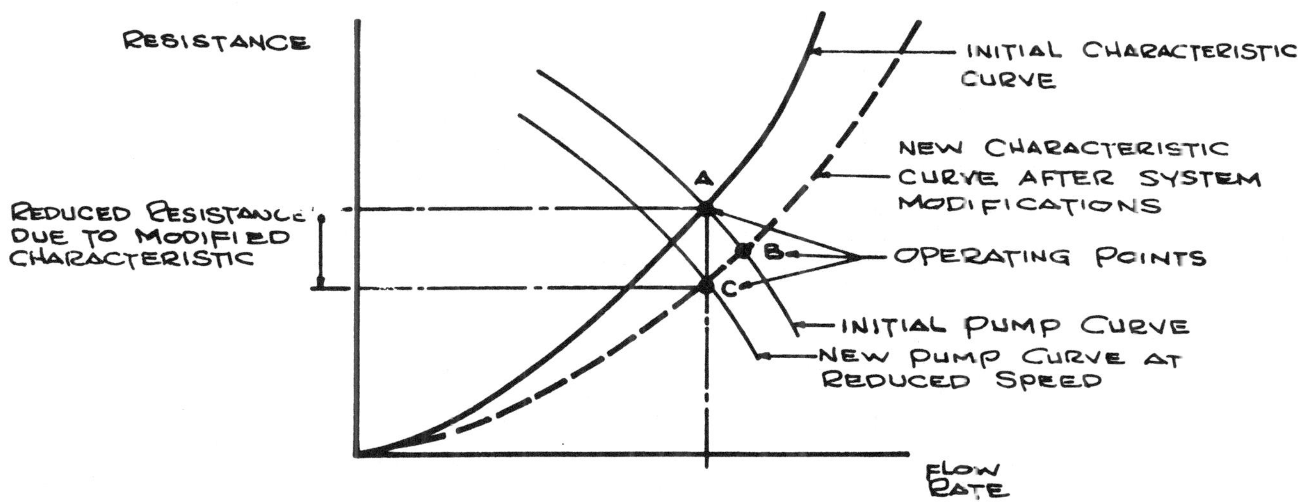

As shown in the diagram, modification of a piping system characteristic will change the operating point from A to B along the pump curve and result in an increased flow rate at a lower resistance. To reduce the flow rate back to its initial point, it is now necessary to reduce the pump speed so that the new pump curve intersects the new characteristic curve at point C, giving the desired flow rate. These principles apply equally to fan/duct systems (ECO #21) and pump/pipe systems.

PROCEDURE

When changing water flow rates and system resistance, sequence the adjustments and modifications in the following order:

(A) Measure initial flow rates and resistance in order to construct the system characteristic curve.

(B) Reduce loads and calculate the new reduced flow rate required to meet them.

(C) Modify the system to reduce resistance to flow and, from measurements, construct the new system characteristic curve.

(D) Determine from the new characteristic curve the degree of resistance at the new reduced flow rate.

(E) Reduce the pump speed or reduce the pump impeller diameter so that the new pump curve crosses the new system characteristic curve at the desired operating point.

The following modifications will further reduce pump energy losses:

Reduce Resistance to Flow in Piping Systems

The resistance to flow exerted by a piping system is the sum of the resistances of all its individual parts in the index circuit. Some of the parts cannot be easily modified, but others are candidates for reducing resistance.

(A) Scale: Heat-exchangers exert a high resistance to flow and are prone to fouling by scale deposits and dirt which further increase their resistance. Disassemble heat exchangers, remove scale mechanically or chemically, and flush out the tubes and shell. Institute a maintenance and water treatment program for heat exchangers based on regular observations of pressure drop and temperature differentials.

Over long periods of time, the formation of advanced scale deposits may result in radically increased resistance. Determine, by inspection, whether such advanced scaling exists and rectify it by chemical cleaning. There are many specialized contractors who have the expertise for this work.

(B) Filters and Strainers: Replace high resistance elements in filters and strainers with low resistance elements or baskets.

(C) Pumps: Pumps must develop sufficient head to overcome the resistance to flow through the longest (or index) circuit, even through this head may exceed the requirements of the other subcircuits. If the index circuit resistance grossly exceeds resistances of the other circuits and cannot be further reduced, consider the installation of a small booster pump for the index circuit only, and reduce the head of the main pump accordingly.

FIGURE9-13 WATER RESISTANCE THROUGH LONG RUNS

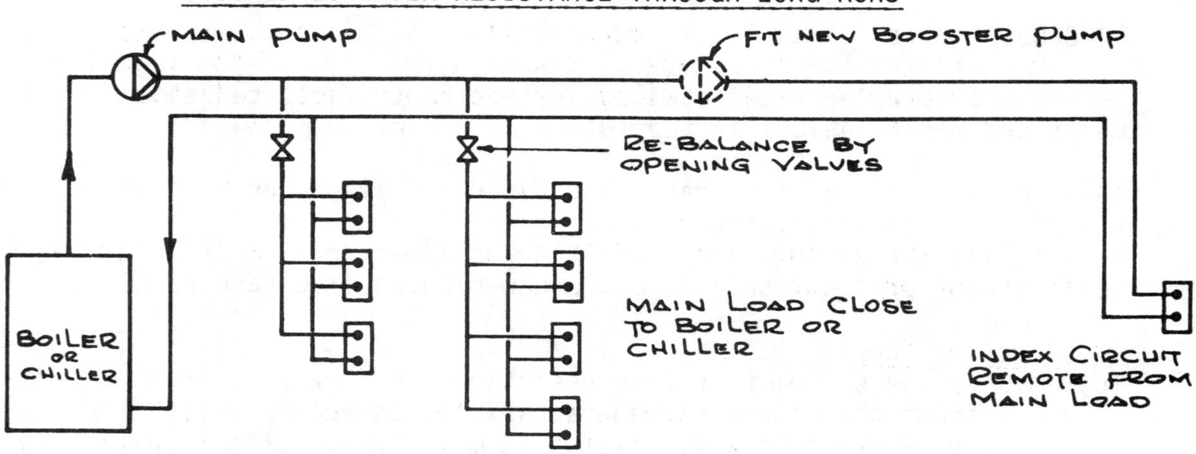

When piping systems are first installed, the installing contractor usually balances flows by trial and error. He often closes balancing valves to a greater extent than is needed, imposing extra head on the pump. Designed safety margins may also result in oversized pumps, and the excess head is absorbed by closing down the valve on the pump discharge. To reduce such resistance to flow, rebalance the system by first opening fully the balancing valve on the index circuit and the pump(s) valve or by removing orifice plates from the pump circuit. Then adjust each circuit balancing valve to achieve proportional flow rates, starting with the next longest circuit and progressing to the shortest circuit. This process is trial and error. Each valve adjustment will affect flow rates in circuits already adjusted, but two or three successive adjustments of the whole system will give good balance.

When all available options to reduce system resistance have been implemented, measure or calculate the new system characteristic and determine the new operating point, taking into account the previous steps to reduce the required flow rate.

Reduce water flow rates

Having reduced resistance to flow to the maximum extend feasible, reduce the flow rate through the following procedures:

(A) Reduce the pump speed in proportion to reductions in flow rate (deriving the appropriate reduction from the pump performance curve available from the manufacturer).

(B) Reduce the speed of indirect drive pumps by changing the size of the motor sheaves.

(C) Reduce the speed of direct drive pumps by exchanging the motor for one of a lower speed. (This is possible only within the range of commercially available motors, and a compromise may be necessary.)

(D) If it is not possible or economically feasible to lower the speed of the pump, reduce the flow rate by changing the size of the impeller. (Seek manufacturer's specifications for each specific application.)

ANALYSIS

To calculate the savings which might be achieved through this ECO, apply the following procedures:

(A) Reducing Pumping Head:

(1) Total all of the steps contemplated to reduce water flow resistance to determine the new system pumping head. This reduction may be calculated by using manufacturer's data and ASHRAE Guide.

(2) Determine the new energy use/yr from Figures 8-44 and 8-45 using the new system pumping head.

(3) Substitute this for the present energy use on Distribution Load Survey Form (Figure 8-10).

(B) Reducing Water Flow:

(1) Calculate the new water flow in gpm based on measures to reduce the building load.

(2) Calculate the new kw required by pumps to satisfy the new water flow requirements using this formula:

$$\text{New kw} = \text{Initial kw} \times \frac{\text{New gpm}}{\text{Initial gpm}}$$

(3) If reduced water flow will be accomplished by throttling valves, calculate the new kw for pumps with this formula:

$$\text{New kw} = \text{Initial kw} \times \frac{\text{New gpm}}{\text{Initial gpm}}$$

(4) Substitute this new kw for the present kw on the Distribution Load Survey Form (Figure 8-10).

Conclusion: All of the figures pertaining to future energy use with regard to this ECO should now be calculated and carried through the appropriate forms as explained in Chpater VII. This will determine whether the costs of implementing and maintaining this ECO are justified in terms of future savings for the building owner.

ECO #23: INSULATE DUCTS

THEORY

Warm air ducts are commonly installed without insulation and routed from the equipment room through unoccupied spaces, shafts, and ceiling voids where their heat loss is unproductive in meeting the occupied space heating load. Although the temperature difference between duct and ambient temperature is relatively small, heat loss in long duct runs can be significant.

Of equal importance is the temperature drop in supply air which accompanies heat loss. In long duct runs serving many rooms in one zone, this will result in the last room having a lower supply air temperature than the first. The tendency in this case is to heat the last room to comfort conditions, resulting in overheating in each preceding room with consequent additional waste of energy over and above the duct heat loss.

PROCEDURE

Warm air ducts may be insulated with rigid fibrous material applied either with adhesive or with special clips or bands. The ducts may also be insulated with flexible mats clipped or wired on (this is particularly applicable for round or oval ducts). Ducts may also be insulated with spray-on foam or fibrous material as described for insulating the undersides of roofs. (It is worth considering insulating roofs and ducts under one contract.)

Insulation applied to ducts supplying only warm air need not be vapor sealed. Insulation applied to ducts supplying warm air in winter and cold air in summer must be vapor sealed to prevent condensation within the insulation. (Check with the local fire department to determine whether the selected insulation is acceptable or whether it is considered a fire and smoke hazard.)

ANALYSIS

To determine the heating energy to be saved by insulating ducts, refer to Figure 8-46 and carry out the following procedure:

(A) Calculate the exposed surface area of uninsulated ductwork.

(B) Determine the temperature differential between the duct and ambient air.

(C) Enter the graph at insulation thickness to be added. Read out the heat loss in Btu/sq ft hour.

(D) Read out the duct heat loss for the existing insulation from the graph.

(E) Subtract the new insulated duct loss from the present duct heat loss and multiply by the surface area of insulated duct to derive the saving in Btuh.

(F) Determine the hours of system operation per year and multiply by the hourly saving to derive the total annual saving in Btu.

(G) Revise the heating system efficiency to reflect the savings on the Annual Heating and Cooling Tabulation Form (Figure 8-11).

If the ambient temperature around the duct varies by more than $5^{o}F$ over the year, determine and use the average temperature for the time considered in the calculation.

In addition to conserving heat-energy, duct insulation will also reduce cooling energy if the same duct is used to supply cooled air in summer. It will also enable a closer control of the single-zone room conditions to be obtained year-round, thus reducing the need to overheat or overcool some rooms. These additional energy savings should be recognized when making an economic analysis. Repeat the above procedure for cooling savings.

Conclusion: All of the figures pertaining to future energy use with regard to this ECO should now be calculated and carried through the appropriate forms as explained in Chapter VII. This will determine whether the costs of implementing and maintaining this ECO are justified in terms of future savings for the building owner.

ECO #24: INSULATE PIPING

THEORY

Energy losses from piping systems, along with frictional resistance, occur in the form of heat gain in chilled water and heat loss from hot water or steam systems. Pipe insulation, which normally keeps these losses to a minimum, frequently becomes damaged and should be repaired or replaced.

PROCEDURE

Bare Piping

Insulate hot pipe to reduce non-productive heat loss. Insulate cold pipe to reduce condensation. Insulation may be a sectional, rigid type of fibrous material, plastic, or glass wool. The material used should be capable of withstanding the surface temperature of the pipe. (Some plastic insulating materials deteriorate or melt at temperatures in excess of 220°F.) An alternative insulating material is plastic magnesia, applied wet; it is particularly suitable where there are many changes of direction, valves, and fittings which make it difficult to apply sectional insulation. All fittings, flanges, etc. should also be insulated against heat loss.

Damaged or Inadequate Insulation

Where there is insulation on existing pipework already, but it is in a bad state of repair, it should be stripped off and replaced. If the worn sections are restricted to a few locations, the insulation might be replaced in those areas only. Where hot pipes are insulated but the insulation is of minimal thickness and effectiveness, it should be supplemented by additional insulation applied over it. (Do not insulate steam traps or the first six feet of the condensate discharge piping from the trap.)

Check with the local fire department to determine whether the selected insulation is acceptable or is considered a fire and smoke hazard. In buildings where food is either stored, processed, or sold, avoid the use of fibrous insulating materials if there is any danger of fiber migration. Check with the local Food and Drug Administration for prohibited materials.

ANALYSIS

To determine the energy saved by insulating pipes, refer to Figures 8-47 and 8-48 for heating and Figure 58 for cooling and carry out the following procedure:

(A) Select the appropriate graph for the surface temperature of the pipe to be insulated.

(B) Tabulate all diameters and lengths of pipe to be insulated.

(C) Enter the graph at the appropriate pipe size, intersect with the existing insulation thickness line, and read out the energy loss in Btu/year.

(D) Repeat this procedure for the same size pipe, but intersect with the added insulation thickness line and read out the annual energy loss in Btu/year.

(E) The difference between these two figures is the energy saved per year for insulating ten feet of the pipe size being treated. Multiply by the $\frac{\text{length of pipe}}{10}$ to be insulated to obtain the total annual saving.

(F) Repeat the above procedure for each different size of pipe.

(G) Total the savings and revise the heating and cooling effeciency to reflect the savings on the Annual Heating and Cooling Tabulation Form (Figure 8-11).

Conclusion: All of the figures pertaining to future energy use with regard to this ECO should now be calculated and carried through the appropriate forms as explained in Chapter VII. This will determine whether the costs of implementing and maintaining this ECO are justified in terms of future savings for the building owner.

ECO #25: REPLACE STEAM TRAPS

THEORY

Steam traps are used to remove condensate, air, and carbon dioxide from the steam-utilizing unit as quickly as they accumulate. With time, internal parts begin to wear and the trap fails to open and close properly, permitting live steam to escape and reducing boiler efficiency, fuel economy, and equipment capacity.

PROCEDURE

Traps should be periodically inspected and all worn or malfunctioning steam traps replaced. Figure 9-14 below indicates the waste caused by steam leaks in a 125 psi system.

FIGURE 9-14 STEAM TRAP WASTE

Size of Orifice	lbs. of Steam Wasted/Month
1/2"	708,750
3/8"	400,00
1/4"	178,000
1/8"	44,200
1/16"	11,200
1/32"	3,000

For steam at 50 lbs. pressure, the waste is about 75 percent of the figures above; at 20 lbs., it is about 50 percent and at 5 lbs., it is about 25 percent.

ANALYSIS

(A) Using Figure 8-50, determine the lbs/hr steam loss through leaking traps.

(B) Total the loss from all traps.

(C) Revise the heating system effeciency to reflect the savings on the Annual Heating and Cooling Tabulation Form (Figure 8-11).

<u>Conclusion</u>: All of the figures pertaining to future energy use with regard to this ECO should now be calculated and carried through the appropriate forms as explained in Chapter VII. This will determine whether the costs of implementing and maintaining this ECO are justified in terms of future savings for the building owner.

F. <u>HVAC Equipment Energy Conservation Opportunities</u> Page

F. HVAC Equipment ECO's: #26 - #43

Introduction

The importance of reducing the annual energy consumption of HVAC equipment can hardly be exaggerated. Due in part to the widespread use of oversized heating and cooling equipment to compensate for energy inefficiencies in building design and construction-- a subject discussed in Chapters II and III --HVAC equipment generally accounts for the major share of a building's annual energy consumption.

The HVAC equipment may appropriately be viewed as the heart and lungs of a typical building. In the complexity and extent of its burners, boilers, chambers, motors, switches, thermostats, tanks, and chillers, it is the force behind virtually every zone and extremity of the building-- just as the human heart lungs condition the blood which is then sent to every zone and extremity of the human body. It is due to this core function coupled with the built-in wastefulness of many existing structures, that HVAC systems tend to offer such a wide variety of energy conservation opportunities, some of them quite dramatic in their favorable ratio of initial investment to energy and dollars saved ("payback").

Those ECO's presented in the following pages have been arranged as nearly as possible in the order in which they ought generally to be investigated-- i.e. from the simplest and cheapest to implement to the more complex and expensive. Where advantage has been taken of immediately implementing operational ECO 5, 6, 7, reassess these ECO's in the light of other ECO's which have reduced building and distribution loads. Every building will impose its exceptions, however, especially in view of the interrelationships between the HVAC equipment and other systems, particularly lighting and distribution. As with all ECO's and groups of ECO's, therefore, those pertaining to HVAC equipment ought to be examined and judged strictly in terms of the actual building under analysis, and certainly in relation to such other controlling factors as the building and distribution load conditions effected by the microclimate, internal space conditions (temperature, humidity, and lighting) building envelope, and distribution system.

Each set of HVAC equipment is of course unique and its particular characteristics can only be identified by on-site inspection and measurement. It will, however, have many characteristics which are common to all core elements of a system, at least in its generic group. As with all elements, however, it must also be understood as part of the system in which it operates. The minimum information required to understand the present operation of a system and to provide a basis for deciding which modificaitons are likely to prove cost-effective is presented below. (Measurement of each of these HVAC functions should be as nearly simultaneous with other measurements as possible.)

(A) Air Flow Rates:

(a) total outdoor air
(b) total return air
(c) total supply air
(d) trunk ducts
(e) terminal units
(f) air cooled condenser

(B) Water Flow Rates:

(a) through boilers
(b) through chillers
(c) through cooling towers
(d) through heat exchangers
(e) through coils and terminal units

(C) Temperatures:

(a) outdoor air dbt and wbt
(b) return air dbt and wbt
(c) mixed air entering coils dbt and wbt
(d) supply air leaving coils dbt and wbt
(e) hot deck and cold deck
(f) air at terminals
(g) conditioned areas dbt and wbt (typical for each functional use)
(h) boiler supply and return
(i) chiller supply and return
(j) condenser supply and return
(k) heat exchanger supply and return
(l) coil supply and return

(D) Refrigerant Temperatures:

(a) hot gas line
(b) suction line

These and othe items pertaining to HVAC ECO's must be approached in a systematic manner rather than considering individual items out of context. As often suggested in preceding chapters, energy systems depend on and react to other systems, so a modificaiton of one will frequently cause a reaction in another which may be either beneficial or counter-productive. (NOTE: For additional information on the characteristics of different types of HVAC systems, see Appendix A.)

ECO #26: IMPROVE CONTROL AND UTILIZATION OF OUTSIDE AIR

THEORY

As discussed in ECO #2, poor control and improper utilization of outside air for ventilation is a major source of wasted energy in buildings. This is especially true during the more intense portions of the heating of the heating and cooling seasons, when outside ventilating air must be either heated or cooled to maintain comfortable indoor temperatures.

Primary cause of "overventilation" in many buildings is the use of poorly installed, poorly sized, or worn out dampers for controlling the admission of outside air for ventilation. These defects occur with particular frequency in roof-top ventilation units, many of which have been fitted with dampers of inferior quality to begin with, while other dampers have simply been damaged or have deteriorated over the years. Such dampers cannot be accurately controlled, and even when fully closed will leak in such quantities as to exceed a building's minimum fresh-air requirements. (A common cause neglect by maintenance crews is the fact that many dampers are "hidden" in the duct-work behind grilles or louvered vents.)

Many building owners, furthermore, are not aware that they can achieve significant savings in ventilation energy through the use of "economizer" cycles during periods of favorable weather. The basic principle in "economizer" cooling is simply the introduction of outdoor air into conditioned spaces to remove internal heat gain without operating the refrigeration compressor. "Economizer" cycles can be utilized in relation to either dry bulb conditions or enthalpy conditions as discussed below. Enthalpy control provides better economy because outside air is used anytime it has a lower heat content than the return air.

PROCEDURE

Dampers

Ill fitting dampers should be replaced by higher quality opposed blade dampers with seals at the edges and ends of blades to minimize air leakage and subsequent energy loss. If economizer cooling is used in summer, the outside air dampers must be sized to pass the full volume of air required by the supply fan. If, on the other hand, economizer cooling is not contemplated, and minimum quantities of outdoor air will be used year round, reduce the size of outdoor air intake louvers and dampers in proportion to reduced outdoor air requirements (calculated in ECO #2). Then blank off the unused portions of out air inlets.

For complete utilization of outdoor air for cooling, it may be necessary to install return air and outdoor air dampers, in addition to providing a means of relieving indoor air pressure. Possible options with regard to the latter include opening windows, operating an exhaust system, or installing propeller-type exhaust fans in the wall (See Figure 9-15 below for relief and return dampers.)

FIGURE 9-15 OUTSIDE AIR DAMPERS

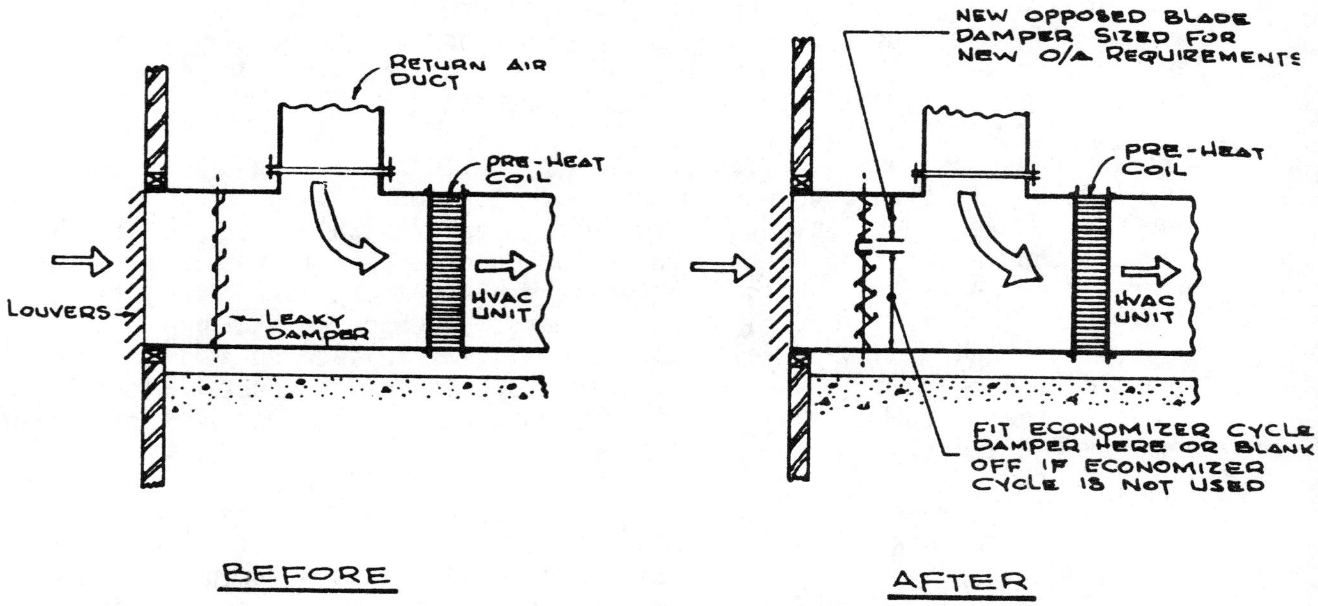

Outside Air-Cooling: Unoccupied Periods:

(A) Shut down compressors and chillers.

(B) Provide a control to sense outdoor conditions and open dampers when the dry-bulb temperature is at least 5° lower than indoor temperature but above 65°.

(C) Open exhaust dampers or exhaust air fans if available. Otherwise partially open windows or install exhaust or return air fans to provide relief from indoor air pressure.

Outside Air Cooling: Occupied Periods:

It is under these conditions that economizer and enthalpy cycle cooling systems may be appropriate. Their respective features and installation procedures are described below.

Economizer cycle cooling

This system monitors and responds to dry-bulb temperature only. It is suitable where wet-bulb degree hours above 65° are fewer than 3000 per year. If this is the desired system for the building in question, then provide controls, dampers, and interlocks to achieve the following controls sequence:

(A) When the outdoor dry-bulb temperature is lower than the supply air dry-bulb temperature required to meet the cooling load, turn off the compressor and chilled water pumps. Then position each of the dampers controlling outdoor air, return air, relief air, and exhaust air to attain the required supply air temperature.

(B) When the outdoor dry-bulb temperature is higher than the supply air temperature required to meet the loads, but is lower than the return air temperature, energize the compressors and chilled water pumps and position dampers for 100 percent outdoor air.

(C) Use minimum outdoor air whenever the outdoor dry-bulb temperature exceeds the return air dry-bulb temperature.

(D) Whenever the relative humidity in the space drops below desired levels and more energy is consumed to raise the rh than is saved by the economizer system, consider using refrigeration in places of economizer cooling. This condition may exist in very cold climates and must be analyzed in detail.

(NOTE: Where there are no automatic controls to operate on economizer cycle, operate dampers and fans manually or consider installing controls.)

<u>Enthalpy Cycle Cooling</u>

This system monitors and responds both to wet-bulb and dry-bulb temperatures. It is most effective in locations which experience more than 3000 wet-bulb degree hours per year. If this is the desired system for the building in question, then provide controls, dampers, and interlocks to achieve the following sequence of controls:

(A) The four conditions listed for the economizer cycle above are also applicable to the system except that enthalpy conditions are measured rather than dry-bulb conditions.

(B) If changes are contemplated in outside air intake, consult all codes bearing on ventilation requirements. Fire and safety codes must also be observed.

<u>Systems Applications</u>

Economizer and enthalpy cycle cooling are applicable to the following types of ventilation systems: (A) Single duct constant volume systems; (B) Variable volume systems; (C) Induction systems; (D) Terminal reheat systems, dual duct systems, and multi-zone systems, assuming appropriate modifications.

ANALYSIS

(A) Operate the system as presently installed and measure the outdoor air volume in cfm which is introduced under various normal damper settings.

(B) Calculate the volume of outdoor air in cfm which would be introduced if new dampers were installed or existing dampers replaced.

(C) Substitute the new cfm figure for the present figure on the Annual Heating and Cooling Tabulation Form (Figure 8-11).

(D) If the building is in a location where there are 3000 or more annual degree hours wbt greater than 66°F, the enthalpy controller could prove to be a good investment.

(E) For areas under 3000 such degree hours, an economizer system controlled by db temperature could prove feasible.

Economizer and enthalpy systems are less effective if used in conjunction with heat recovery systems. Trade-offs must be analyzed. It is not possible in this manual to provide blanket rules for determining energy saved or costs for economizers which affect each individual application.

The degree of cooling which can be achieved with outdoor air cooling can be assessed. Generally, the following procedure is used:

1. Determine the building heat gain at various outdoor temperatures and plot to obtain building energy curve.
2. Add all internal loads.
3. Determine outside temperature where heating is required.
4. Establish maximum outside temperature that will satisfy cooling requirements.
5. Using weather data for the area, determine hours per year when conditions exist between Nos 3 and 4 above.
6. Calculate Btu/Yr of free cooling available.
7. Subtract this figure from the annual cooling load calculated on the Annual Heating and Cooling Tabulation Form, Figure 8-11.

The local Temperature Control manufacturers and Air Conditioning equipment manufacturers can be helpful in establishing this savings.

Conclusion: All of the figures pertaining to future energy use with regard to this ECO should now be calculated and carried through the appropriate forms as explained in Chapters, VII. This will determine whether the costs of implementing and maintaining this ECO are justified in terms of future savings for the building owner.

ECO #27: RECIRCULATE EXHAUST AIR USING ACTIVATED CARBON FILTERS

THEORY

The quantity of air exhausted from a building often exceeds the minimum outdoor air requirements and requires excess outdoor air to be introduced to make up the difference and maintain the pressure balance of the building. To balance the minimum outdoor air requirements and the total air exhausted from the building, the quantity of exhaust air may be reduced either by reducing the rate of exhaust or by retaining the same rate of exhaust and recirculating a proportion of the air back into the building.

Recirculating untreated exhaust air is often unacceptable because it contains unpleasant odors. This air, however, can be treated by passing it through activated carbon filters which will remove the odors. Candidates for such treatment are central exhaust systems from toilet areas, dining rooms, lounges, etc.

Activated carbon filters remove odors from the air by absorption. After a period of time they become saturated and their performance falls off. At this point, the filters must be removed and either regenerated in-house or replaced with freshly regenerated filters supplied by the manufacturer. Heat energy is required for regeneration but can be derived from a waste heat source.

PROCEDURE

To recirculate the air, ducting must be installed from the discharge side of the exhaust system to the return air side of the HVAC system. This ductwork should contain the activated carbon filters. Recirculate that portion of the total exhaust air for the building which exceeds 90 percent of minimum outdoor air requirements. The length of time the filters can remain in service depends on the circumstances of each particular case; manufacturer's advice should be sought on this point. The cost of installing the recirculating ductwork and filters will vary for each particular case depending on the disposition of exhaust and HVAC systems and the quantity of air handled. To obtain an accurate assessment, designs should be prepared and quotations solicited from local contractors.

ANALYSIS

To calculate the heating energy which might be saved by reducing the volume of outside air, refer to Figure 8-31. Then:

(A) determine the new outside air requirement in cfm based on the above procedure;

(B) substitute the new cfm figure for the present figure on Annual Heating and Cooling Tabulation Form (Figure 8-11).

Conclusion: All of the figures pertaining to future energy use with regard to this ECO should now be calculated and carried through the appropriate forms as explained in Chapter VII. This will determine whether the costs of implemting and maintaining this ECO justified in terms of future savings for the building owner.

FIGURE 9-16 CARBON FILTER APPLICATION

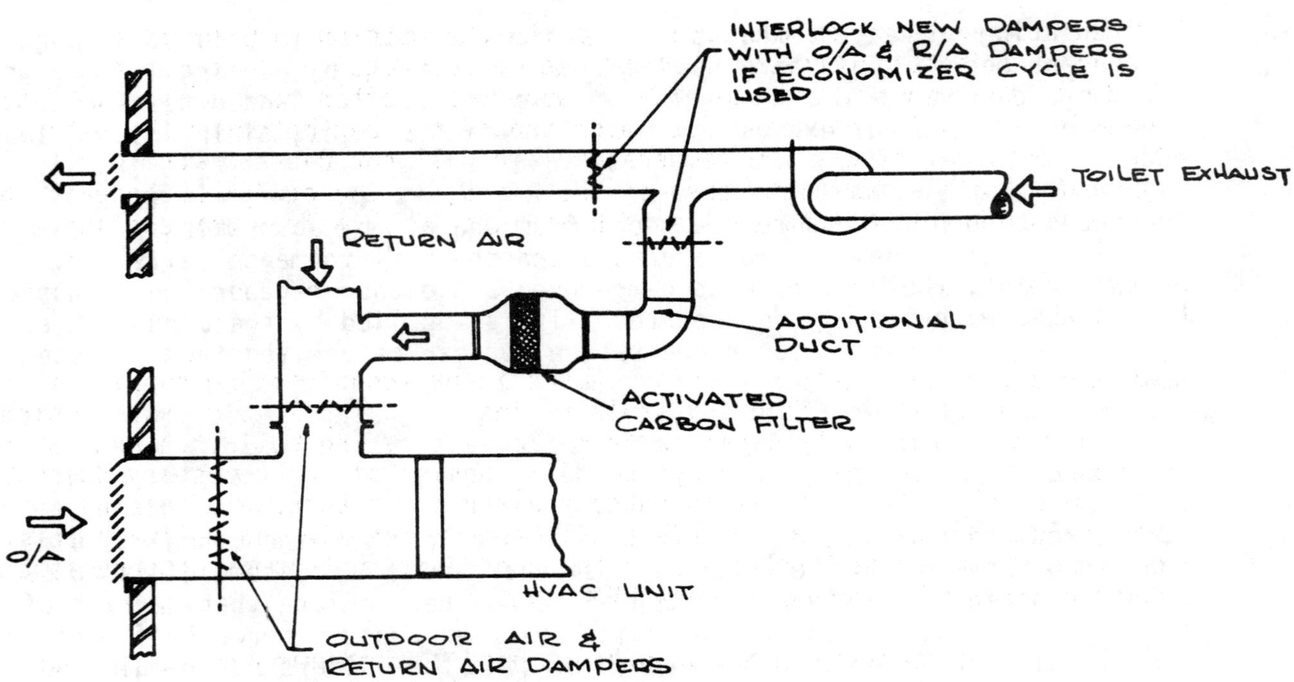

ECO # 28 USE SEPARATE MAKE-UP AIR FOR EXHAUST HOODS

THEORY

Kitchen equipment exhaust hoods and other process equipment hoods frequently remove large quantities of hot air from a building which must be replaced by introducing outdoor air. This make-up air is often introduced through the building HVAC system which heats it in winter and cools and dehumidifies it in summer to the level required to maintain occupant comfort within the building.

It is not necessary in most buildings for the make-up air supplied to exhaust hoods to be treated to the same degree required by occupants. In fact, the temperature the make-up supply air often need only be 50^{0}-55^{0}, assuming it is introduced close to or around the perimeter of the exhaust hood. This relatively low-temperature air will not cause discomfort to occupants in the immediate area of the hoods, as the equipment beneath the hoods radiates large quantities of heat which will offset any cooling effect.

PROCEDURE

To reduce the quantitiy of energy wasted in pre-conditioning supply-air to exhaust hoods, it is often advisable to install a supply-air system which is completely or partially separate from the rest of the buildings HVAC system. In most cases, this separate system will consist of the necessary ducts, fans, dampers, etc. to draw in outdoor air pass it through a heating coil enroute to the exhasut hood. This will temper the air and maintain tolerable conditions immediately adjacent to the hood while not imposing unnecessary demands on the buildings primary HVAC system.

(NOTE: The effectiveness of an exhaust hood in capturing heated air and fumes, smoke, steam, etc. is a function of the face velocity at the edge of the hood. To maintain a satisfactory capture velocity, large open hoods require large volumes of supply air. But face or capture velocities can be maintained and even increased while decreasing the volume of exhaust air by installing baffles or false hood within and existing hood. See diagram below.)

ANALYSIS

To calculate the energy which might be saved through this ECO, refer to Figure 8-31. Then:

(A) determine the new outside air requirements in cfm based on the above procedure;

(B) substitute the new figure for the existing figure on the Annual Heating and Cooling Tabulation Form (Figure 8-11).

Conclusion: All of the figures pertaining to future energy use with regard to this ECO should now be calculated and carried through the appropriate forms as explained in Chapter VII. This will determine whether the costs of implementing and maintaining this ECO are justified in terms of future savings for the building owner.

FIGURE 9-17 EXHAUST HOOD MAKE-UP AIR

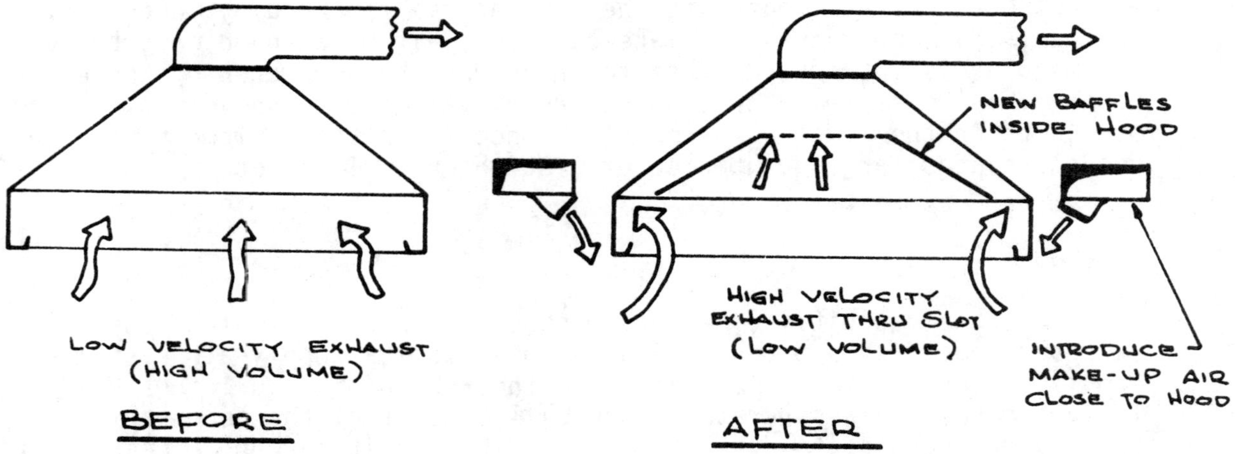

ECO #29: EMPLOY EVAPORATIVE COOLING OF OUTDOOR AIR

THEORY

In geographic areas which experience high dry-bulb temperatures but low coincidental wet-bulb temperatures (65°F wet-bulb or less), evaporative cooling may be used to reduce the temperature of the outdoor air. For instance, outdoor air at 95° dbt - 65° wbt can be adiabatically cooled by water sprays to 77° dbt - 65° wbt giving an rh of 50 percent. This effectively eliminates the ventilation cooling load and under certain conditions will allow the use of economizer cooling, even though the outdoor dry-bulb temperature exceeds the design dry-bulb temperature within the building. Even in relatively humid climates, there are a significant number of hours per year that adiabatic cooling can be used, allowing the operation of refrigeration equipment to be minimized.

PROCEDURE

Evaporative cooling can be achieved by installing sprays within the main HVAC units or in duct extensions together with drain pan eliminators. If the design of the cooling coil is suitable, it is often effective to spray the water directly onto the upstream face of the coil and use the existing downstream eliminators.

Spray water can be continuously recirculated from the drain pan through the spray headers and back to the drain pan by a small pump, but continuous make-up and blow-down should be provided to limit precipitation of solids. If low temperature well water is available, this can first be used in the cooling coil and then used as feed to the evaporative cooler sprays.

ANALYSIS

To determine the cost-effectiveness of evaporative cooling, an analysis must be made of climatic data for each particular installation to assess the number of hours of possible use and the quantity of energy to be saved. The volume of water required and its cost must be considered, as well as the cost of pump operation. The "bottom line," of course, is the volume of energy in Btu which might be saved on an annual basis through the utilization of evaporative cooling for all or part of the building's present cooling requirements. The energy saved will be a reduction in the outside air cooling load and should be subtracted from this load on the Annual Heating and Cooling Tabulation Form (Figure 8-11).

Conclusion: All of the figures pertaining to future energy use with regard to this ECO should now be calculated and carried through the appropriate froms as explained in Chapter VII. This will determine whether the costs of implementing and maintaining this ECO are justified in terms of future savings for the building owner.

ECO #30: EMPLOY DESICCANT DEHUMIDIFICATION

THEORY

Basically, desiccant dehumidification converts the latent cooling load to a high-temperature sensible cooling load, a proportion of which can easily be handled without resorting to mechanical refrigeration. Dessiccant dehumidification is particularly applicable in geographic locations experiencing long periods of high wet-bulb temperatures.

Desiccants such as Silica Gel can be used as adsorbents to reduce the moisture content of the air. Dehumidification is achieved adiabatically, and the air dry-bulb temperature increases proportionally with the quantity of moisture removed. For example, air at 85° dbt - 75° wbt would change adiabatically to 122° dbt - 75° wbt with a corresponding reduction in the moisture content to 56 grain/lb. Ordinarily air can be reduced to 60° at 50% rh by sensible heat removal only with a relatively high-temperature cooling medium, other than chilled water, in the cooling coils. Cooling tower, well water, or waste cool water can be utilized in the coils.

PROCEDURE

Apply desiccant dehumidification to buildings or zones of buildings subject to high internal latent heat gains or to zones requiring a high percentage of outdoor air. Both fixed and rotary bed, factory-assembled units are available, up to capacities of 30,000 cfm. Single bed fixed units must be taken off line periodically for rejuvenation. Rotary units operate in the same manner as thermal wheels (ECO #42) with the air to be conditioned on one side of the wheel and regeneration air on the other side.

It is important to remember that when dessicant material has been used for awhile, it becomes charged with moisture which must be driven out through a regeneration process requiring heat. If possible, recaptured waste heat or solar energy should be used for regeneration.

ANALYSIS

To determine the economic feasibility of desiccant dehumidification for a particular building, analyze local weather data and existing system operating conditions to ascertain the hours of effective use. Refer to the chart below for savings in refrigeration load. The reduction should be subtracted from the calculated cooling load on the Annual Heating and Cooling Tabulation Form (Figure 8-11).

<u>Conclusion</u>: All of the figures pertaining to future energy use with regard to this ECO should now be calculated and carried through the appropriate

forms as explained in Chapter VII. This will determine whether the costs of implementing and maintaining this ECO are justified in terms of future savings for the building owner.

FIGURE 9-18 DESICCANT DEHUMIDIFICATION

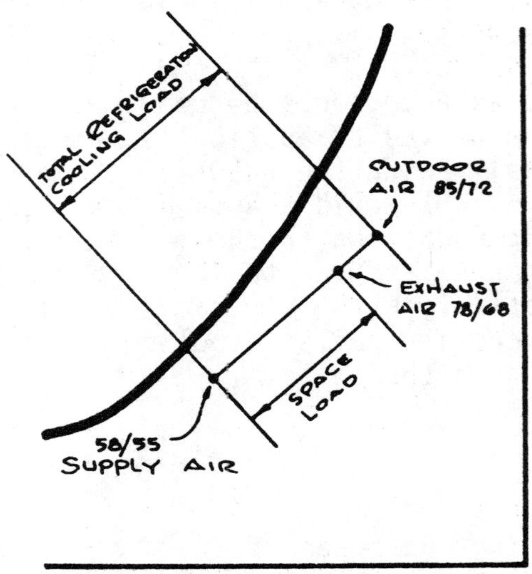

PSYCHROMETRIC CYCLE FOR 100% REFRIGERATION COOLING

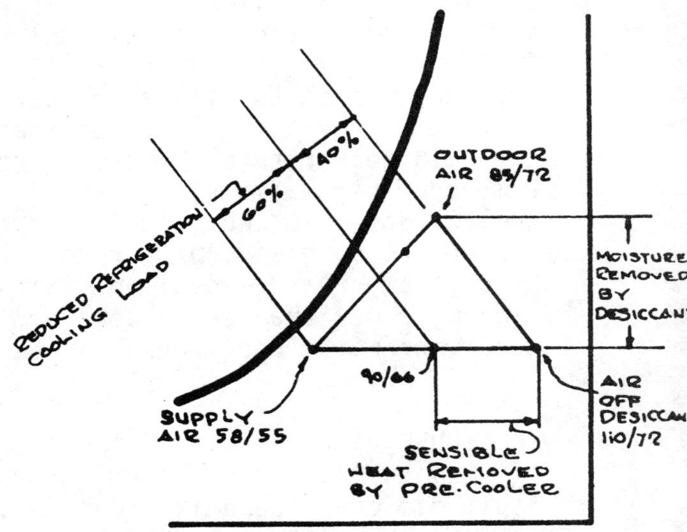

PSYCHROMETRIC CYCLE FOR DESICCANT DEHUMIDIFICATION

ECO #31: REDUCE ENERGY CONSUMED BY REHEAT SYSTEMS

THEORY

"Reheat" refers to a system of air-conditioning which avoids the overcooling of a space not by reducing the volume of chilled air but by tempering it with warm air to achieve the desired levels of temperature and relative humidity. The heat so employed is called "reheat." It is a popular system because of its relatively low initial cost and its high degree of controllability. A variant of "reheat" is often employed in double-duct and multizone HVAC systems.

While convenient to use, these systems are extremely exorbitant in their energy consumption. In fact, the reheat function alone can easily consume as much energy per year as the HVAC system to which it is applied. Thus, in buildings which employ reheat systems, the energy consultant will inevitably find a prime opportunity for energy conservation.

PROCEDURE

Since reheat systems are used primarily to offset excessive cooling, the majority of waste due to reheat often occurs during milder weather and-- if the system is left operating --during the night. The degree to which this occurs will vary with each system and the the methods by which it is operated and controlled. For example, if a terminal reheat system (discussed below) produces a constant 55° temperature off the cooling coil whenever it is operating, then the volume of reheat energy consumed at night will generally be excessive. On the other hand, if the air-chilling function is controlled by space thermostats to deliver only enough chilled air to balance those zones demanding the greatest volume, then night-time consumption of reheat energy will be greatly reduced as compared with day-time consumption. It is therefore quite essential that the consultant be very familiar with the controls and operation of of the reheat system in the particular building being studied. Only thus may an accurate determination be made as to the specific procedures necessary to reduce reheat energy.

The four major HVAC systems which typically employ reheat are briefly discussed below, along with procedures for reducing the energy they consume.

<u>Terminal reheat systems</u>

Terminal reheat allows each zone to be individually controlled but wastes energy in the cooling season, as all of the supply air must be cooled to a low enough temperature to meet the most critical load zone and then reheated for zones of lesser loads to avoid overcooling.

As with single-duct systems, energy can be conserved by reducing the supply air volume. Many reheat systems are controlled at a fixed supply temperature of around 55°dbt/55°wbt. To conserve energy, fit controls to reschedule the supply air temperature upward according to demands of the zone with the greatest cooling load. If one zone has cooling loads grossly in excess of all others, the controlling thermostat should be located in that space.

The greatest quantity of energy can be saved by adding variable air volume boxes to each of the major branch ducts. Each VAV box should be controlled by a space thermostat located in its particular zone, and its associated reheat coil should be provided with controls to prevent reheat until the VAV box has reduced the zone supply air volume to 50%.

In buildings where a reheat system supplies zones of different occupancy, add dampers and control valves to enable water to be shut off during unoccupied times. In addition:

(A) Reduce the water flow and temperature of hot water to reheat coils.

(B) Use waste heat from condensate, incinerators, diesel or gas engines, or solar energy for reheat; but in all cases, raise the cold duct temperature to reduce refrigeration load.

(C) Modify controls to operate terminal reheat systems on a termperature demand cycle only. Caution: Control of humidity will thus be eliminated and zone control modified.

(D) Add or adjust controls to schedule supply air temperature according to the number of reheat coils in operation. If 90% of the coils are in operation, reset supply air temperatures to a higher level until the number of reheat coils falls to 10%.

(E) Install an interlock between the two valves to prevent simultaneous heating and cooling.

(F) If air conditioning is needed during unoccupied hours or in very lightly occupied areas, install conrtrols to de-energize the reheat coils, raise the cold duct temperature, and operate on a demand cycle.

(G) Under an alteration or expansion program, install variable volume rather than terminal reheat or other systems.

(H) De-energize or shut off terminal reheat coils, raise the chilled water and supply air temperature of the central system, and add re-cooling coils to ducts in areas where lower temperatures are needed.

<u>Multi-zone systems</u>

Most existing multi-zone units have a single heating coil serving the hot deck and a single cooling coil serving the cold deck. Each zone supply temperature is adjusted by mixing the required quantities of hot and cold air from these coils. With these types of units, the hot deck temperature must be sufficiently high to meet the heating demands of the coldest zone, and cold deck air must be sufficiently low to meet the demands of the hottest zone. All intermediate zones are supplied with a mixture of hot and cold air wasting energy in a manner similar to reheat systems.

Newer multi-zone units are now available with individual heating and cooling coils for each zone supply duct, so the supply air is heated or cooled only to meet the zone load demand. These new units use far less energy than units with common coils, and where renovations are contemplated or the existing multizone unit is at or near the end of its useful life, replacement should be considered, using a multi-zone unit with individual zone coils. In addition:

(A) Analyze multi-zone systems carefully, treating each zone as a single zone system, and adjust air volumes and temperature accordingly.

(B) Hot deck and cold deck dampers are often of poor quality and allow considerable leakage even where fully closed. To conserve energy, check these dampers and modify to avoid leakage.

(C) Install controls or adjust existing controls to give the minimum hot deck temperature and maximum cold deck temperature consistent with the loads of critical zones.

(D) Where a multi-zone unit serves interior zones that require cooling all summer and most of the winter, energy can be conserved by converting to VAV reheat. To achieve this, blank off the hot deck and add low pressure VAV dump boxes with new reheat coils in the branch duct after the VAV box. This system works well in conjunction with an economizer cycle, and reheat energy is minimized. But careful analysis is required of the existing system and zone requirements to make the correct selection of equipment.

(E) Arrange the controls so that when all hot duct dampers are partially closed, the hot deck temperatures will progressively reduce until one or more zone damper is fully opened and, when all of the cold duct dampers are partially closed, the cold duct temperature will progressively increase until one or more of the zone dampers is fully opened. In addition:

    (a) Install controls to shut off the fan and all heating control valves during unoccupied periods in the cooling season, and shut off the cooling valve during unoccupied periods in the heating season.

    (b) Convert system entirely or in part to variable volume by adding terminal units and pressure bypass and/or add fan coil units in specific areas requiring constant air volume.

Dual-duct lowvelocity systems

Dual-duct lowvelocity systems supply hot and cold air in individual ducts to the various zones of the building which are supplied with a mixture of hot and cold air to maintain the desired zone conditions. Control of mixing is normally achieved by automatic dampers in the branch ducts, each damper being positioned according to the dictates of the space thermostat. Each duct should be analyzed as a single-zone system to determine whether the air volume or its temperature set-point can be advantageously changed to conserve energy. Reduce the temperature of the hot duct and increase the temperature of the cold duct to that point where the heating and cooling loads of the most critical zone can just be met.

Return air is a mixture of all zones and reflects the average building temperature. In some designs of central station equipment, it is possible to stratify the return air and the outdoor air by installing splitters so that the hottest air favors the hot deck and coldest air favors the cold deck. This will reduce both heating and cooling loads.

Dualduct high-velocity systems

Dual-duct high-velocity systems operate in the same manner as low-velocity systems, except that the supply fan runs at a high pressure and each zone requires a mixing box with sound attenuation.

Considerable quantities of energy are required to operate the fan at high pressure, and a close analysis of pressure drops within the system should be made, along with a reduction in fan pressure to the minimum required to operate the mixing boxes.

To reduce system pressure, replace existing high pressure mixing boxes with lower pressure types. In addition:

(A) When there is no cooling load, install controls to close off cold air duct; de-energize chillers and cold water pumps and operate as a single-duct system, rescheduling the warmer air duct temperatures according to heating loads only.

(B) When there are no heating loads, install controls to close off the warm air ducts; shut off hot water, steam or electricity to the warm duct and operate the system with cold duct air only, rescheduling supply air temperature according to cooling loads.

(C) Replace obsolete or defective mixing boxes to eliminate leakage of hot or cold air when the respective damper is closed.

(D) Provide volume control for the supply air fan and reduce capacity, preferably by speed reduction when both the hot deck and cold deck air quantities can be reduced to meet peak loads. Reducing the heat loss and heat gain provides an opportunity to reduce the amount of air circulated.

(E) When there is more than one air handling unit in a dual-air system, modify duct work if possible so that each unit supplies a separate zone to provide an opportunity to reduce hot and cold duct temperatures according to shifting loads.

(F) Change dual-duct systems to variable-volume systems when the savings from such an investment are not overcome by adding VAV boxes and fan control.

ANALYSIS

The amount of reheat energy being used has been established by the method described under "Reheat Use Tabulation". The amount that this energy can be reduced by implementing the procedures described above is a judgment on the part of the consultant. There is no way of actually calculating the amount of residual reheat, short of computer simulation. Nonetheless, a basic understanding of when and how the reheat system is employed in a building will enable the consultant to make reasonably accurate judgment as to the amount of energy reduction which might be achieved by this ECO.

Reheat is used to off-set excessive cooling, therefore the majority of waste due to reheat occurs during milder weather, and during the night if the system is in operation. The degree to which this occurs will vary with each system and the methods by which it is operated and controlled. For example, if a terminal reheat system has a constant 55° temperature off the cooling coil whenever it is in operation, then the amount of reheat use at night will be quite excessive. On the other hand, if the cold air off of the cooling coil is controlled by the space thermostats to only deliver air cold enough to satisfy the zone calling for the most cooling, then night-time consumption. It is therefore quite essential that the consultant be very familiar with the controls and operation of a reheat system in order to accurately determine the procedures necessary to reduce the amount of energy reduction that will result.

Make a judgment as to the reduction of reheat and substitute the new reheat energy for the present energy on the Energy Tabulation Form (Figure 8-12).

Conclusion: All of the figures pertaining to future energy use with regard to this ECO should now be calculated and carried through the appropriate forms as explained in Chapter VII. This will determine whether the costs of implementing and maintaining this ECO are justified in terms of future savings for the building owner.

ECO #32: ADJUST FUEL/AIR RATIOS OF FIRING SYSTEM

THEORY

Since the boiler system in the structure being modified (if it has one) was originally sized for an unmodified heating load, the implementation of any number of the above ECO's will require adjustments in the boiler system itself, creating a new ECO. Especially important is the fuel/air ratio in the firing system, which will have to be adjusted to correspond with the modified heating load of the building. Below is a typical heating system time/load curve:

FIGURE 9-19 HEATING SYSTEM TIME/LOAD CURVE

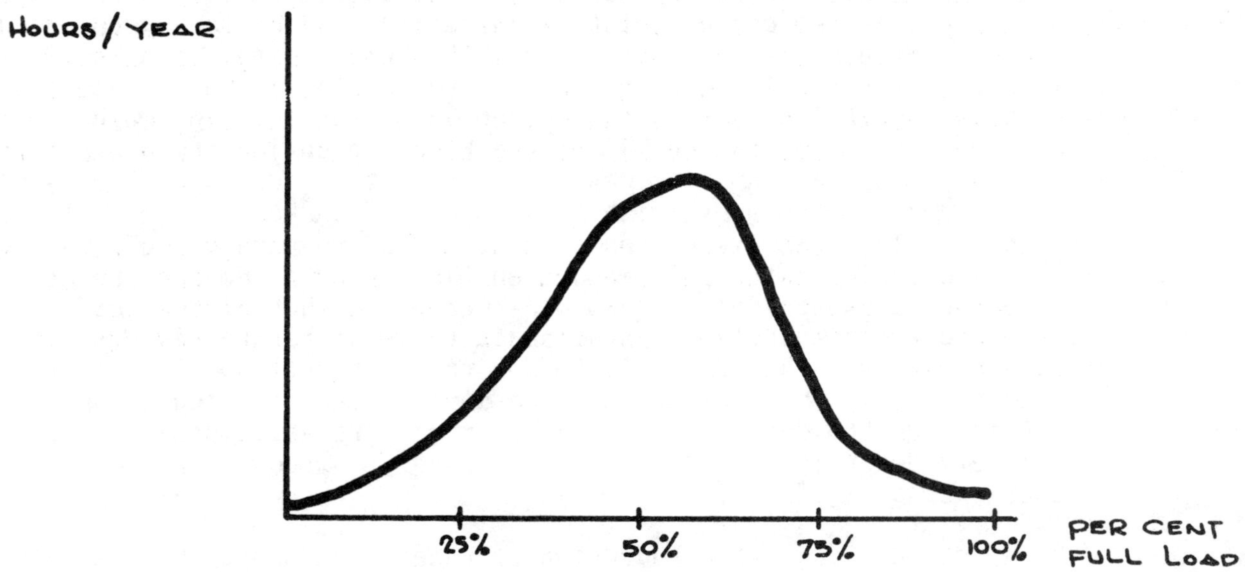

The heating system operates between 50% and 60% full load for more hours/year than at higher or lower loads, and the new optimum operating point of the heating boilers should be selected accordingly to give the highest efficiencies over this range.

PROCEDURE

Primary and secondary air should be allowed to enter the combustion chamber only in regulated quantities and at the correct place. Defective gaskets, cracked brickwork, broken casings, etc. will allow uncontrolled and varying quantities of air to enter the boiler and will prevent accurate fuel/air ratio adjustment. If spurious stack temperature and/or oxygen content readings are obtained, inspect the boiler for air leaks and repair all defects before a final adjustement of the fuel/air ratio.

When substantial reductions in heating load have been achieved, the firing rate of the boiler may be excessive and should be reduced. Consult the firing equipment manufacturer for specific recommendations. (A reduced firing rate in gas and oil burners may require additional bricking to reduce the size and shape of the combustion chamber.)

ANALYSIS

Use Figure 8-34 to determine the optimum fuel/air ratio for any given combination of circumstances. Indicators of maximum combustion efficiency are stack temperature, percentage CO_2 and percentage O_2.

The cost of fuel/air ratio adjustments and firing equipment modifications will vary with each application, and quotations should be obtained from qualified contractors. These costs, added to the projected annual fuel costs of the modfied boiler and/or its firing system, may then be compared with the annual energy consumption of the unmodified boiler system to determine the economic feasibility of this ECO. This annual energy consumption may be obtained from the Annual Heating and Cooling Tabulation Form (Figure 9-12) and, if domestic water is heated by the boiler, from the Building Load Survey Form (Figure 8-9). Substitute the new consumption figure for the present figure on both these forms.

Conclusion: All of the figures pertaining to future energy use with regard to this ECO should now be calculated and carried through the appropriate forms as explained in Chapter VII. This will determine whether the costs of implementing and maintaining this ECO are justified in terms of future savings for the building owner.

ECO #33: INSTALL FLUE GAS ANALYZER

THEORY

The efficient combustion of fuel in a boiler requires an optimum fuel/air ratio providing for a percentage of total air sufficient to insure complete combustion of the fuel without overdiluting the mixture and thereby lowering boiler burner efficiency. Optimum combustion efficiency, which varies constantly with changing loads and stack draft, can be achieved only through a continuing analysis of flue gases. The information required for continuous analysis and adjustment of fuel/air ratios is (A) flue gas temperature and (B) flue gas CO_2 or O_2 content.

Devices are available which continuously measure CO_2 and stack temperature to produce a direct reading of boiler efficiency. These indicators provide boiler operators with the requisite information for manual adjustment of boiler fuel/air ratio. They are suitable for smaller installations or buildings where money for investment in capital improvements is limited. A more accurate measure of combustion efficiency, however, is obtained by an analysis of oxygen content rather than of other gases such as carbon dioxide and carbon monoxide. As shown in Figure 8-34, the cross checking of O_2 concentrations is useful in judging burner performance more precisely. Due to the increasing utilization of multifuel biolers, however, O_2 analysis is the single most useful criterion for all fuels since the O_2 total air ratio varies only within narrow limits.

PROCEDURE

For larger boiler plants, consider the installation of an automatic continuous oxygen analyzer with "trim" output that will adjust the fuel/air ratio to meet changing stack draft and load conditions. Most boilers can be modified to accept an automatic fuel/air mixture control by a flue gas analyzer, but a gas analyzer manufacturer should be consulted for each particular installation to be sure that all other boiler controls are compatible with the analyzers.

It is important to note that some environmental protection laws might place a higher priority on visible stack emissions than on efficiency and optimization of fuel combustion, especially where fuel oil is burned. The effect of percent total air on smoke density might prove to be an overriding consideration and limit the approach to minimum excess air. All applicable codes and environmental statutes should be checked for compliance.

ANALYSIS

Refer to Figure 8-34 to calculate the degree of improved efficiency to be obtained by optimizing boiler fuel/air ratios through the addition of a flue gas analyzer. Substitute the new fuel efficiency figure for the present figure on the Annual Heating and Cooling Tabulation Form (Figure 8-11).

Conclusion: All of the figures pertaining to future energy use with regard to this ECO should now be calculated and carried through the appropriate forms as explained in Chapter VII. This will determine whether the costs of implementing and maintaining this ECO are justified in terms of future savings for the building owner.

ECO #34: REPLACE EXISTING BOILERS WITH MODULAR BOILERS

THEORY

Heating boilers are usually designed to operate at maximum efficiency only when producing their rated output of Btu. As shown in the graph below, however, most boilers operate at 60 percent or less of capacity for 90 percent of the heating season, resulting in significant boiler inefficiencies and wasted fuel.

FIGURE 9-20 EFFECTS OF BOILER CYCLING

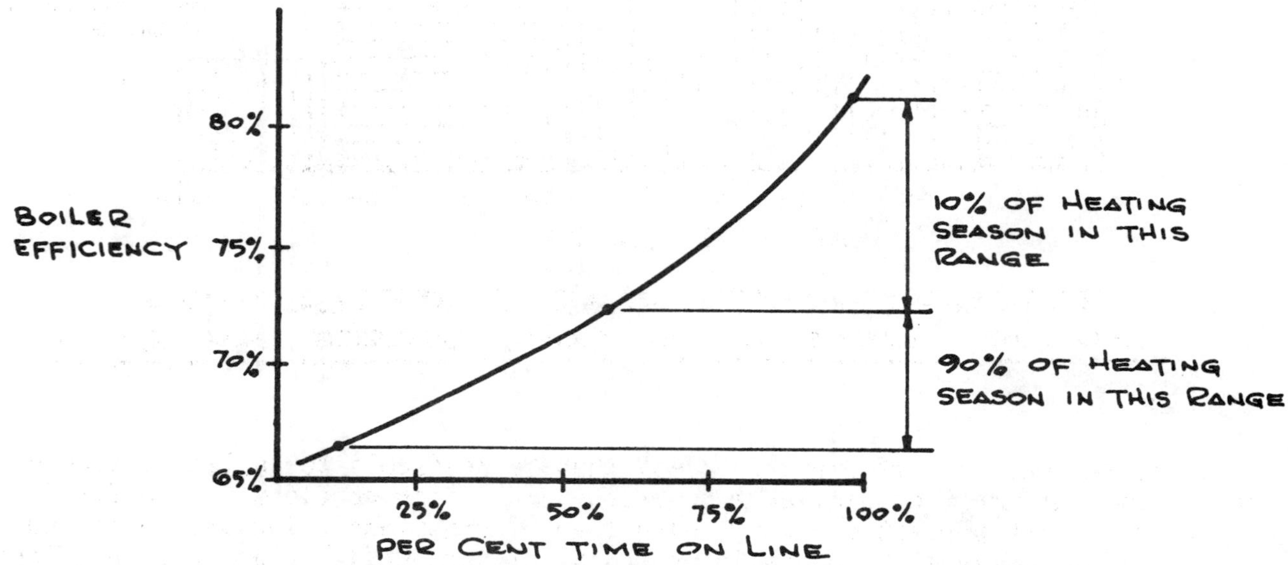

EFFECT OF CYCLING TO MEET PART LOADS

This waste can be diminished but not eliminated by high-low firing systems in large capacity boilers, as shown in the graph below.

A generally superior means of meeting a fluctuating boiler load demand is a system of modular boilers which can be fired independently. Each small-capacity unit has a relatively low thermal inertia (giving rapid response and low heat-up and cool-down losses) and will either be firing at a maximum efficiency can be improved from 68 to 75 percent in a typical installation where a single unit, large-capacity boilers are replaced by any modular boilers. This represents a saving of approximately 9 percent of the present yearly fuel consumption of most commercial buildings.

FIGURE 9-21 SEASONAL HEAT LOAD DISTRIBUTION

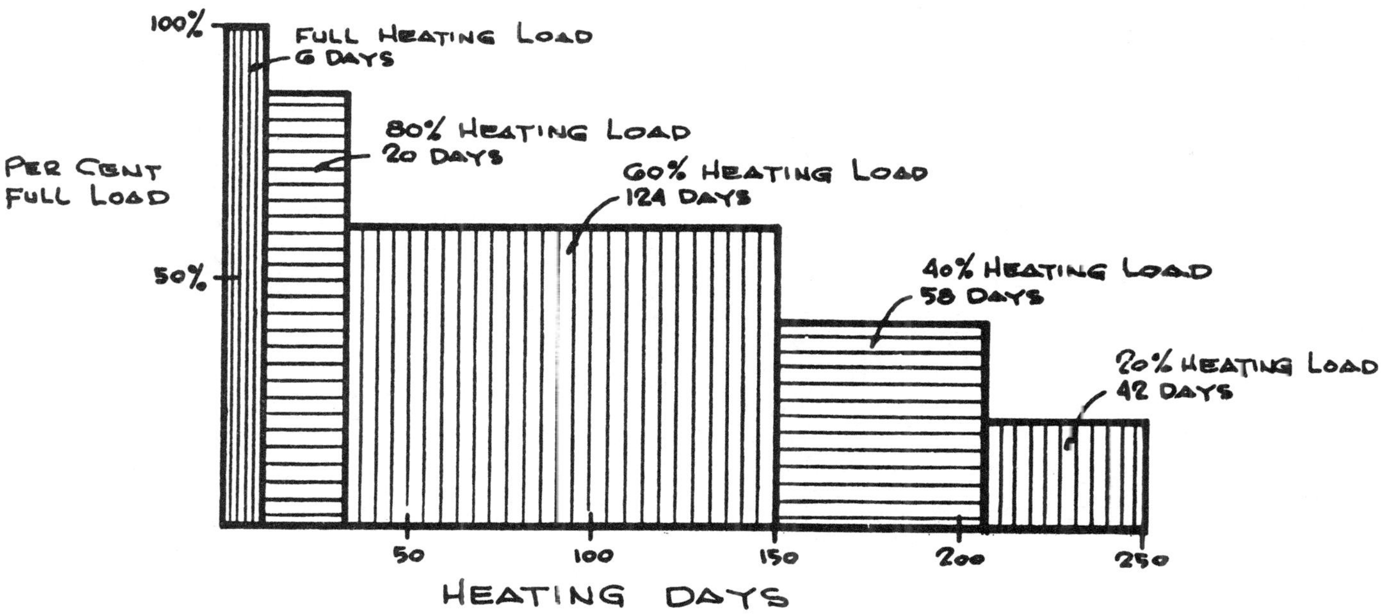

TYPICAL HEAT LOAD DISTRIBUTION
250 DAY SEASON IN 6000 DEGREE DAY ZONE

PROCEDURE

To increase seasonal efficiency, it may thus be advisable to replace a single boiler with an array of smaller modular boilers. Each module would be fired on demand at 100 percent capacity, with load fluctuations being met by firing more or fewer boilers. Especially where the present boiler plant has deteriorated to the point where it is at or near the end of its useful life, it is often worthwhile to consider replacement with modular boilers sized to meet the reduced heating load resulting from other ECO's.

Note: Modular boilers are particularly effective in buildings with intermittent short-time occupancy, such as churches. They provide rapid warm-up for occupied periods and low standby losses during extended unoccupied periods.

ANALYSIS

To calculate the potential energy savings due to this ECO, refer to the graphs above and (A) determine the seasonal efficiency of the proposed modular boiler installation; (B) Substitute this new efficiency figure for the present figure on the Annual Heating and Cooling and Tabulation Form (Figure 8-11).

Conclusion: All of the figures pertaining to future energy use with regard to this ECO should now be calculated and carried through the appropriate forms as explained in Chapter VII. This will determine whether the costs of implementing and maintaining the ECO are justified in terms of future savings for the building owner.

ECO #35: PREHEAT COMBUSTION AIR AND HEAVY FUEL OIL TO INCREASE BOILER EFFICIENCY

THEORY

Preheating the primary and secondary air will reduce its cooling effect when it enters the boiler combustion chamber, thus increasing the efficiency of the boiler. It will also promote a more intimate mixing of fuel and air which will further improve efficiency. Waste heat from flue gases, blowdown, condensate, hot wells, etc., or from solar energy may be used to preheat combustion air and/or oil, either in the storage tanks (low sulfur oil requires continuous heating to prevent wax deposits) or at the burner nozzle.

FIGURE 9-22 EFFICENCY INCREASE WITH PRE-HEATED AIR

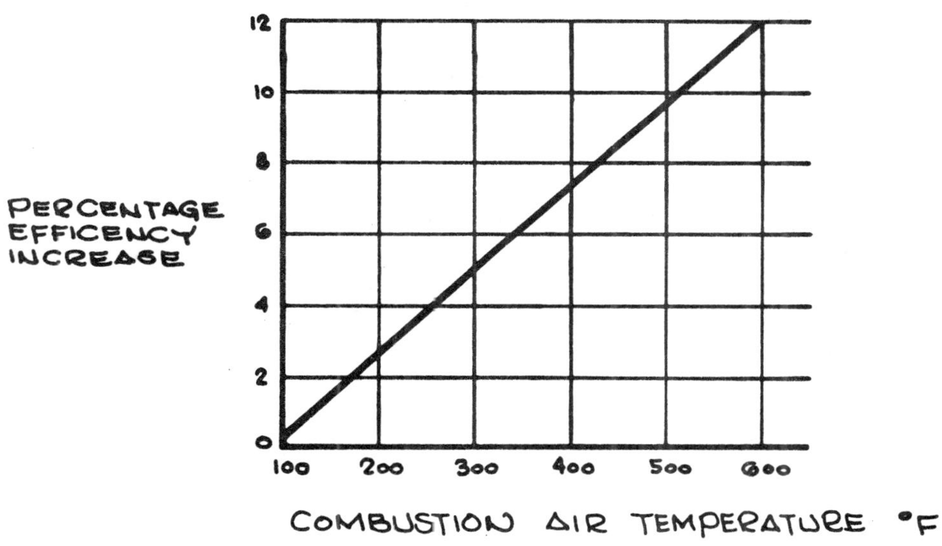

PROCEDURE

Make a heat exchange directly from flue gases to combustion air using static tubular, plate, or rotary exchangers. Heat exchange may also be made indirectly through run-around coils in the stack and combustion air duct. In most boiler rooms, air is heated incidentally by hot boiler and pipe surfaces and rises to collect below the ceiling. Use this air directly as preheated combustion air by ducting it down to the firing level and directing it into the primary and secondary air inlets.

As indicated by the graph above, boiler efficiency will increase by approximately 2% for each 100°F added to combustion air temperature. Oil must be preheated to at least the following temperatures to obtain complete atomizatio

No. 4 oil - 135°F
No. 5 oil - 185°F
No. 6 oil - 210°F

Heating beyond these temperatures will increase efficiency, but care must be taken not to overheat, as vapor-locking could cause flame-outs. The increased efficiency obtained by preheating oil could be as high as 3% but depends on the particular consituents of the oil.

ANALYSIS

In approaching the analysis of this ECO, obtain the manufacturer's recommendations on preheated fuel and air for the particular equipment being considered. Obtain the fuel oil dealer's recommendations on the preheating levels most appropriate for the fuel to be used. (Combustion air can be preheated up to 600°F for pulverized fuels and up to 350°F for stoker-fired coal, oil, and gas. The upper temperature limit is determined by the design and materials of the firing equipment.)

With these recommendations in hand: (A) analyze the effectiveness or the proposed boiler modifications in terms of increased seasonal boiler plant efficiency; (B) enter subsequent changes in heating conversion efficiency on the Annual Heating and Cooling Tabulation Form (Figure 8-11).

Conclusion: All of the figures pertaining to future energy use with regard to this ECO should now be calculated and carried through the appropriate forms as explained in Chapter VII. This will determine whether the costs of implementing and maintaining this ECO are justified in terms of future savings for the building owner.

ECO #36: MAINTAIN FUEL BURNING EQUIPMENT AND HEAT TRANSFER SURFACES

THEORY

Fuel burning equipment allowed to become dirty and out of adjustment becomes increasingly inefficient with continued usage. Likewise, both fire-side and water-side heat transfer surfaces become less and less effective if allowed to become fouled by products of combustion, scaling, and other impurities. All heat not properly transferred is discharged through the stack.

PROCEDURE

After reducing the building and distribution heating load, clean and/or replace dirty oil nozzles, oversized or undersized nozzles, fouled gas parts, and improperly sized combustion chamnbers. Reduce nozzle sizes and modify combustion chambers for proper combustion.

The condition of the heat transfer surface directly affects heat transfer from the combustion chamber and/or hot gasses. Keep the fire-side of the heat transfer surface clean and free from soot or other deposits and the air and water-sides clean and free of scale deposits. Remove deposits by scraping where they are accessible, by chemical treatment, or by a combination. In the case of steam boilers, once the water-side of the boiler is clean, institute correct water treatment and blow-down to maintain optimum heat transfer conditions.

ANALYSIS

Using data from the ASHRAE Guide, calculate the seasonal heat and fuel waste due to fouled boiler equipment. Then: (A) determine the projected efficiency of the modified eqipment; (B) enter subsequent changes in heating conversion efficiency on the Annual Heating and Cooling Tabulation Form (Figure 8-11).

Conclusion: All of the figures pertaining to future energy use with regard to this ECO should now be calculated and carried through the appropriate forms as explained in Chapter VII. This will determine whether the costs of implementing and maintaining this ECO are justified in terms of future savings for the building owner.

ECO #37: REPLACE STEAM ATOMIZING BURNERS WITH AIR ATOMIZING BURNERS

THEORY

Air-atomizing burners are considerably more efficient than steam-atomizing burners, due primarily to the relatively higher O_2 content of the fuel at the instant of combustion.

PROCEDURE

Replace obsolete or defective steam-atomizing burners with new air-atomizing burners. Also consider replacing the existing steam-atomizing burners in satisfactory condition with new air-atomizing burners.

ANALYSIS

(A) Compare the seasonal efficiencies of the existing steam-atomizing burners with those of the proposed air-atomizing burners.

(B) If the proposed new burners are more efficient than the existing burners, enter subsequent changes in heating conversion efficiency on the Annual Heating and Cooling and Tabulation Form (Figure 8-11).

Conclusion: All of the figures pertaining to future energy use with regard to this ECO should now be calculated and carried through the appropriate forms as explained in Chapter VII. This will determine whether the costs of implementing and maintaining this ECO are justified in terms of future savings for the building owner.

ECO #38: REDUCE BLOWDOWN LOSSES

THEORY

The purpose of blowing down a boiler is to maintain a low concentration of dissolved and suspended solids in the boiler water and to remove sludge in the boiler to avoid priming and carryover. There are two principal types of blowdown intermittent-- manual blowdown and continuous blowdown. Manual blowdown (or sludge blowdown) is necessary for the operation of the boiler regardless of whether continuous blowdown is being used. The frequency of manual blowdown will depend on the volume of solids in the boiler makeup water and the type of water treatment used. While continuous blowdown requires a steady supply of additional energy (because the makeup water must be heated), these losses can be minimized with automatic blowdowr control and heat recovery systems.

PROCEDURE

Install automatic blowdown controls to monitor the conductivity and pH of the boiler water allowing the boiler to blow down only when required to maintain acceptable water quality. Further savings can be realized by piping the blowdown water through a heat exchanger or through a flash tank with a heat exchanger.

ANALYSIS

(A) Determine the blow-down rate and calculate the total heat available from blowdown.

(B) Compute the heat to be recovered by using a heat exchanger and/or by adding a flash tank.

(C) Enter subsequent changes in heating conversion efficiency on the Annual Heating and Cooling Tabulation Form (Figure 8-11).

(Note: The percent of heat recovery will change with boiler operating conditions. The heat recovery will change with boiler operating conditions. The heat rovery range is 78 percent for a 15 psig boiler to a 98 percent for a 300 psig boiler.)

Conclusion: All of the figures pertaining to future energy use with regard to this ECO should now be calculated and carried through the appropriate forms as explained in Chapter VII. This will determine whether the costs of implementing and maintaining this ECO are justified in terms of future savings for the building owner.

ECO #39: INCREASE EVAPORATOR AND/OR DECREASE CONDENSER WATER TEMPERATURES AND MODIFY CONTROLS

THEORY

When all applicable ECO's have been implemented to reduce both the load and thermal losses in the distribution system, the quantity of cooling required by the primary equipment will most probably have been reduced. Reductions in the cooling load will in turn permit changes in the equipment operation conditions. The performance, or seasonal efficiency, of an existing refrigeration system can be increased by (A) changing the mode of operation and the operating conditions to conform more closely to the part load conditions which are the rule, rather than the exception, in vir tually all buildings; (B) improving maintenance and service procedures.

The mode of operation can be changed in two ways:

(A) <u>Raising Evaporator Temperatures</u>

The evaporator temperature is a function of operating mode and maintenance. All measures to reduce loads provide opportunities to raise the temperature for peak loads, thus presenting even greater opportunities for part-load conditions. Adjustments for load often permit an increase in the supply air temperatures even with a reduction in the total amount of air circulated. This will increase the evaporating temperature and suction temperature of a direct expansion system and conserve power for the compressor. Raising supply air temperatures with chilled water systems permits higher chilled water temperatures, higher evaporator temperatures, higher suction temperatures, and a reduction in power used by the chiller.

Systems with an inadequate charge or refrigerant and fouled evaporator surfaces have a greatly reduced refrigerating capacity and waste energy.

(B) <u>Lower Condenser Temperatures</u>

The condenser temperature is a function of the outdoor or ambient dry-bulb conditions (for operation of the air-cooled condensers) and wet-bulb conditions (for operation of the cooling tower or evaporative condensers) and of the condition and operating mode of the air cooled condensers, cooling towers and shell and tube water-cooled condensers.

Cooling towers, air-cooled condensers, and evaporative condensers are usually selected initially to provide a given condensing temperature for maximum expected outdoor conditions; consequently, they will provide lower condensing temperatures when outdoor conditions are below the maximum expected level.

Both of the above modifications will increase the coefficient of performance (COP) of the refrigeration equipment. The COP is related directly to evaporating and condensing temperatures which, for a water chiller with cooling tower, are typically 40°F and 100°F respectively. If the evaporating temperature can be raised (by using chilled water at a higher temperature of, for instance, 50°F instead of 40°F) and/or if condensing temperatures can be reduced, then the COP will increase and a greater cooling effect will result from the same power input.

In general, raising the chilled water temperature 10° and reducing the condensing water temperature 10° will result in an increase in efficiency or approximately 20 to 25 percent. Under part-load conditions, this increase in efficiency is more marked; on a seasonal basis it will have a greater effect in conserving energy than would be indicated by consideration of full-load operating conditions only.

PROCEDURE

To increase the evaporator temperature, provide controls to achieve the following:

(A) Raise the chilled water temperature to "follow the load". Install a limit switch in each modulating or diversion valve to measure whether the valve is fully open or partially open. Arrange the control circuits so that when all coil control valves are either closed or in a partially open position (indicating light load conditions), the chilled-water temperature supply set point should be raised until one or more coil control valves return to the fully open position.

(B) Raise supply air temperature to follow the load.

(C) Permit relative humidity to rise in the conditioned air space (to 60 percent or more).

(D) Decrease the superheat on the DX coil (consult manufacturer).

(E) Fully load one compressor before starting the next refrigeration unit module.

To decrease the condenser water temperature:

(A) If low temperature well water is available or can be made available, consider its use for condenser water rather than using cooling towers. The energy required to pump the well water is likely to be approximately the same as for a condenser water pump and cooling tower fans, but energy savings in compressor horsepower will be worthwhile.

(B) Where existing cooling installations operate at constant condenser water temperature and cooling tower fans are cycled on and off, modify the controls to operate cooling tower fans continuously whenever there is a machine on-line. Allow the condenser water temperature to drop until it reaches a pre-determined lower limit, at which point the cooling tower fans can be cycled on and off.

(C) To reduce condensing temperatures and maintain low fouling factors, install automatic tube cleaners. One such device comprises a cylindrical brush in each tube which is periodically forced from one end of the condenser tube bundle to the other by a reversal in the direction of water flow. It has been demonstrated that this type of system can be very effective in keeping condenser tubes clean and, depending on the exisitng level of condenser tube fouling, can yield substantial energy savings.

ANALYSIS

(A) To calculate the volume of energy which might be saved by increasing evaporator temperature as described above, refer to Figure 8-52 and determine the percent increase in COP based on the higher temperature.

(B) To calculate the volume of energy which might be saved by reducing condenser temperature as descrived above, refer to Figure 8-51 and determine the percent increase in COP based on the lower temperature.

(C) Enter subsequent changes in cooling conversion efficiency on the Annual Heating and Cooling Tabulation Form (Figure 8-51).

Conclusion: All of the figures pertaining to future energy use with regard to this ECO should now be calculated and carried through the appropriate forms as explained in Chapter VII. This will determine whether the costs of implementing and maintaining this ECO are justfied in terms of future savings for the building owner.

ECO #40: ISOLATE OFF-LINE CHILLERS AND COOLING TOWERS

THEORY

Light cooling-load demands on a multiple chiller installation are often met by circulating chilled water through all the chillers, even though only one chiller is operating. This not only wastes pump energy by maintaining an unnecessarily high flow-rate of water through the system, but, due to the bypass effect of offline chillers, it forces the remaining on-line chiller to produce chilled water at lower temperatures to offset the mixing effect of the bypass. This in turn causes the on-line chiller to operate at lower evaporator temperatures, reducing the COP of the cooling unit. Pump and fan energy are also wasted by circulating water through idle cooling towers corresponding with idling chillers.

PROCEDURE

Under light-load conditions isolate the off-line chillers and reduce the chilled water flow-rate. Shut off unnecessary chilled water pumps when multiple pumps are installed. If multiple chilled water pumps are not installed, consider the installation of additional pumps or install a multi-speed drive on the present pump.

In conjunction with the isolation of off-line chillers, unneeded units of the cooling towers should also be shut down and the flow of condenser water correspondingly reduced. In cases where existing modules are all oversized for light loads, consider installing new smaller refrigerating and cooling tower units.

ANALYSIS

Determine whether isolating the existing units without adding equipment will meet the minimum load demands. If not, consider adding a low-volume pump and isolating valves. (Only if the load on the chiller is still considerably higher than the system load should new refrigeration equipment be considered.) Then: (A) substitute the improved COP of the modified cooling system for the present COP on Form AHCT; (B) substitute the new horsepower of the low-volume pump(s) for the present horsepower on the Annual Heating and Cooling and Tabulation Form (Figure 8-10).

Conclusion: All of the figures pertaining to future energy use with regard to this ECO should now be calculated and carried through the appropriate forms as explained in Chapter VII. This will determine whether the costs of implementing and maintaining this ECO are justified in terms of future savings for the building owner.

FIGURE 9-23 EFFECT OF BYPASS THROUGH OFF-LINE CHILLERS

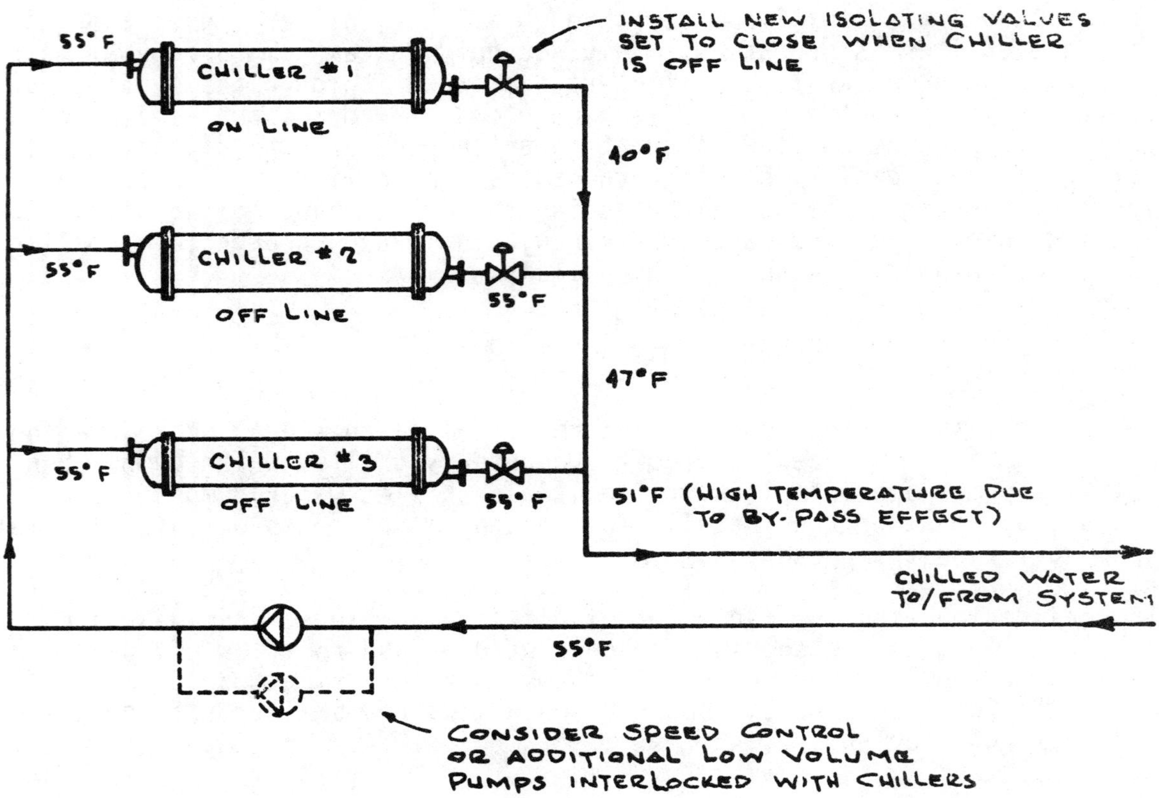

ECO #41: REPLACE AIR-COOLED CONDENSERS WITH COOLING TOWERS

THEORY

The condensing temperature of a refrigeration machine may be significantly lowered by replacing air-cooled condensers with cooling towers in climates of fewer than 15,000 wet-bulb degree hours. Air-cooled condensers frequently operate at temperature differences of 20°F between air and refrigerant; thus, if the outdoor air temperature is 95°F dbt, the condensing temperature will be 115°.

Based on the above discusson, it follows that if an air-cooled condenser used with a reciprocating refrigeration machine were replaced by a cooling tower in an area that experienced 95°F dbt - 72°F wbt design conditions, then the condensing temperature would be reduced from 115°F to 97°F. This reduction in condensing temperature will increase the COP of the refrigeration machine.

Cooling tower performance is related to wet-bulb temperature. Most towers commonly operate at temperature differences of 10°F between the air (wet-bulb) and condenser water. Thus, if the outdoor air is 72°F wbt, the condenser water temperature would be 82°F off the tower and, at 2 gpm/ton, would be 97°F, giving a mean condensing temperature of 90°F.

PROCEDURE

Install cooling towers on a one-to-one basis if the air-cooled condensers are large, or install one large cooling tower to replace a number of small air-cooled condensers. Whichever method is adopted, each refrigeration machine will require the addition of a water/refrigerant heat exchanger (shell and tube condenser).

ANALYSIS

To calculate the future energy which might be saved through this ECO, refer to Figure 8-51. Then: (A) determine the percent increase in COP based on the modified cooling system; (B) change the present COP on the Annual Heating and Cooling Tabulation Form (Figure 8-11) to reflect the above improvement in perfomance.

Conclusion: All of the figures pertaining to future energy use with regard to this ECO should now be calculated and carried through the appropriate forms as explained in Chapter VII. This will determine whether the costs of implementing and maintaining this ECO are justified in terms of future savings for the building owner.

ECO #42: USE A PIGGYBACK ABSORPTION SYSTEM

THEORY

Where existing centrifugal chillers are driven by steam turbines and the exhaust steam is condensed before being returned as feed water to the boiler, an absorption chiller can be substituted for the condenser to produce additional chilled water from the waste heat.

PROCEDURE

Install a steam absorption piggyback chiller for use as a condenser. Under part-load conditions, it will be more efficient to use the piggyback absorption machine to meet as much of the load as possible and to modulate the turbine driven machines to meet the remainder.

Piggyback absorption is particularly effective in buildings using purchased steam when condensate is not returned to the supplier (or where no credit is given for the temperature of the returned condensate). In such cases, it is worth considering the purchase of a larger machine and derating it by operating at outlet temperatures down to 180°F. This will reclaim the maximum possible heat from exhaust steam and condensate. Consider also using lower temperature condensate from the absorption machine as a heat source for low-grade systems such as domestic hot water, ventilating air pre-heaters, etc. (Refer to ECO #43.)

Low-temperature absorption machines are currently being developed to operate at inlet temperatures of 190°F or lower and yet retain a reasonably high COP. Consider using a low-temperature absorption machine in buildings that have large quantities of low-grade waste heat (e.g. cooling water from engine/generator sets).

Low temperature absorption machines are also capable of operating in conjunction with solar collectors and can extend the use of a solar system from winter only to year round.

ANALYSIS

To calculate the savings which might be achieved through this ECO:

(A) Use the following formula to determine the amount of cooling available through absorption:

$$\text{Absorption Cooling} = \frac{S_t \times H}{S_a}$$

Absorption cooling = tons of absorption cooling per ton of turbine cooling.

S_t = lbs of steam per ton required by the absorber in Btu.

S_a = heat input per ton required by the absorber in Btu.

H = heat content of the turbine exhaust steam in Btu per lb.

(B) Determine the percentage reduction in output from the turbine.

(C) Reduce the calculated cooling on the Annual Heating and Cooling Tabulation Form (Figure 8-11).

Example

As a demonstration, a low-pressure turbine driving a centrifugal chiller requires 12 lbs. of dry steam at 145 psig for each ton of refrigeration. Exhaust steam from the turbine at 15 psig and 20% moisture content will yield approximately 780 Btu's/lb when cooled and condensed to 220°F, giving a total usable heat of 12x780 = 9360 Btu/ton. An absorption chiller used as a steam condenser requires a heat input of 18,000 Btu/ton and will yield 9360/18,000 = 0.52 tons for each ton of refrigeration produced by the centrifugal chiller. (Newer units under development will require only 13,000 Btu/ton.) In this case, the addition of an absorption piggyback chiller to condense waste steam will decrease the steam/ ton of refrigeration by 30%.

Conclusion: All of the figures pertaining to future energy use with regard to this ECO should now be calculated and carried through the appropriate forms as explained in Chapter VII. This will determine whether the costs of implementing and maintaining this ECO are justified in terms of future savings for the building owner.

ECO #43: UTILIZATION OF HEAT RECLAMATION SYSTEM

THEORY

Heat reclamation is the recovery and utilization of heat energy that is otherwise rejected as waste. Waste heat is a substitute for a portion of the heat energy that would normally be required for heating, cooling and domestic hot water systems. Heat recovery conserves energy, reduces operating costs, and reduces peak loads.

The performance of any heat recovery system depends upon the following factors:

(A) The temperature difference between the heat source and heat sink.

(B) The latent heat difference (where applicable) between the heat source and sink.

(C) The mass flow multiplied by specific heat of each source and sink.

(D) The efficiency of the heat transfer device.

(E) The extra energy input required to operate the heat recovery device.

(F) The fan or pump energy absorbed as heat by the heat transfer device, which can enhance or detract from the performance.

Although the basic principles in heat recovery are the same, there are many methods to implement it by using different devices applicable to different systems or situations. Heat recovery devices reduce the peak heating and cooling loads when used with outdoor air systems and many reduce or completely eliminate the requirements for heaing and/or cooling equipment in major building expansions. The following are some of the most frequently used methods for heat recovery.

(A) Thermal Wheels:

A thermal or heat wheel is a rotating heat exhanger driven by an electric motor with a high-thermal inertia core. Such wheels are capable of transferring energy from one air stream to another and, in very large boiler plants, from flue gas to air. The hot and cold air streams must be immediately adjacent and parallel to permit installation of the heat wheel. Duct modifications may be necessary. Two types of thermal wheels are available. One which transfers sensible heat only and another which transfers both sensible and latent heat.

(B) Run-Around System:

This system comprises two or more extended surface fin coils installed in air ducts and interconnected by piping. The heat exchanger fluid, consisting of ethylene glycol and water, is circulated through the system by a pump, removing heat from the hot air stream and rejecting it into the cold air stream. A run-around-coil system may be used in winter to recover heat from warm exhaust air for use in pre-heating cold outdoor air, and in summer to cool hot outdoor air by rejecting heat into cooker exhaust air.

(C) Heat Pipe Sytems:

Heat pipes comprise extended surface finned tubes extending between adjacent air ducts. The tubes are continuous from one duct to the other on the same horizontal plane. Each tube contains liquid refrigerant which evaporates at the warm end, absorbing heat from the water air stream, and migrates as a gas to the cold end where it condenses and releases heat into the cold air stream. The condensed liquid then runs back to the hot end of the tube to complete the cycle.

(D) Air-to-Air Heat Exchangers:

Air-to-air heat exchangers transfer heat directly from one air stream to another through direct contact on either side of a metal heat transfer surface. This surface may be either convoluted plate (more common for low-temperature use in an HVAC system) or tube (more common for boiler flue gas heat transfer).

(E) Heat Pump as Heat Exchanger:

Heat pumps are actually heat transfer devices and, unlike those previously described, upgrade the temperature by as much as a factor of 3:1. This feature makes them particularly attractive for use with low temperature heat sources. They also have the capacity to transfer latent heat as well as sensible heat.

(F) Shell and Tube Heat Exchangers:

Shell and tube heat exchangers consist of tubular shell with a flange, in which a tube bundle of "U" bend construction is inserted. This device transfers heat between two physically separated fluids, one circulating through the tubes while the other passes through the shell. The difference in the leaving temperatures of the two liquids is called the "approach".

(G) Incinerators burning solid waste produce 6,000 Btu/lb of common trash. Special heat recovery incinerators are now available in sizes down to approximately 450°, which can be used in a heat exchanger as a source of high and low temperature heat.

(H) Heat-of-Light Systems:

The major advantage of a "heat-of-light" system lies in its reduction of heating, cooling and HVAC system and distribution loads, rather than in savings in electrical energy for lighting. However, slightly higher lamp efficiencies will result as the cooling effect on the lamps increases their output. (Refer to Figure 8-53 for the effect of heat transfer on light output.) The two types of heat-of-light systems, "dry" and "wet", provide the following advantages.

(1) Excess heat from interior areas of the building can be collected and distributed to perimeter areas.

(2) The sensible room heat component of the cooling load is decreased, permitting a reduction in the quantity of air required for cooling (thus saving fan horsepower).

(3) In the case of wet heat-of-light systems, the cooling load is reduced and less power is required for the refrigeration units. (Refer to Figures 8-53 and 8-54 for the thermal performance of wet heat-of-light systems).

PROCEDURE

Due to the numerous types of heat exchange devices, it is imperative that a selection be first made of the one most probably worth investigating for any given application. To facilitate the reader in making such selection, this section is divided into two parts. Part I, consisting of a matrix called "Applications of Heat Reclamation Devices" and a narrative describing some possibilities for "heat recovery applications," and Part II, consisting of procedural description of the "heat recovery devices" described under the Theory Section of this ECO.

Part I - Heat Recovery Applications:
(Refer to Figure 9-24)

(A) Transfer energy between exhaust and outdoor air ducts when there is more than 4,000 cfm being exhausted.

When there are more than 3,500 heating degree days and more than 8,000 cooling degree hours above 78°F dbt, consider thermal wheels, heat pipes and other devices listed in the matrix. Where supply and exhaust ducts are remote from each other and cannot be brought together, consider systems other than heat pipes and thermal wheels.

Install a thermal wheel or heat pump to recover both sensible and latent heat in locations with more than 12,000 wet-bulb degree hours above 66°F, wbt.

When justified for the heating mode only, install an air-to-water-to-air heat pump to transfer energy from the exhaust air stream to the fresh air stream.

Utilize exhaust-air heat energy to temperature make-up air and preheat combustion air, or use this system for space heating via heat pumps.

(B) Recover waste heat from the boiler flue gases whenever the stack temperature is greater than 350°F.

FIGURE 9-24 APPLICATIONS OF HEAT RECLAMATION DEVICES

ENERGY SOURCE	TEMPER VENTILATION AIR	PREHEAT DOMESTIC HOT WATER	SPACE HEATING	TERMINAL REHEAT	TEMPER MAKE-UP AIR	PREHEAT COMBUSTION AIR	HEAVY OIL HEATING	INTERNAL TO EXTERNAL ZONE HEAT TRANSFER
Exhaust Air	1a, 1b, 1c, 2,3,4,5				1a,1b,1c, 2,3,4,5	Direct		5, Direct
Flue Gas	1b, 3, 4		3, 4		1b, 3, 4	1b,2,3,4		
Hot Condensate	3, 6	3, 6	3, 5, 6	3, 6	3, 6	3, 6	6	
Refrigerant Hot Gas	6	6		6	6			
Hot Condenser Water	3, 6	3, 6	3, 5, 6	6	3, 5, 6	3, 5, 6	6	5
Hot Water Drains		6						
Solid Waste	7	7	7	7	7	7	7	
Engine Exhaust & Cooling Systems	6, 8	6	6, 8	6	6, 8	6, 8	6	
Lights	5, 9		5, 9	5, 9	5, 9			5, 9

Device

1. Thermal Wheel
   a) Latent
   b) Sensible
   c) Combination
2. Run-around coil
3. Heat pipe
4. Air-to-Air Heat Exchanger
5. Heat Pump
6. Shell/Tube Heat Exchanger
7. Incinerator
8. Waste Heat Boiler
9. Heat-of-Light

Install a heat pipe or an air-to-air exchanger to transfer energy from the hot flue gas to temper ventilation air, pre-heat domestic hot water, heat space, or pre-heating combustion air.

Take into account the corrosive effect of flue gas when selecting materials.

Allow for change-of-draft conditions caused by a heat exchanger.

Provide an alternative source of combustion air when heat exchanger dampers are closed for cleaning.

(C) Recover heat from laundry and/or kitchen waste water:

When more than 30,000 gal/week of water at temperatures above 120°F is discharged to waste, use as a heat source for heat pump or other HVAC system requirements.

Consideration must be given to the characteristics of the waste water, particularly the soap/detergent content of laundry waste water and the grease content of kitchen waste water. Piping and/or material modifications may be necessary to enable the heat exchanger to handle water with high concentrations of these impurities. In addition a holding tank may be required to maintain a steady flow rate through the heat exchanger when water is being sporadically discharged.

Waste heat thus recovered may be used by any system requiring hot water, such as domestic hot water and heating systems.

(D) Recover heat from engine or combustion turbine exhaust and cooling systems:

On engines larger than 50 hp.

Exhaust gas heat recovery is restricted by the partical limitations of the heat exchanger plus the prevention of flue gas condensation. the recommended minimum exhaust temperature is approximately 250°F. Depending on the initial exhaust temperature 50% to 60% of the available exhaust heat can be removed.

(E) Recover heat from incinerators:

If the quantity of solid waste exceeds 1,000 lbs/day.

(F) Recover heat from condensate return systems:

(1) When district heat steam condensate is discharged to waste.
(2) When steam condensate from equipment supplied by on-site boilers is at a temperature of 180°F or greater.

(G) Recover heat from refrigeration system hot gas:

Where there is a steady and concurrent demand for refrigeration and waste heat, and where the refrigeration systems operate 1,000 or more per year.

Do not reduce superheat to the point where liquid slugging occurs.

The heat exchanger must be located after the hot gas bypass or other unloading devices. If located outdoors, drains must be provided to prevent freezing.

(H) Recover heat from condenser water systems:

Install a heat exchanger or heat pipe in the hot condenser water line to temper outdoor air, pre-heat domestic hot water, or modify the piping in air handling units to utilize hot condenser water to heat air.

Install a coil to extract heat from the hot condenser water line to heat intake air in an air-cycle heat pump which can then transfer its condenser heat to the space requiring it.

Generally, it is not economical to replace existing condensers by double-bundle condensers; however, in the event that replacement is being contemplated due to age or the installation of new refrigeration equipment, give consideration to a double-bundle condenser.

(I) Recover heat from air-cooled condensers:

In supermarkets, grocery stores, and any building with 15 hp or more of refrigerated display cases or storage boxes, and heating degree days exceeding 3,000.

When air-cooled condensers are located away from spaces requiring heat, condenser heat can be recovered by adding insulated ducts.

(J) Utilize heat from internal spaces for heating perimeter areas:

When replacing an existing chiller, install a double-bundle condenser, or cascade condensers, to recover heat that is otherwise dissipated by the cooling tower.

Install tanks to store hot condenser water from daytime cooling for nighttime heating. This water can be used directly for space heating or as a heat source for a heat pump system.

FIGURE 9-25 WASTE HEAT RECOVERY FROM CENTRAL CHILLER

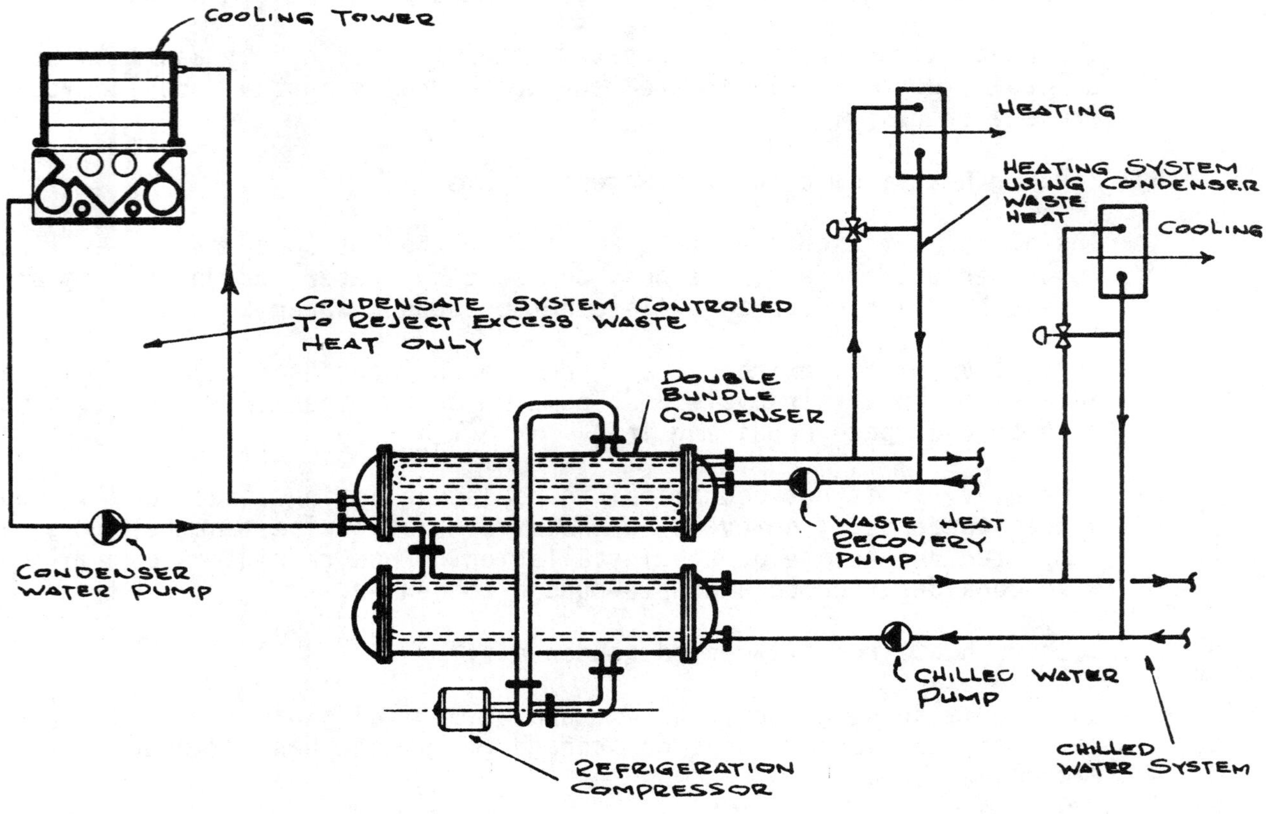

Part II - Heating Recovery Devices

(A) Thermal Wheels:

Exhaust air (from the building or space) and make-up air (from outside) flowing in the opposite direction are introduces to each half of the thermal wheel through separate adjacent ducts. A thin cylinder containing heat transfer media slowly rotates between the two air streams within the thermal wheel. Energy absorbed by the media is transferred from one air stream to the other. During a heating cycle, incoming low-temperature fresh air is heated and humidified by heat transferred from the warm, moist exhaust air. During summer months, the cooling and dehumidifying of moist higher-temperature outside air is accomplished in a similar manner by giving up heat to the cooler exhaust air stream.

Thermal wheels are commercially available in sizes ranging from 3,000 cfm to 60,000 cfm. They can be provided with purge sections and can also be arranged for vertical and horizontal mounting. Larger thermal wheels can be provided with variable speed, SCR controls.

FIGURE 9-26 THERMAL WHEEL SECTIONS

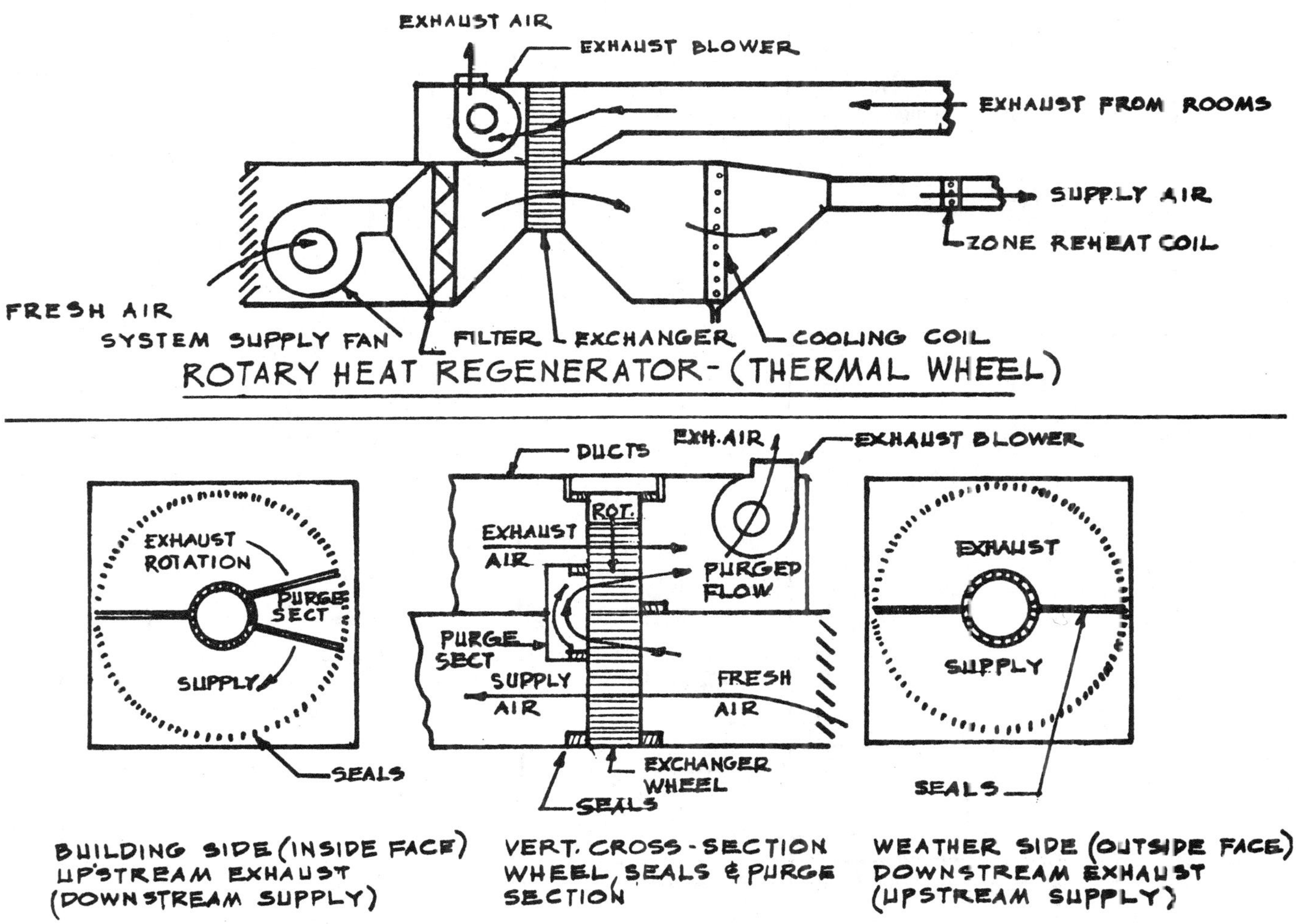

When applying thermal wheels to ventilation systems, install a roughing filter and/or insect screen to protect the core. Icing, which can possibly occur under some outdoor conditions with enthalpy wheels, can be eliminated by pre-heating 10°F. When used to transfer heat from flue gasses, the construction of the thermal wheel must be suitable for the temperatures encountered.

The sketches in Figure 9-26 outline the possible arrangement of a thermal wheel between an exhaust and fresh air system.

(B) Run-around coil system:

The coil type exchanger has the advantage of permitting heat transfer between supply and exhaust systems in widely separated portions of the building, since coils are installed in each of the intake and exhaust ducts, and are connected by piping only. The liquid heat transfer medium is pumped from coil to coil. It is relatively easy to run piping long distances through the structure, interconecting a number of separate exhaust and fresh air systems to form one complete system. A possible arrangement for an application of run-around coils is shown below:

FIGURE 9-27 RUN AROUND HEAT RECOVERY SYSTEM WITH MULTIPLE COILS

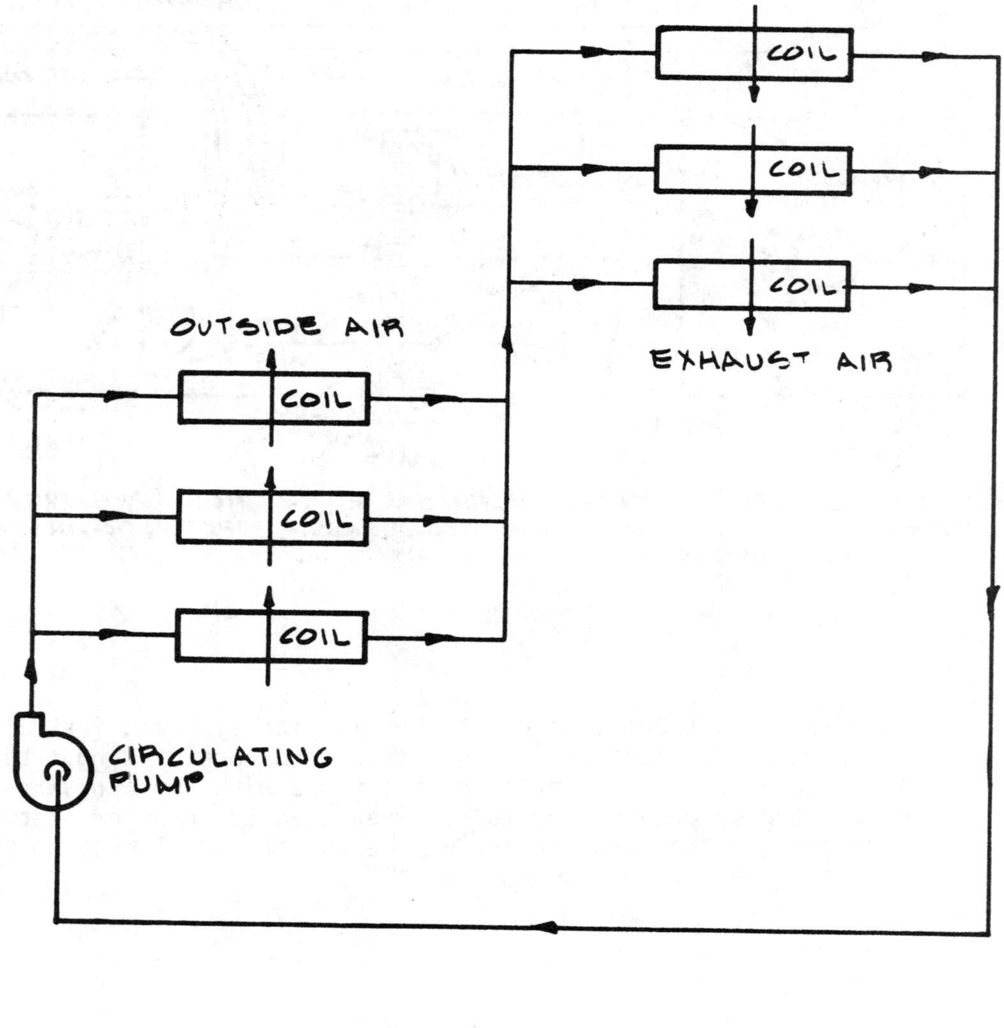

(C) Heat pipe systems:

Heat pipes between hot and cold air ducts can only be used when these are immediately adjacent to each to each other, and some modifcation of existing ductwork may be necessary for installation. To operate effectively, the heat tubes must slope down from the cold end to the hot end. Accurate control of the slope must be provided. A possible arrangement of a heat pipe recovery system is indicated below:

FIGURE 9-28 HEAT PIPE EXCHANGER

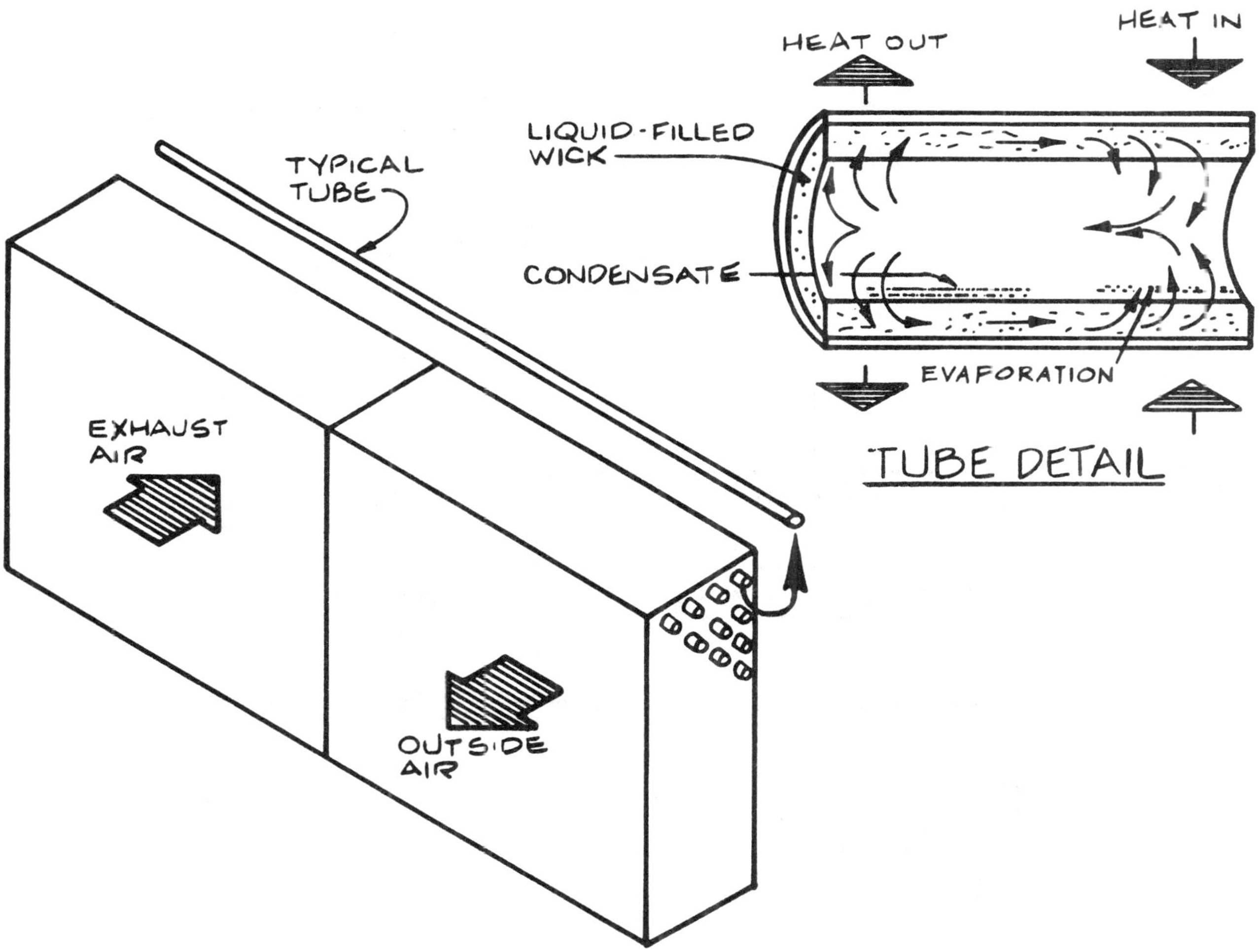

(D) Air-to-air heat exchangers:

The most common application of air-to-air heat exchangers, as it will apply to office buildings, involves recovering of wasted heat through the boiler flue gas.

The heat exchangers may be purchased as packaged units or can be custom made. They transfer sensible heat only and are not designed for cooling applications. Size is limited only by the physical dimensions of the space available. Applications for industrial applicatons are numerous. A typical application for recovery heat to cold combustion air or other warm air requirements is indicated below:

FIGURE 9-29 FLUE GAS HEAT EXCHANGER

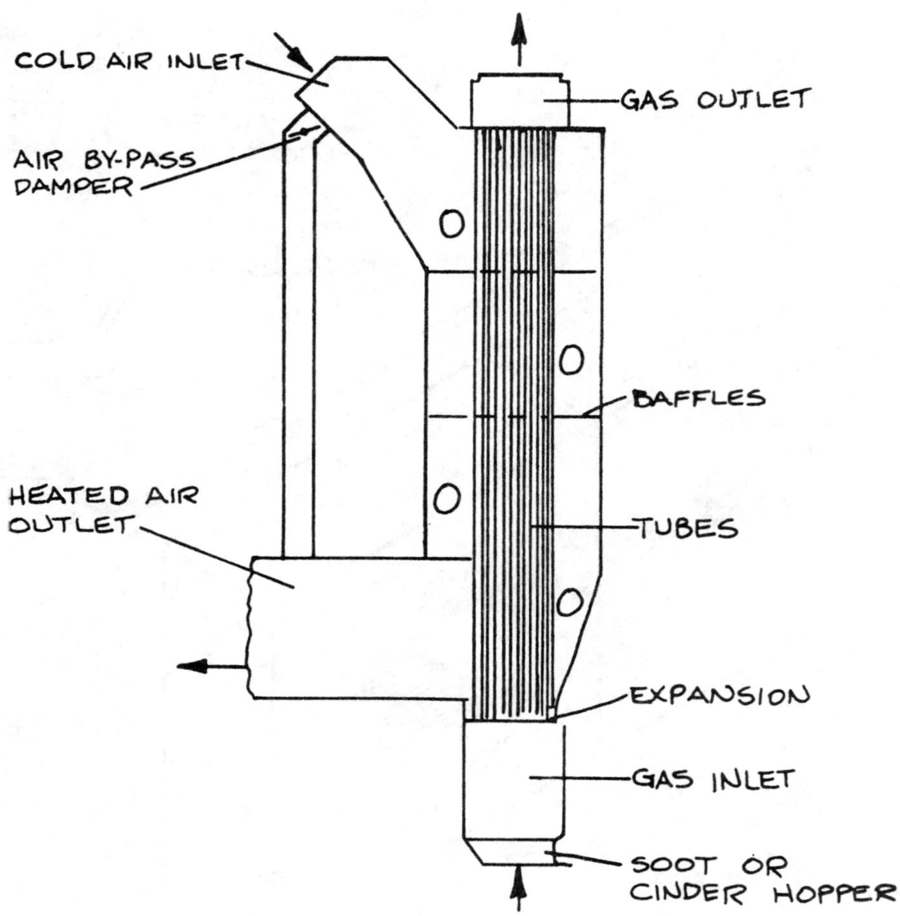

(E) Heat pumps:

Heat pumps are commonly used to heat and cool spaces. Employed in the heating mode, they have a high COP and can be used to extract more heat from a given source than any other heat exchange device. In the heating mode the power input is approximately 1/3 to 1/5 of the heat output. Heat pumps are far more efficient than electric resistance heating and compare well with oil or gas fired heating systems. When used in the cooling mode, performance decreases, and they should not be considered for this purpose alone unless simultaneous heating and cooling will occur.

Heat pumps are available in the following configurations, each having particular application:

(1) Air-to-air (very limited applications)
(2) Water-to-air
(3) Water-to-water
(4) Air-to-water

Sizes range from 1/2 to 25 tons as commercially available package units. Larger sizes can be obtained by modifications to conventional refrigeration machines, however, not using readily available equipment which has been tested and rated by reputable manufacturers is not recommended.

Heat pumps make heat available from low-temperature sources. For example, heat can be extracted from cold drain water at approximately 50°F with rejection at temperatures high enough for use by domestic hot water system, air pre-heating and other similar applications. When selecting heat pumps, choose machines that have the higher COP over the range of expected operating conditions.

(F) Shell and tube heat exchangers:

Shell and tube heat exchangers as a means of transferring and converting heat from one form to another they can be used to capture energy from hot condensate, hot refrigerant gas, condenser water, and hot drain lines from kitchens and laundries and use it by transferring it to heating domestic hot water or convert it into other useful forms of energy.

Heat exchangers should be insulated to prevent unnecessary heat loss and should be constructed of materials to suite the application.

Shell and tube heat exchangers can be used to exchange heat in the following configurations:

(1) liquid-to-liquid
(2) Steam-to-liquid
(3) Gas-to-liquid

All three configurations are commercially available in a wide range of sizes and outputs with reliable heat exchange data.

(G) Incinerators:

In employing incinerators for heat recovery the heat should be utilized at the highest practicable temperature. Typical applications are absorption refrigeration, heating, domestic hot water, etc.

(H) Heat of lights:

Both systems require special fixtures; in addition, the wet systems require a completely new water circulating system and connections to a cooling towe Wet heat of light systems are thus not particularly suitable for retrofitti unless a complete remodeling of the existing HVAC and Electricical Systems a building is contemplated.

(1) "Dry" heat-of-light systems:

For a dry heat-of-light, room air is extracted through the lighting fixture, is passed over the lamp and ballast, and is either ducted to a fan or drawn into the ceiling plenum space. The heated air from the lamp and fixture can be supplied to cooler perimeter zones during the heating season or recirculated back to an existing air handling unit or discharged outdoors during the cooling season. If a ceiling plenum system is used, a separate fan is necessary to draw the warm air from the ceiling plenum and deliver it to cooler zones or to discharge it outdoors.

When ceiling plenums are used as collection chambers, each zone must be isolated from the other by a vertical barrier, and ceilings over conditioned areas should be insulated to a "U" valve of .1 or better to limit reradiation to the occupied spaces. (Consult all codes to co with fire protection requirements.)

(2) "Wet" heat-of-light systems: The special lighting fixtures accompanyi these systems include built in water passages and air extract inlets. Air is extracted through the fixture, but in addition, water is circulated through passages to the cooling tower where the heat is remove and the cool water piped back to the fixtures. The heat removed from fixtures can be used for reheat purposes in the cooling system instead of being rejected to the outdoors. The circulating water system can also be used to pick up heat from special water-cooled louvered veneti blinds to reduce solar heat gain at the windows, as well as to make th heat available for other uses.

ANALYSIS

Yearly energy requirements have been established on the various forms and these energies represent the potential for heat reclaim. Listed below are the locatio of these energies:

(1) Outside Air - Annual Heating and Cooling Tabulation Form (Figure 8-11).

(2) Exhaust Air - Building Load Survey Form (Figure 8-9).

(3) Laundry and Kitchen Water Use - Building Load Survey Form (Figure 8-9).

(4) Cooling Systems - Annual Heating and Cooling Tabulation Form (Figure 8-11).

(5) Condensater Return - Annual Heating and Cooling Tabulation Form (Figure 8-11) for Heating and Building Load Survey Form (Figure 8-9) for domestic water.

(6) Refrigeration System - Distribution Load Survey Form (Figure 8-10).

The efficiency of the various heat reclaim systems must be applied to the above energies as this is discussed below for each type system. When the energy reduction due to reclaim has been determined the energy affected should be reduced on the appropriate form.

The potential benefits of the systems described are as follows:

(A) Thermal Wheels:

The maximum potential heat that can be exchanged between a warm exhaust air stream and a cold outdoor air stream can be derived from the following formula:

$Q = (Q_e - Q_o) \times M_e$ or M_o (whichever is least), where:

Qe = heat content of exhaust air in Btu/lb

Qo = heat content of outdoor air in Btu/lb

Me = exhaust air flow rate in lbs/hr

Mo = outdoor air flow rate in lbs/hr

Thermal wheels typically have an efficiency of 60% to 80% for sensible heat transfer and 20% to 60% for latent heat transfer. These efficiencies should be applied to the theoretical maximum heat transfer Q to obtain the actual heat transfer rate in Btu per hour. Check with the manufacturer for specific efficiencies.

The total system efficiency is equal to the heat transfer efficiency plus the heat of the driving motors in the heating mode. During the cooling cycle, the energy required to rotate the wheel, plus the additional energy required to overcome the air resistance, must be deducted from the gross load deduction due to heat transfer.

The preceding formula provides the load reduction for one hour under a specific set of conditions. To determine annual load reduction, analyze the specific conditions for each hour of operation per year. The above procedure may be used either for sensible heat recovery or for enthalpy recovery, provided the units are in Btu/lb for each respective condition. It may also be used in the same manner for the cooling where the direction of heat flow is reversed.

(B) Run-Around Coils:

The formula for determining system efficiency described for thermal wheels should also be applied to run-around coil systems.

When calculating reductions in load due to run-around coils, the effect of coil depth on the rate of heat transfer and the resistance to air flow must be recognized. A trade-off analysis between the extra heat transfer obtained by deep coils against the extra resistance to air flow should be made to determine the optimum coil size. Where there is no possiblility of freezing, water may be used as the heat transfer medium. Where frost protection is necessary and ethyle glycol is used as the heat transfer fluid, the different specific heats and rates of heat transfer must be accounted for in the calculations.

The initial cost of run-around coils is less than the rotary type heat exchangers by about 40%. They are most economical for small systems or for heating systems only.

(C) Heat Pipe System:

To calculate the effectiveness and annual load reduction of a heat pipe system, use the same procedure described for thermal wheels. Heat pipe coils are available in sizes ranging from 2,500 cfm to 20,000 cfm.

(D) Air-to-Air:

The efficiency of each installation and its cost must be calculated on the basis of the particular circumstances that apply. While efficiencies generally are lower than 50%, air-to-air heat exchangers are relatively inexpensive, have low resistance to air flow, require no motive power input, and are trouble-free and durable.

(E) Heat Pumps:

To estimate the energy savings from a heat pump operation is not a simple task due to the numerous variables, some of which are as follows:

(1) Type of heat pump - air-to-air, water-to-air, water-to-water.

(2) Spaces of application - relative proportion of interior to exterior area.

(3) Climate considerations - heat pump is lossing capacity at the heating mode as the outdoor temperature drops under certain levels.

(4) Type of auxiliary heat to be used in conjunction with the heat pump - dropping of capacity under certain temperatures often requires providing additional means of heating at these conditions. The type of heat (electric, oil, gas and the time of its operation will affect the overall heat pump

energy consumption considerably. Therefore, only an engineer considering the above conditions, along with others that may be applicable to a specific project, can make the necessary evaluations for selecting such a system. However, it is not recommended that a replacement of an existing heating/cooling system with a heat pump be considered unless:

a) It involves a complete modernization of an existing building that requires replacement of the existing system, or;

b) The old HVAC system is obsolete and a replacement of it is unavoidable, or;

c) An addition to the building is contemplated for which heat pumps could be added instead of another system.

(F) Shell and Tube Heat Exchangers:

To determine the energy savings by the use of heat exchanger the following points shall be considered:

(1) Btu of heat wasted.

(2) Btu of heat to be recovered.

(3) Cost benefit analysis considering the first investment cost versus the savings in dollars due to the energy recovered.

It should be pointed out that first costs rise rapidly as the approach temperature is decreased, and this should be considered when making an economic evaluation. (Do not strive to achieve the last degree of caputred heat.)

(G) Incinerators:

The energy saved by using heat recovery incinerators shall be evaluated by analyzing the cost effectiveness of the considerable first cost investment for their employment. In an office building that this manual is addressing itself primarily, is doubtful that this item can be justified.

(H) Heat of Lights:

Although "heat-of-light" systems can conserve energy and reduce operating cost, the installation of such systems will be quite costly unless renovations to the building, requiring completely new lighting and duct work systems, are contemplated. If lower lighting levels for the facility are adopted, as recommended in the Lighting Section, total heat-of-light systems will not be economically feasible for most existing buildings - even on a life cycle cost basis; accordingly, the priority of such systems should be low compared to other energy conservation measures. An exception would be where high lighting levels are required over large areas to meet specific task requirements. In evaluating the heat of light savings, refer to Figures 8-53 and 8-54.

G. Domestic Hot Water System Energy Conservation Opportunites — Page

G. Domestic Water System ECO's: #44 - #46

Introduction

In general, the heating and distribution of domestic water in typical U.S. buildings consumes less energy on an annual basis than other major building systems, particularly HVAC and lighting systems. But there are significant exceptions to this rule and numerous opportunities to save energy and dollars through improvements in domestic hot water systems-- often at nominal cost to the building owner.

The volume of energy required for hot water is particularly significant in hospitals, apartment complexes, restaurants and kitchens, motels, laundries, and athletic facilities in high schools and colleges. In many parts of the northern U.S., the energy consumed by hot water systems in multi-family dwellings is second only to heating systems. It is second only to cooling systems in parts of the southern and southwestern U.S. Domestic hot water for hospitals often demands more energy than lighting, and the high hot water consumption rates of motels and laundries help explain the increasingly widespread installation of solar hot water systems in such facilities.

Thus, while savings due to improved hot water systems may occasionally seem less dramatic than savings due to improvements in HVAC systems or building envelopes, they would never be overlooked in a serious energy management program. The following ECO's offer suggestions for profitable reductions in the energy consumption of domestic hot water systems.

(NOTE: These ECO's, as well as ECO #4, are restricted solely to the generation and distribution of hot water. In practice, these ECO's will often be evaluated and implemented in conjunction with other ECO's nominally pertaining to such other "energy-users" as boilers and manufacturing or production processes utilizing heat and/or steam.)

The order in which these ECO's are examined are similar to the order of all other ECO's, namely:

1. Load Reduction (ECO #2)
2. Distribution Losses (ECO #4)
3. Primary Equipment (ECO's #45 and 46)

ECO #44: INSULATE HOT WATER STORAGE TANK AND PIPING

THEORY

The loss of heat from domestic hot water storage tanks must be continuously offset by the addition of heat to maintain a ready supply of hot water. This heat loss occurs 24 hours a day, whether the building is occupied or not. Heat losses from uninsulated hot water pipes, as discussed in ECO #24, can also be substantial.

PROCEDURE

Storage tanks should be covered with insulation having a minimum conductivity of 0.3. Insulate bare tanks and apply additional insulation to tanks having less than three inches of insulation. Replace or repair all missing or torn insulation as required. Applicable codes should of course be checked to determine the acceptability of various insulation materials. Obtain quotations for accurate cost assessment. (For information on insulating hot water pipes, see ECO #24.)

ANALYSIS

Calculate the savings from this ECO by comparing the heat loss from the tank before and after insulation (assume that water temperature and ambient air temperature are constant). To assist in these calculations, Figure 9-30 indicates the annual energy loss in Btu for various sizes of tanks located in a space with an ambient air temperature of 65°F, with an insulation thickness as indicated.

Multiply the savings in Btuh by the number of operating hours per year to obtain the total annual energy savings due to storage tank insulation. (For benefits resulting from insulating hot water pipes, see ECO #24.) Revise the efficiency of the domestic water system to reflect the reduced energy consumption on the Building Load Survey Form (Figure 8-9).

Conclusion: all of the figures pertaining to future energy use with regard to this ECO should now be calculated and carried throgh the appropriate forms as explained in Chapter VII. This will determine whether the costs of implementing and maintaining this ECO are justified in terms of future savings for the building owner.

FIGURE 9-30 RATIOS OF INSULATION THICKNESS AND TANK SIZE TO HOURLY HEAT LOSS

Insulation k=0.3	Tank Size in Gallons	Btuh lost at water temperature of		
		100^0F	140^0F	180^0F
Bare	50	1301	3024	5100
	100	2190	5039	8584
	250	3840	8923	15051
	500	6292	14620	24660
1"	50	206	441	678
	100	346	743	1142
	250	607	1303	2002
	500	994	2135	3280
2"	50	109	236	362
	100	184	397	609
	250	323	696	1069
	500	528	1140	1751
3"	50	74	160	245
	100	125	270	412
	250	219	473	722
	500	359	775	1184

ECO #45: USE HEAT RECOVERY SYSTEMS

THEORY

Heat reclamation is the recovery and utilization of heat energy that is otherwise rejected as waste. The subject has been discussed at length in ECO #43 ("HEAT RECLAMATION"), and the principal means of heat recovery are listed there. This ECO is basically an extension of ECO #43. It presents the three most common methods of reclaiming heat for specific application to domestic hot water. Those methods are:

(A) Hot gas heat exchanger: A typical refrigeration machine with a water-cooled condenser rejects approximately 15,00 Btuh for each 12,000 Btuh of refrigeration. An air-cooled condenser rejects up to 17,000 Btuh. Up to 5,000 Btuh of the heat rejected from either system can be recaptured.

(B) Hot drain heat exchanger: Buildings with kitchens, laundries, and other service facilities which utilize large quantities of hot water in many cases discharge hot waste water to drains, from which the waste heat can be recaptured.

FIGURE 9-31 HOT GAS HEAT EXCHANGER

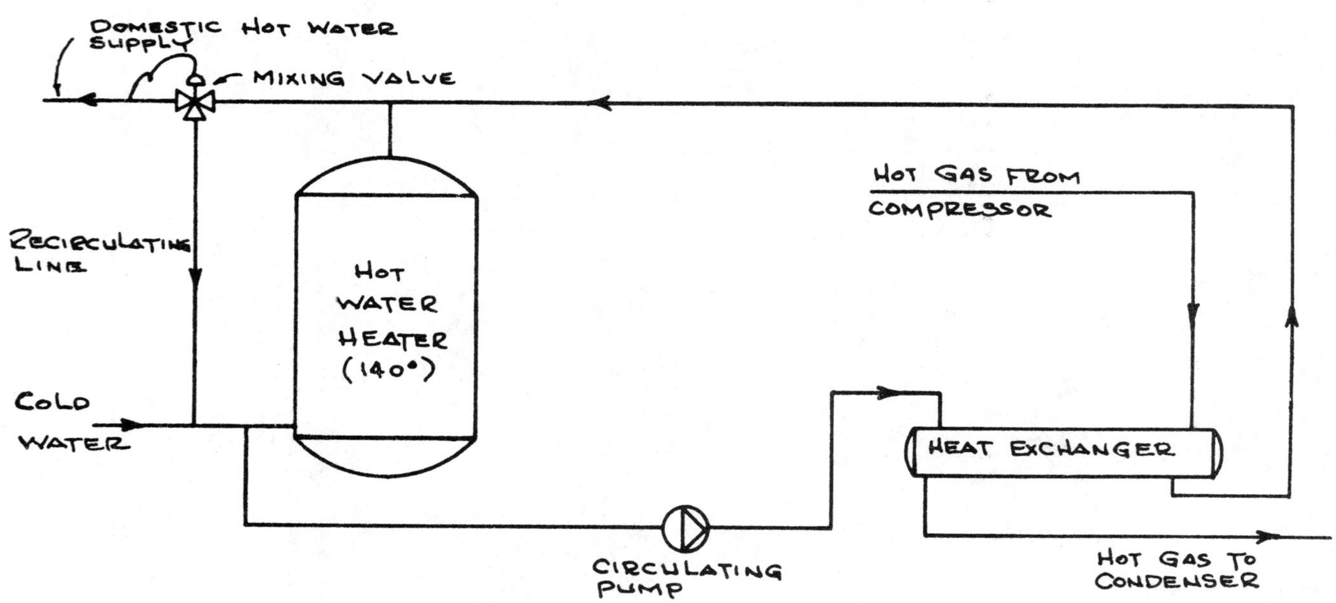

(C) Hot condensate heat exchanger: The condensate return portion of many steam systems exhausts large quantities of heat in the form of flash steam when the hot condensate is reduced to atmospheric pressure in the condensate receiver. Much of this heat, too, can be reclaimed.

PROCEDURE

Hot gas heat exchanger: To recover waste heat from the compressor, install a heat exchanger in the hot gas line between the compressor and condenser of the chiller (see ECO #43). A typical arrangement in conjunction with a domestic hot water system is shown in Figure 9-30 above. (Hot gas temperature depends on head pressure but is usually on the order of 120-130°F.)

Cold water is circulated through the heat exchanger by the circulating pump. When hot water is not being used, water is pumped back to the heat exchanger through the recirculating line. When hot water is needed, it is fed from either the heat exchanger, the hot water heater, or both. A mixing valve is provided to maintain the desired temperature.

Hot condensate heat exchanger: Recover waste heat by installing a heat exchanger in the condensate return main before the receiver to reduce condensate temperature to aproximately 180°F. Use the heat recovered to preheat water as required. Figure 9-31 shows the schematic installation of a hot condensate heat exchanger. (The quantity of heat recovered depends on the pressure and temperature characteristics of the boiler.)

FIGURE 9-32 HOT CONDENSATE HEAT EXCHANGER

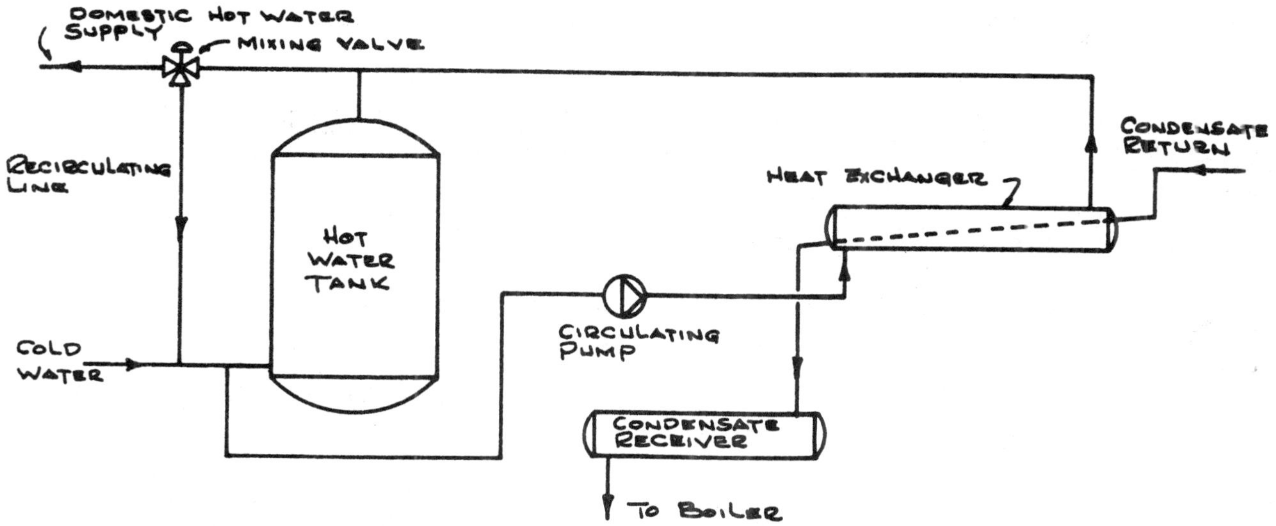

Hot drain heat exchanger: Recover waste heat by installing a heat exchanger. The extra heat can then be used to preheat domestic hot water (among other applications). In general, it is economical to preheat water from 50°F to 105°F without excessive equipment cost. The hot water at 105°F can then be fed into the domestic hot water tank for further heating to utilization temperature if required.

FIGURE 9-33 HOT DRAIN HEAT EXCHANGER

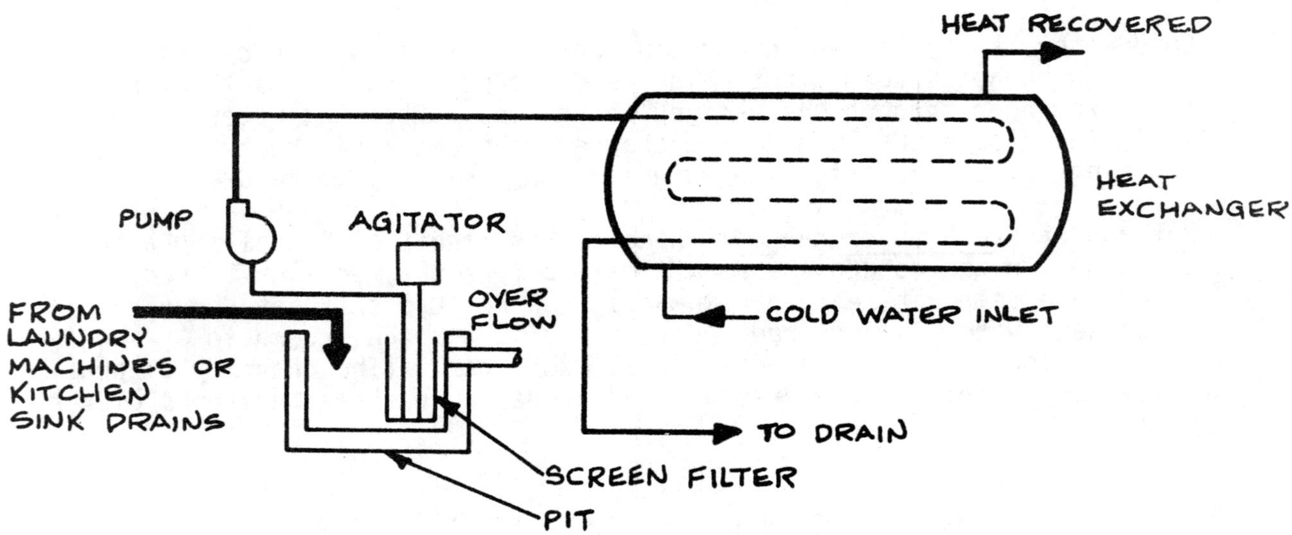

ANALYSIS

First calculate the heat to be recovered from any of the above methods and convert it to Btu/year. Then, using the unit cost of fuel for the application in question, convert fuel savings to dollar savings and compare these savings with the cost of implementing the ECO, plus recurring maintenance expenses, to arrive at a cost-benefit analysis.

To calculate the savings which might be achieved through reclamation of waste heat from domestic hot water systems:

(A) Calculate the heat to be recovered form any of the above methods and convert it to Btu/yr.

(B) Subtract the above figure from the present figure for Btu/yr under "DOMESTIC WATER" on the Building Load Survey Form (Figure 8-10).

Conclusion: All of the figures pertaining to future energy use with regard to this ECO should now be calculated and carried through the appropriate forms as explained in Chapter VII. This will determine whether the costs of implementing and maintaining this ECO are justified in terms of future savings for the building owner.

ECO #46: REPLACE CENTRAL SYSTEM BY LOCAL HEATING UNITS AND/OR SEPARATE SUMMER GENERATION OF HOT WATER

THEORY

Domestic hot water for any given building may be generated either centrally by a single unit and distributed through the buiding, or it may be generated locally by a number of smaller units located nearer the various zones where it is used. While a centralized system offers certain obvious advantages, including ease of maintenance and control, it may not be the most efficient or desirable system in buildings where one or more of the following conditions occur:

(A) The distribution system is so long as to produce excessive heat loss through the piping.

(B) The need for high-temperature hot water (above 140°F) is not uniform throughout the building; i.e., there are some spaces requiring high-temperature water, while the majority of the building could suffice with water at 90°F to 100°F. This generally means that the entire building is supplied with high-temperature water at a considerable energy penalty.

(C) The space-heating boiler is also used to generate domestic hot water. This usually means that the boiler must be operated at low efficiency during the cooling season merely to supply domestic hot water, again incurring a heavy penalty in terms of wasted energy.

(D) Domestic hot water is required for short periods of heavy use at various points throughout the building.

When one or more of the above conditions exist in the building under study, the possibility of replacing the central unit with an array of smaller units should be considered, along with other related improvements in system efficiency.

PROCEDURE

Depending on the actual conditions of the building in question, consider implementing one or more of the following modifications.

(A) When the bulk of the hot water is centrally generated and centrally consumed, excepting a few remote areas which require extensive piping, isolate those remote areas from the central system and provide them with smaller local heating units.

(B) When multiple temperature requirements are met by a central domestic hot water system, the minimum temperature water is determined by the maximum usage temperature, with lower temperatures attained by mixing in cold water at the tap. Where higher temperatures are required only at a few specific locations, lower the temperature of the central boiler and install separate booster heaters units for the high-temperature areas.

(C) When the space-heating boilers also provide heat for domestic hot water during the cooling season, causing the boiler to fire at an efficiency under 60 percent, install a smaller hot-water boiler for use at maximum efficiency during the cooling season, allowing the central boiler to be shut down.

(D) When the building system requires hot water for short periods of heavy use at various locations in the buildings, replace the central system and its long runs of piping with smaller hot water heaters located closer to the points of use.

(NOTE: Before installing local hot water units, review applicable codes for restrictions on their locations and/or required modifications to the building, such as fire-walls around fuel-burning equipment.)

ANALYSIS

(A) Analyze the patterns of hot water use within the building to determine the feasibility of replacing the central system with local water heaters and/or of adding a summer-only boiler.

(B) Measure the energy losses in the existing system and calculate the savings to be derived from the modifications discussed above. (The energy saved is the sum of the reduced distribution losses, plus the improved generation efficiency of local and summer-only units as opposed to a central boiler. Refer to ECO's #24 and 44.

(C) Revise the efficiency of the domestic water system to reflect the increase on the Building Load Survey Form (Figure 8-9).

<u>Conclusion</u>: All of the figures pertaining to future energy use with regard to this ECO should now be calculated and carried through the appropriate forms as explained in Chapter VII. This will determine whether the costs of implementing and maintaining this ECO are justified in terms of future savings for the building owner.

H. Power Systems Energy Conservation Opportunities

H. Power System ECO's: #47 - #52

INTRODUCTION

The volume of electric power consumed by a building-- measured in kilowatt hours (kwh) - depends upon (1) the demands of the systems consuming power to supply the building load (lighting, heating, ventilating, cooling, domestic hot water, commercial refrigeration, elevators, escalators, business machines, communications systems, cooking, snow melting, etc.); (2) power losses through conversion and distribution systems which supply those loads; (3) characteristics of the electric service and distribution systems.

Specific suggestions for reducing a building's "parasitic" distribution loads are detailed in some of the earlier Energy Conservation Opportunities. For example, when the quantity of air or water circulated in an HVAC system is reduced, less power is needed to drive fan pumps, thus reducing annual energy consumption. Improper distribution voltage, low power-factors, transformers at inefficient loading points, losses from standby transformers, excessive operation of motors and equipment, improper control and low motor-load factors, and excessive voltage drops in undersized conductors or wiring are also major causes of wasted electricity.

Opportunities to reduce electrical energy consumption include the replacement of obsolete with more efficient equipment, such as SCR controllers for elevators and transformers with low-loss factors. Reducing the kilowatt hours of electricity consumed by the building reduces the consumption of raw source energy (the amount of energy required to generate electricity at the power plant) by a factor of more than three. Boiler and generator losses at the power plant, coupled with transmission and distribution losses from the plant to the building, account for about 70 percent of the energy used in generating electricity.)

Reducing energy losses due to inefficient or improperly utilized power systems, while saving a building owner relatively substantial amounts of money in utility bills, saves even more in the form of fossil fuels and other materials consumed in generating electrical power.

ECO #47: REDUCE ENERGY CONSUMPTION OF EQUIPMENT AND MACHINES

THEORY

Most buildings contain many electrically driven machines which are left switched on and idling all day and even all night, despite the fact that those machines are used only for short periods while the building is occupied. It is natural to think primarily of large motors and large loads when embarking upon an energy conservation program, but the aggregate horse power and energy consumption of many small inefficient motors can easily exceed the horsepower of one large motor. In addition to motors, such other devices as resistance heating equipment, electric cooking equipment, meat preparation rooms in supermarkets and restaurants, electric signs, business machines, and mechanical and electrical equipment can be turned off for a substantial portion of the week, resulting in significant fuel economies.

PROCEDURE

The object of this ECO is to reduce the energy consumption of equipment and machines to the absolute minimum required by their productive operation. For the most part, this means effective and continuous control over the time such equipment is running, thus ensuring that when the equipment is not performing a necessary function it is not running at all.

There are of course any number of methods for improving control over building power systems. A complete investigation of these methods will require consultation with systems and controls manufacturers, especially in view of the rate at which new control devices are introduced in the market. Among the more common methods of control are the following:

(A) Train or retrain building engineers and maintenance personnel to use every existing means of reducing the operating time of the equipment under their surveillance.

(B) Where circuiting permits, control the operation of small motor-driven equipment with a single timer.

(C) Investigate recircuiting possibilities where multiple circuits supply power to building equipment.

(D) Larger motors and groups of motors and panelboards may be controlled by a remote manual switch, time switch, photo-electric cell, relay, contactor, controller, or by a combination of such controls.

ANALYSIS

To calculate the savings which might be achieved through reductions in consumption of energy by equipment and machaines: (A) determine the new (reduced) kwh/yr consumed by power systems following implementation of the proposed modifications; (B) substitute this new figure for the present figure on the Building Load Survey Form (Figure 8-9).

Conclusion: All of the figures pertaining to future energy use with regard to this ECO should now be calculated and carried through the appropriate forms as explained in Chapter VII. This will determine whether the costs of implementing and maintaining this ECO are justified in terms of future savings for the building owner.

ECO #48: REDUCE PEAK LOADS

THEORY

Some utility rate structures are based not only on a building's total consumption of electricity, but also on its peak demand, which may occur for a short period per month or per year, but which establishes demand charges for the whole month or year. In order to reduce the electrical demand charge, a load-shedding program should be introduced. Since many of the largest loads are deactivated for considerable periods of time, load-shedding, or leveling, conserves energy while reducing power losses in electrical equipment and transmission lines.

PROCEDURE

Consult with the utility which serves the building to analyze the rate structure and to determine when the peaks occur and how they might best be reduced. (The total monthly and yearly demands of the building's "energy-users" have already been tabulated and accounted for on the Energy Tabulation Forms (Figure 8-12) "Electric Demand," and the Utility Consumption /Cost Survey Form (Figure 8-8). Determine periods when the energy-users <u>must</u> be in operation, as well as the periods when they can be disconnected without impairing the safety of occupants or equipment. Establish a load-shedding program to shut down those loads. This program may be: (A) manual, where cost and demand upon personnel are not obstacles; (B) automatic, where control devices prevent large motors from operating simultaneously.

Peak loads which are not predictable in advance can be monitored, and equipment operation can be automatically programmed to de-energize selected loads at peak demand, reactivating them when reduced demand permits. Available automatic load-limiting devices range in complexity from a simple thermal sensor (which works like a circuit-breaker to switch off a low-priority load when the building load reaches a preset point) to a unit that reads current and provides for shedding and restoring loads. Automatic load shedding is also commonly accomplished with a watt-hour meter, combined with time switches and load controllers. (The peak demand controller in large and complex building systems is usually part of a computerized energy management system.)

ANALYSIS

To calculate the energy savaings which might be achieved by reducing peak loads: (A) determine the new (reduced) annual load demand in kwh based on proposed modifications of the building's power systems; (B) substitute this new figure for the present figure on Energy Tabulation Form, Figure 8-12, ("Electric Demand").

Conclusion: All of the figures pertaining to future energy use with regard to this ECO should now be calculated and carried through the appropriate forms as explained in Chapter VII. This will determine whether the costs of implementing and maintaining the ECO are justified in terms of future savings for the building owner.

ECO #49: UTILIZE EFFICIENT TRANSFORMERS

THEORY

Most dry-type transformers range from 93 to 98 percent efficient; the losses derive from the core (magnetizing) and coils (resistance and impedance). Even when equipment served by the transformer is inoperative, some energy is lost unless primary power to the transformer is switched off. Dry-type transformers are available in many temperature-rise classifications. The lower the temperature-rise rating, the lower will be the coil losses due to the larger conductors used in the winding. The most common temperature-rise ratings are 80°C, 115°C and 150°C. Select the 80°C rise transformer for greater efficiency. Transformers with lower heat-rise ratings also have longer life expectancy, which may offset the higher cost (about 10 percent).

PROCEDURE

To reduce the energy lost in transformers:

(A) Select the most efficient dry-type transformer when replacement is necessary due to load changes or breakdowns.

(B) If larger liquid-filled transformers are required for new additions, select those with the highest efficiency.

(C) Completely disconnect from primary power the transformers that do not serve any active loads. Avoid disconnecting transformers feeding clocks, heating control circuits, fire alarms, and aother critical equipment.

ANALYSIS

To calculate the savings which might be achieved through more efficient transformers: (A) determine the new (reduced) annual electric consumption mand in kwh based on proposed modifications; (B) substitute this new figure for the present figure on the Energy Tabulation Form "Electrical" (Figure 8-12).

Conclusion: All of the figures pertaining to future energy use with regard to this ECO should now be calculated and carried through the appropriate forms as explained in Chapter VII. This will determine whether the costs of implementing and maintaining this ECO are justified in terms of future savings for the building owner.

ECO #50: REPLACE OVERSIZED MOTORS

THEORY

Because original load estimates for a building's mechanical equipment are usually conservative, most motors are oversized for the equipment load they serve. This is especially true when building loads have been reduced through conservation methods. If the ratio of the motor's load to the motor's horsepower rating is small, the power factor and load factor will not correspond, and the motor will operate inefficiently, thus wasting energy. See Figure 8-55 ("Variations in Efficiency With Load For Motors").

PROCEDURE

Undertake a comprehensive study of all the motors in the building to determine their load factors. Gather and record the following information:

(A) The equipment served by the motor.

(B) The nameplate information for each motor, including: brand, type, frame, horsepower, speed or speeds, voltage phases and frecuency, mounting, full load rating.

(C) Measure with an ammeter and record the full-load running current of each phase leg.

(D) Measure with voltmeter and record the voltage at the motor when measuring running current.

(E) Record pulley size and type, i.e. V belt, chain, etc.

(F) Find motor loading by dividing the current draw, as recorded in item (C) above, by nameplate current drawn. Convert this to percent and use in Figure 8-55.

(G) Motors that are not loaded to at least 60 percent of their potential are inefficient and reduce the power factor of the entire electrical system. Exchange underloaded motors with others on the premises to achieve as close to full loading on each motor as possible. If a motor needs replacement, consider interchanging motors then, too.

(H) For equipment which cycles on and off at short intervals, heat build-up is a less critical problem than in other cases. The motor service factor establishes the maximum overload possible without exceeding the motor's temperature rating. Utilize this information (available through the manufacturer) when in the process of exchanging and interchanging motors currently in use. For cycling equipment, select smaller motors, rated slightly below maximum load requirements, and allow some overloading to occur.

(I) Small split-phase or shaded-pole motors, often used in perimeter fan coil units, have low power factors and are inefficient. Turn off these motors during mild winter nights and permit the fan coil to operate by natural convection.

Factors to consider when installing new motors:

(A) Match the phase and frequency with electrical distribution system characteristics.

(B) When available motors do not exactly match system voltage, select the following:

(1) For motors loaded to 75 percent or less of their capacity motors rated at slightly more than system voltage.

(2) For motors loaded to more than 75 percent of their capacity, motors which are rated slightly under the system voltage. (Problems may arise if the nameplate rating of the motor selected is in excess of system voltage by more than 5 percent-- a likely situation in these days of brownouts and voltage cutbacks which actually increase the difference.)

ANALYSIS

To calculate the energy savings which might be achieved through replacement of oversized motors, refer to Figure 8-55. Then:

(A) determine the new (reduced) annual consumption in kwh based on proposed modifications (total motor load equals the sum of the load of each motor which is modified);

(B) substitute these new figures and reduced total consumption for the present figures on the Distribution Load Survey (Figure 8-10).

Conclusion: All of the figures pertaining to future energy use with regard to this ECO shoud now be calculated and carried through the appropriate forms as explained in Chapter VII. This will determine whether the costs of implementing and maintaining this ECO are justified in terms of future savings for the building owner.

ECO #51: CORRECT THE POWER FACTOR

THEORY

A low power-factor (pf) increases losses in electrical distribution and utilization equipment such as wiring, motors, and transformers. It reduces the load-handling capability and voltage regulation of the building electrical system. At a unity power-factor, losses are at a minimum. When pf is below a designated level, utilities often have a provision for a separate additional penalty charge. Although the methods of computing the charge vary, the magnitude is about the same. As an example, one utility company charges a penalty of $0.25 per kilowatt-ampere reactive (kvar) over 50 percent of total kw demand. Some utility companies register their demand in kva (apparent power), others in kw (actual power) and kvar (reactive power), and others in kw where there is no charge or penalty for lower-factor loads.

A poor power-factor also reduces the capacity of the electrical service. Even with a pf of 80%, electrical service spare capacity is decreased by an additional 25% of the load, and transmission losses are increased by 56% when compared to unity pf.

The power-factor is the ratio of actual power (kw) to apparent (kva); it may

FIGURE 9-34 POWER FACTOR CORRECTION

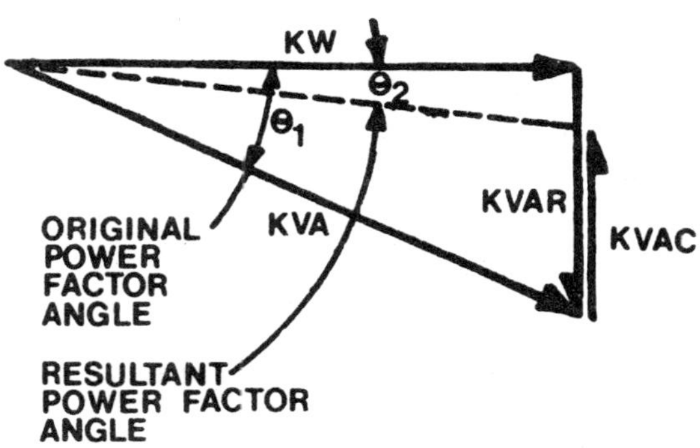

$$\text{Power Factor} = \frac{\text{Kw}}{\text{Kva}}$$

Power Factor = Cosine of the Angle θ

Kw = Kilowatt Meter Reading

3 Phase Average System

$$= \frac{\text{A mmeter Reading} \times \text{Voltage*}}{1000}$$

Kvar = Sine of the Angle** θ x Kva

= Kw x Tangent of the P.F. Angle θ

Kvar of Corrective Capacitance = Kvar x % Correction (Divided by 3 To Obtain Capacitance Per Phase Leg For A Balanced 3 Phase Load)

---

* For 3 phase voltage use 1.732 x system voltage.

** A power factor reading of 75% or 0.75 for instance, is equal to the cosine of a 41.5° angle. The sine of that same angle equals 0.66, and tangent equals 0.88.

also be expressed as the cosine of the phase angle between the impressed voltage and the current. The following figure shows the relationship between pf, kw, kvar and kva, and indicates the effect of corrective capacitance.

PROCEDURE

There are a number of devices available for the purposes of power-factor correction. Three of the more common types are capacitors, synchronous motors and synchronous condensers. Synchronous condensers are the most expensive and are not a practical solution for building services applications.

Capacitor correction is relatively inexpensive both in material and installation costs. Capacitors can be installed at any point in the electrical system, and will improve the power-factor between the point of application and the power source. However, the power-factor between the utilization equipment and the capacitor will remain unchanged. Capacitors are usually added at each piece of offending equipment, ahead of groups of small motors (ahead of motor control centers or distribution panels) or at main services. (Refer to National Electric Code 1975, article 460 for installation code requirements.) The advantages and disadvantages of each type of installation are listed below:

A. Individual Equipment

Advantages:

1. increased loan capabilities of distribution system;
2. can be switched with equipment thus no additional switching is required (for suggested maximum capacitor ratings in connection with this option, see Figure 9-35 below);
3. better voltage regulation because capacitor use follows load;
4. capacitor sizing is simplified;
5. capacitors are coupled with equipment thus would move with equipment if rearrangements are instituted.

Disadvantages:

Small capacitors cost more per kvar than larger units (economic break-even point for individual correction is generally at 10 hp).

B. Grouped Equipment

Advantages:

1. increased load capabilities of the service;
2. reduced material costs relative to individual correction;
3. reduced installation costs relative to individual correction.

Disadvantage:

Switching means may be required to control amount of capacitance used.

C. Main Service

Advantage:

Low material installations cost.

Disadvantages:

1. switching devices will usually be required to control the amount of capacitance used;
2. does not improve the load capabilities of the distribution system.

Where loads contributing to low power factor are relatively constant, and system load capabilities are not relevant, correcting at the main service could provide a cost advantage. When the low power factor is derived from a few selected pieces of equipment, individual equipment correction would be cost effective. Most capacitors used for pf correction have built in fusing; if not, fusing must be provided.

D. In summary, the power factor may be corrected by:

1. Providing a capacitor where the situation dictates. This is the least expensive method of reducing the reactive power of the system. For 1 to 60 kvac capacitance, one capacitor may be installed. If more than 60 kvac is required, use banks of capacitors.

2. Providing synchronous motors, keeping in mind, however, that small synchronous motors are not practical in most commercial buildings.

3. Providing synchronous condensers. As mentioned above these represent the most expensive solution to the problem.

ANALYSIS

The cost of installation depends on the location and type of mounting. In order to determine the capacitance required to correct pf, follow one of these procedures:

(A) When billings include a pf penalty, a kvar meter demand reading can be obtained from the utility. Determine the required corrective capacitance (kvac) by multiplying kvar by percent correction desired.

(B) As an alternative, use a pf meter to determine the pf for individual equipment, then use an ammeter for an ampere reading (at normal loading). Use the pf and ampere readings to calculate kvar and required kvac.

(C) When a pf meter is not available, the power-factor can be determined by dividing the wattage found via wattmeter (if available) by the product of the average of the ammeter readings of the phases and the system voltage:

$$pf = \frac{w}{\frac{amp \times V}{3}}$$

where,

amp = amperes (phase A + phase B + phase C) from ammeter readings
pf = power factor
V = system voltage
w = wattage from wattmeter reading

(NOTE: The minimum power-factor for which the power company does not include a penalty charge is approximately .85. Any further increase in the power-factor may prove to be futile and uneconomical. The utility charges, along with the initial costs, must be carefully evaluated before proceeding with the installation of the most appropriate capacitors.)

To calculate the savings which might be achieved through correction of the power factor: (A) determine the annual cost of utility company penalties due to low power factor; (B) subtract this annual figure from the annual cost of electricity on Utility Consumption/Cost Survey Form (Figure 8-8); (C) substitute the new total for the present total on the Utility Consumption/Cost Survey Form.

<u>Conclusion</u>: All of the figures pertaining to future energy use with regard to this ECO should now be calculated and carried through the appropriate forms as explained in Chapter VII. This will determine whether the costs of implementing and maintaining this ECO are justified in terms of future savings for the building owner.

FIGURE 9-35 PARTIAL LISTING OF SUGGESTED MAXIMUM CAPACITOR RATING WHEN MOTOR AND CAPACITOR ARE SWITCHED AS A UNIT*

	Nominal Motor Speed (Revolutions Per Minute)							
	3600		1800		1200		900	
Motor Horsepower Rating	Kvar	Percent Amperes Reactive**	Kvar	Percent Amperes Reactive**	Kvar	Percent Amperes Reactive**	Kvar	Percent Amperes Reactive**
3	1.5	14	1.5	15	1.5	20	2	27
5	2	12	2	13	2	17	3	25
7.5	2.5	11	2.5	12	3	15	4	22
10	3	10	3	11	3.5	14	5	21
15	4	9	4	10	5	13	6.5	18
20	5	9	5	10	6.5	12	7.5	16
25	6	9	6	10	7.5	11	9	15
30	7	8	7	9	9	11	10	14
40	9	8	9	9	11	10	12	13
50	12	8	11	9	13	10	15	12
60	14	8	14	8	15	10	18	11
75	17	8	16	8	18	10	21	10
100	22	8	21	8	25	9	27	10
125	27	8	26	8	30	9	32.5	10
150	32.5	8	30	8	35	9	37.5	10
200	40	8	37.5	8	42.5	9	47.5	10
250	50	8	45	7	52.5	8	57.5	9
300	57.5	8	52.5	7	60	8	65	9
350	65	8	60	7	67.5	8	75	9
400	70	8	65	6	75	8	85	9
450	75	8	67.5	6	80	8	92.5	9
500	77.5	8	72.5	6	82.5	8	97.5	9

* For use with 3-phase, 60-hertz NEMA Classification B motors to raise full load power factor to approximately 95%.

** Percent amperes reactive (AR) = the percent reduction in line current due to capacitors. If a capacitor of low kvar is to be used, the new percent AR will be approximately proportional to the actual capacitor rating divided by the kvar value in the tables. Selection of the motor overload relay should be based on the nameplate full load current of the motor reduced by the percent AR value.

ECO #52: UTILIZE A CENTRAL POWER CONTROL SYSTEM

THEORY

The purpose of a central control system is to provide the Physical Plant Chief Engineer and Energy Management Team with a management tool to give constant surveillance of the building and to make the most efficient and economical use of the physical plant systems and personnel.

System types: Proprietary central computerized control systems are marketed by each of the major temperature control manufacturers. These systems all have common features which accomplish a similar range of tasks. Each manufacturer, however, uses coding and computer languages which are unique to this system and cannot be decoded by any other system. Once a basic system has been selected and installed, all subsequent additions must be obtained from the original manufacturer. It is important, therefore, when selecting a system, to investigate its expansion potential in relation to future requirements.

Central computer control systems are usually modular in design, so that additional hardware and software can be added later if the necessary provisions have been made in the basic system. Each manufacturer's system comprises standard "off-the-shelf" hardware, but the application is always tailored to the specific project. Basically, any system is composed of four major parts.

(A) Interface panels which are located at strategic points throughout the building (usually equipment rooms). These panels are the focal point of all signals to and from a particular area.

(B) Transmission system between the Central Console and all interface panels. This transmission system can be a simple single-core cable for digital transmission or a multi-core cable for multiplex transmission.

(C) Central control console and associated hardware located in a control room (usually close to the boiler room and chief engineer's office). The console, tied to the computer and associated hardware, is the point at which the operator enters all instructions and retrieves all data. In addition, the computer generates routine instructions according to the program contents.

(D) Software (program) generated by the manufacturer in conjunction with the prospective user. The software input - via magnetic tapes, paper tapes, or cards - contains the basic operating instructions for the computer and is stored in the form of "bits" of information either in core memory or in bulk memory (cheaper than core for large quantities of data).

FIGLRE 9-36 CENTRAL CONTROL SYSTEM

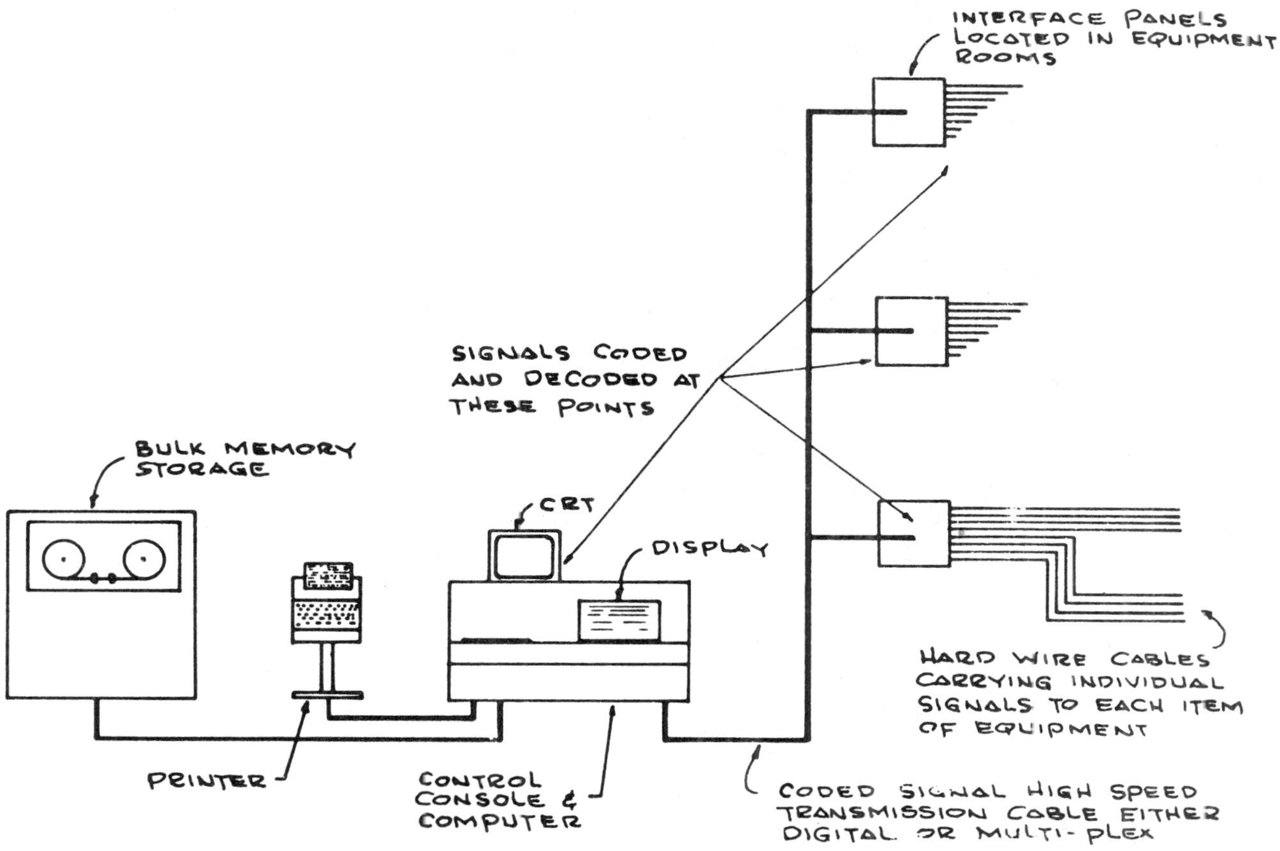

Interface requirements:

To obtain maximum benefit from a central console system it is necessary to measure, monitor and control many different items of equipment. The central control system must be interfaced with the existing equipment to obtain this information in rational form, and to exert its control functions.

Figure 9-37 shows a table cf desirable interface points and the types of functions required for typical building systems. Interface of two types

FIGURE 9-37 TABLE OF INTERFACE POINTS

ITEM	START/STOP LEAD/LAG	READ OUT	RESET	ALARM 1	ALARM 2
Ventilation					
Outside air temperature (dry bulb)		X			
Outside air temperature (wet bulb)		X			
Air temperature on pre-heat coil (dry bulb)		X		X	
Air temperature on pre-heat coil (wet bulb)		X			
Air temperature on cooling coil (dry bulb)		X			
Air temperature off cooling coil (dry bulb)		X			
Air temperature off humidifier (dry bulb)		X			
Air temperature off humidifier (wet bulb)		X			
Air temperature off re-heat coil (dry bulb)		X			
Critical area temperature (dry bulb)		X		X	
Critical area humidity		X		X	
Return air temperature (dry bulb)		X			
Return air temperature (wet bulb)		X			
Fresh air/exhaust air/recirculation air dampers		X	X		
Air velocity beyond filters					X
Supply fan	X	X		X	X
Humidifier spray pump	X	X			X
Supply air flow switch		X		X	
Exhaust fan	X	X		X	X
Heating coil control set-point			X		
Cooling coil control set-point			X		
Humidifier control set-point			X		
Hot Water Heating					
Boiler fuel input	X	X	X		
Boiler supply temperature		X	X	X	
Boiler return temperature		X			
Boiler high limit temperature				X	
Hot water pumps	X	X		X	X
Steam					
Boiler fuel input	X	X	X		
Boiler steam pressure		X	X	X	
Boiler condensate feed temperature		X			
Boiler high steam pressure limit				X	
Boiler low water level limit				X	
Boiler feed pumps	X	X		X	X
Chilled Water					
Chiller power/fuel input	X	X	X		
Chiller supply temperature		X	X	X	
Chiller return temperature		X		X	

FIGURE 9-37 (con't)

ITEM	START/STOP LEAD/LAG	READ OUT	RESET	ALARM 1	ALARM 2
Chilled Water (continued)					
Condenser supply temperature		X	X		
Condenser return temperature		X	X		
Chilled water pumps	X	X		X	X
Chilled water flow rate		X			
Condenser water pump	X	X		X	X
Condenser water flow rate		X			
Cooling tower fan	X	X			X
Air cooled condenser fan	X	X			X
Domestic Hot Water					
Domestic hot water heater fuel/power input	X	X	X		
Domestic hot water storage temperature		X	X		
Domestic hot water high temperature limit				X	
Domestic hot water return temperature		X			
Domestic hot water circulating pumps	X	X			X
Cold Water					
Cold water pressure boost pumps	X	X		X	X
Cold water boosted pressure		X		X	
Other Services					
Control system air pressure		X		X	
Electrical					
Total building power input KVA & KW		X		X	
Primary isolation switch		X			
Load interrupter switch		X			
Transformer winding temperature		X		X	
Transformer secondary voltage fixture		X		X	
Emergency generator	X	X		X	X
Parking lot lights	X	X			
Selected lighting circuits	X	X			

NOTE: Alarm 1 denotes Critical Alarm
Alarm 2 denotes Maintenance Alarm

of signal is required; binary signals, which comprise only two alternatives (on/off, open/close, etc.), and analog signals, which measure a value against a particular scale (temperature, pressure, etc.).

Binary signals are used as instructions to start and stop equipment, open and close valves and dampers, etc. They are also used for retrieving data such as on/off, open/closed, etc. Analog signals are used as instructions to raise or lower temperature set-points, adjust damper positions, raise or lower pressure set points, etc. They are also used for retrieving data such as temperature, pressure, humidity, etc.

To select a central control system, it is first necessary to assemble a complete list of all desired interface points under the two categories of binary and analog, and then arrange them in groups served by individual interface panels.

Signal transmission:

To make effective use of the computer's capabilities, information between the computer console and the interface panels must be transmitted at high speed in a format easily handled by the computer. The most convenient method of transmission is digital, where information is represented by pulses arranged serially and transmitted through a single core conductor. Digital cable is commonly coaxial, although some systems use a twisted pair of insulated wires. (Voice intercom is carried by a separate screened cable.)

Multiplex cable is multi-core, signals being multiplexed and transmitted in parallel. Typically the cable will comprise between 50 and 100 separate wires and have an overall diameter of one inch. It is more difficult to install this cable than coaxial, particularly in existing buildings where empty conduits and throughways are not available.

Some systems still use multiplex cable on the grounds that analog signals can be transmitted in unmodified form, whereas they must be converted from analog to digital form (with a small loss of accuracy at the interface panel) for digital transmission. Conversely, analog signals are sensitive to interference from "spikes" and other spurious signals induced by adjacent building wiring; digital signals are not subject to interference.

When selecting a central control system, the different characteristics of the transmission methods must be analyzed and a choice made based on the particular circumstances. Generally, digital transmission systems provide more options for later additions, as multiplex systems have a definite limit to the number of points that can be connected.

Control console and hardware:

Most central control systems now on the market offer comparable perfomance with regard to the control console and hardware items such as printers, graphic display, CRT, and input/output keyboards. Individual options are available on request to suit the prospective user's particular requirements. For instance, different levels of access into the computer from the keyboard are available and may comprise from three to five levels. Access would range from restriction to normal operation for the lowest level up to reprogramming at the highest level, and would be controlled by key-switch. (It is vital to protect memory from accidental erasure.)

The central processor or computer will be supplied with integral core memory. Capacities typically vary from 16 K words, each word containing 16 bits of information. Core memory is expensive compared with bulk memory, and if the core capacity is likely to be exceeded, then the addition of a bulk memory unit should be considered. (First be sure that the computer selected can be interfaced with external memory.)

Bulk memory can be either disc or tape, both with random access. Disc memory is preferred for control applications and typically has a capacity of 192 K words.

Software:

Software programs for common applications are available from the manufacturer. Each of the major controls manufacturers has a library of application programs which have been developed over a number of years. The cost of these programs is low compared to custom-written software. When selecting a central control system, investigate the library of programs available, as the extent and range of the library are good indices of a particular manufacturer's commitment and ability in the computer control field. Programs and their applications include:

Executive program - This program is essential to the function of the computer itself as it controls the priority of signal processing, directs information in and out of memory, controls the operation of all associated hardware and encodes/decodes incoming and outgoing signals. An essential function is the orderly shut-down of the computer on power failure to ensure that all information is safely stored in permanent memory before all power is lost. Beware of systems that store information in volatile memory which evaporates on power failure.

Start-stop program - This is the simplest program and yet with modifications becomes the most effective energy saver, particularly when used in conjunction with other programs and routines. In its simplest form the program is arranged to start and stop equipment according to predetermined times allowing for weekends and holidays. The program can also open and close dampers and turn lights on and off. It can, in addition, reschedule temperature from occupied level to unoccupied level, as all of these functions are binary in nature.

Manual override can be provided with automatic reversion to program control at the end of an occupied or unoccupied period. Interface with other programs allows the basic program to be overriden and equipment turned on and off to meet some other criteria. Start sequences can be arranged in cascade to avoid simultaneous starting and the cumulative effect of inrush currents causing a high peak electrical demand.

Load-shedding program - This program is used in conjunction with the start/stop program to override normal operation and turn off equipment as a predetermined peak load value is approached. In its simplest form the program shuts down selected equipment according to the measurement of peak load. The program can be modified to include a rolling priority feature whereby selected equipment is turned off according to assigned priorities for an adjustable period of time, after which it is returned to service and a lower priority piece of equipment turned off. The rolling priority function is activated as a pre-determined peak load value is approached and equipment turned off is limited to that necessary to maintain the load below a given level.

Interface with a profile routine allows a continuous profile of electrical demand to be made and immediate future load to be predicted. This predicted load then initiates the rolling priority to alter the slope of the prediction curve. The current maximum peak load experienced in the previous eleven months is held in memory and used as the criteria for load shedding. This value is constantly modified on the basis of monitored information.

Optimizing programs - These programs tend to be more sophisticated, requiring interface with an arithemtic routine to allow optimum settings to be calculated. The program interfaces with the start/stop program, load shedding program, and rest program to modify operation of equipment. Optimizing programs can provide economizer cycle control of outdoor air according to enthalpy and loads. Chilled water and boiler water temperatures can be modified to provide optimum equipment operation according to load. Some optimizing programs are available from the manufacturer's library but others may have to be custom-written.

Reset programs - These programs reset the control points of control thermostats and pressure controllers, and interface with the load-shedding and optimizing programs. In place of, or in addition to, turning off equipment for load shedding and energy conservation, loads can be trimmed; i.e., if a peak load is approaching, space conditions can be reset to reduce cooling or heating loads; outdoor air dampers can also be reset.

Alarm programs - These programs report alarms generated either outside the computer (fire alarm pull boxes, boiler high pressure, etc.) or in computer software. Upper and lower limits can be set in software for values of any monitored point and arranged to report an alarm when exceeded.

Alarm programs can also be interfaced with the start/stop program to monitor hours run and report a maintenance alarm when a designated number of running hours is exceeded. Information on particular maintenance required can be stored in memory and be printed out when the alarm is reported.

Figure 9-38 shows typical relationships between programs and the interface between one and another. The program functions described are an indication only of what can be achieved with a central computer control system and are intended to stimulate investigation into all possibilities, particularly in reference to energy conservation.

PROCEDURE

Computerized central control systems are generally most applicable to larger, more complex buildings or groups of buildings. They should certainly be considered where net floor area exceeds 100,000 sq ft and where energy use is high due to extended occupancy or use of special equipment. Central control systems vary from the relatively simple type performing a few functions to the progressively more complex types performing more and more complicated functions.

A central control system should be tailored to the present requirments of the building and the systems to be controlled. If money for investment is limited, attention should first be given to those functions which will show the quickest and greatest return in energy saved. An overview of the central control system capacity should, however, be made at the time of the initial installation and provisions be included for expanding the system hardware and its software capacity as and when investment money permits. For instance, an initial system may well comprise a control console and central processor unit with one single printer output programmed only to perform start/stop and simple reset functions and to report alarms. Optimization of system operation and load-shedding in this case would be achieved manually at the central console based on decisions made by the operating staff and achieved by overriding the programmed start/stop times and reset points for the various systems.

Future hardware additions to this initial system might well include an extra printer to act as a back up and to handle alarm signals separately from system status reports; a cathode ray tube output to show in English language the status of selected systems and current modifications made to the normal operating program; a bulk storage memory unit to increase the initial core memory included in the central processing unit so that more complex programs can be used and a data file assembled of operating conditions; a high-speed tape input/output device to enable new programs to be fed into the computer and to extract from the computer banks historical data of systems operation; complex custom written software programs to optimize the operation of all systems and to activate automatic load shedding on a selected rolling priority basis.

FIGURE 9-38 INTERFACE BETWEEN PROGRAMS

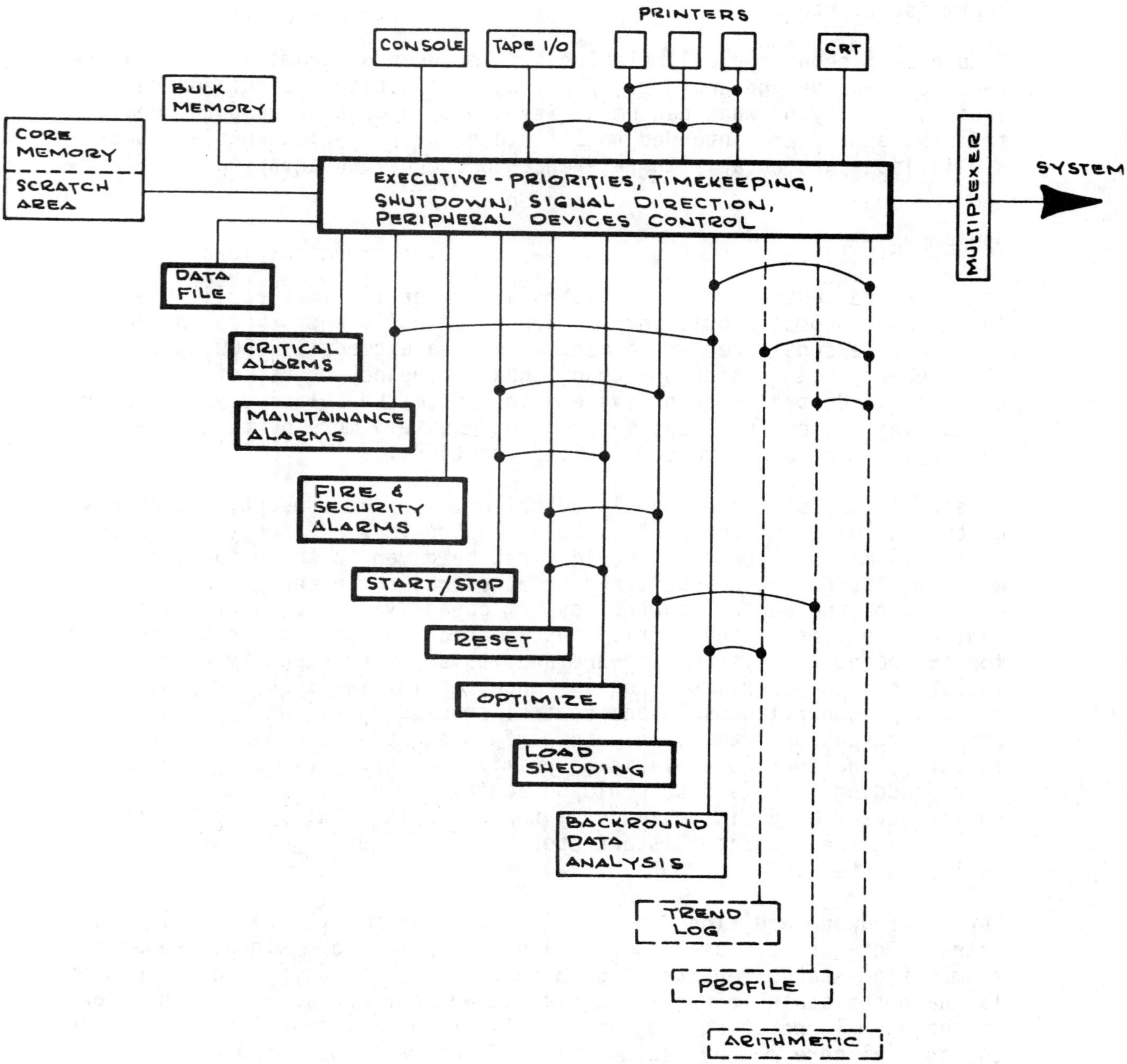

In summary, a central control sytem shoud be capable of some or all of the following functions:

(A) Monitor all systems for off-normal conditions.

(B) Monitor all fire alarm and security devices.

(C) Monitor operating conditions of all systems and reschedule set-points to optimize energy use.

(D) Monitor on a continuous basis selected portions of any systems and store this information in bulk memory for later retrieval and use in updating software. This information can also be used to monitor the effectiveness of the energy management program and indicate changes or modifications which need to be made.

(E) Limit peak electrical demand values by predicting trends of loads and shedding non-essential services according to programmed priorities.

(F) Optimize the operation of all systems to obtain the maximum effect for the minimum expenditure of energy.

(G) Optimize maintenance tasks to effect maximum equipment life with minimum labor and costs.

(H) Provide inventory control of spare parts, materials, and tools used for maintenance.

ANALYSIS

While of course the installation of a computerized central control system is a highly technical and expensive undertaking, it is also one of the most effective methods available to conserve energy in larger buildings and complexes on a consistent and significant basis. Depending on the skill with which it is used, such a system will generally save enough money in utility bills and personnel costs to pay for itself within a few years. As with all ECO's, however, a thoughtful cost-benefit analysis will have to be made to justify the expense of implementation.

The Distribution Load Survey Form (Figure 8-10) identifies the operating time of all pumps, fans, etc. A centralized control will reduce this time. Substitute the new operating time on the form Load Shedding reduction in demand should be reflected on the Energy Tabulation, "Electric Demand", Form (Figure 8-12).

Conclusion: All of the figures pertaining to future energy use with regard to this ECO should now be calculated and carried through the appropriate forms as explained in Chapter VII. This will determine whether the costs of implementing and maintaining this ECO are justified in terms of future savings for the building owner.

SECTION X
MODEL BUILDING ENERGY STUDY

X. MODEL BUILDING ENERGY STUDY

A. Introduction

It is the purpose of this concluding chapter to provide the reader with a detailed prototype of a typical building energy study. Most of the principles, methods, and procedures discussed in the foregoing chapters are here applied to a hypothetical office building in the midwestern U.S. First the building is surveyed in terms of its present energy use as determined from (A) an analysis of fuel and utility bills representing a "base year" of energy consumption; (B) a "physical" study and energy-efficiency analysis of the building itself, including tabulations of energy consumption for each of the buildings major "energy-users".

Based on the results of this two-fold analysis of present energy use, seven Project-ECO's (PECO's) are analyzed in terms of reducing the building's future energy use in such a manner as to provide the owner with a favorable cost-benefit ratio in connection with the implementation of those PECO's. (While additional PECO's might have been applied to the model building, the seven here presented are more than adequate to demonstrate the full range of methods and procedures involved in the analysis of promising ECO's for a particular building. Such detailed treatment of additional PECO's would merely subject the reader to unnecessary repetition.)

(NOTE: While the model building is described as being located in St. Louis, only the climatic data for the St. Louis area has actually been used in the energy study. Such other data as fuel and utility costs are hypothetical, as is the building itself. Otherwise, the "Model Building Energy Study" is based on actual principles of energy use and conservation as discussed in previous chapters of the manual.)

B. Building Data

This building is called the HB&S Building. Located in St. Louis, Missouri, it is 100 percent occupied and provides all utilites for the tenants. The gross square footage of the building is 92,700 sq ft equally divided among 9 floors of 10,300 sq ft each. There is also a smaller basement mechanical equipment room.

The east side of the first floor contains a drug store which is open from 7 a.m. to 7 p.m. Monday through Saturday, and from 8 a.m. to 1 p.m. on Sunday. The west half of the first floor contains an office area which is occupied from 8 a.m. to 5 p.m. Monday through Friday. Some of the tenants occasionally return to their offices for night or weekend work, but this is not a normal occurence.

1. Building Envelope

The outside face of the building is of lightcolored brick, with 30 percent of its surface composed of metal casement windows, single-glazed. The only exception to the metal casement windows is the drug store which has fixed glass panels.

Custodial duties require an additional 12 hours per work week for the drug store and 20 hours per week for the office areas.

2. Mechanical Equipment

The basement equipment room contains two centrifugal chillers and two gas-fired steam boilers. A system of chilled water piping and steam piping distributes heating and cooling throughout the building. Pumps circulate chilled water to air handling units located on each floor. The first floor mechanical room has two air handling units, one serving the drug store and one serving the lobby, hall and office area. Steam reheat coils provide zoning for the office lobby and hall areas.

Floors two through nine have two air handling units, one serving the interior zones and one the exterior zones. The interior is further divided into two zones and the exterior into four zones by the use of steam reheat coils.

3. Lighting - Interior and Exterior

Basic interior lighting consists of fluorescent fixtures with the exception of the halls and lobbies, which are lighted by 150-watt incandescent spots.

The drug store feature 96" slimline fluorescents, and all lights are controlled by one switch. The office areas are lighted with 2 x 4 fluorescents controlled by one switch on each floor. The hall and lobby lighting on each floor are separately switched.

The exterior lighting for the building consists of 250-watt H.I.D. floodlights spaced every 25 ft. around the exterior of the building, and the three-story parking garage is lighted by 96" fluorescent strips. All of the exterior lighting for the building and parking is operated by a time clock.

4. Utility Rates

a. Electrical
Demand Rates: $5/kw/mo. Highest kw reading during 15 minute intervals sets demand for next 12 months.

Consumption:

5.8¢/kwh for first - 50 kwh
3.8¢/kwh for next - 550 kwh
3.1¢/kwh for next - 2400 kwh
1.2¢/kwh for additional - kwh
Fuel adj. charge - 1.5¢ kwh

b. Natural Gas

$1.75/Mcf

FIGURE 10-1 HB&S BUILDING FIRST FLOOR PLAN

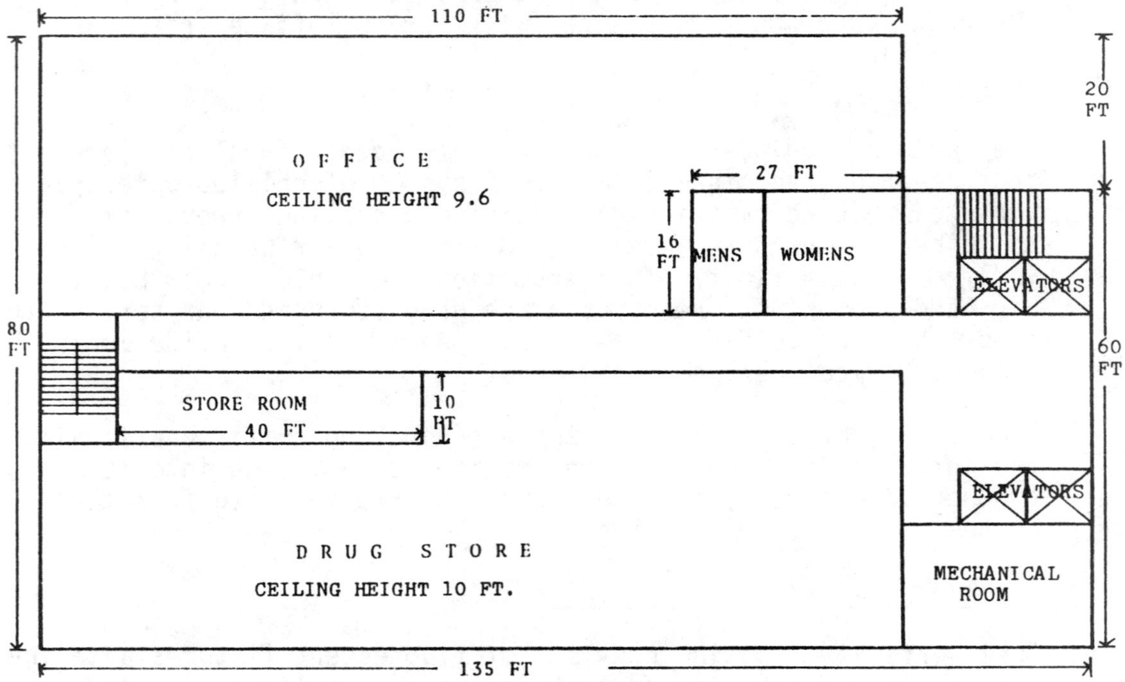

FIRST FLOOR
FLOOR TO FLOOR HEIGHT - 15.0 FT.

FIGURE 10-2 HB&S BUILDING 2ND THRU 9TH FLOOR PLAN

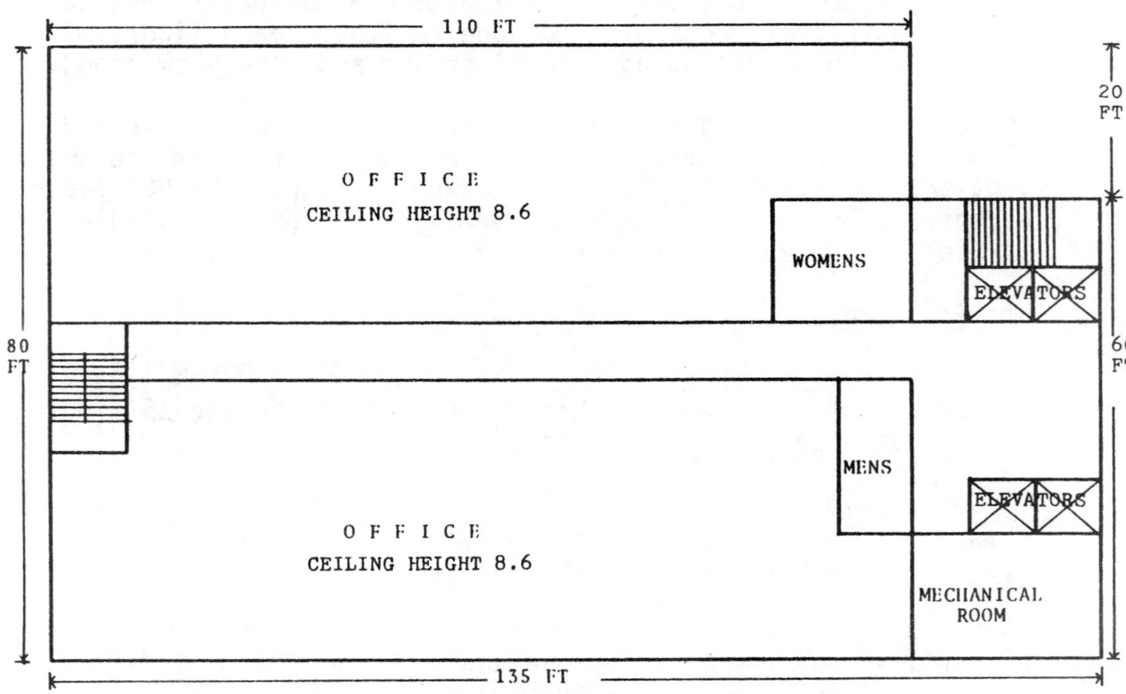

TYPICAL - 2ND THRU 9TH FLOORS
FLOOR TO FLOOR HEIGHT - 13.0 FT.

C. Utility Consumption/Cost Survey

The completed form and the following graph have been derived from study of monthly gas and electric bills representing a "base Year" in the operation of the model building. The graph itself, of course, is simply a visual interpretation of the Utility Consumption/Cost Survey Form, as discussed in Chapter IV (Step Four). The construction of such a graph, based on the principle that a "picture is worth a thousand words", can be an extremely important step in the preparation of a convincing energy study.

FIGURE 10-3
EXAMPLE OF COMPLETED UTILITY CONSUMPTION/COST SURVEY FORM

UTILITY CONSUMPTION/COST SURVEY FORM

JOB: HB&S BUILDING DATE: 9-[illegible]-78

Month Year	Elec Kwh	Elec Demand	Elec $	Fuel Gas mcf	Fuel Gas $	Fuel	Fuel $	Degree Days-Htg
Jan 77	218671	623.2	9935	2061	3607			1006
Feb	225262	546.8	10130	2030	3553			894
Mar	307901	680.7	12749	1883	3295			679
Apr	339913	696.1	13613	1740	3045			132
May	385220	730.0	14837	1207	2112			0
June	403900	840.[illegible]	15341	1061	1857			0
July	456283	887.2	16756	990	1733			0
Aug	461200	887.2	15388	1363	2385			0
Sept	445450	865.0	16463	1526	2670			181
Oct	385500	810.5	14845	1892	3311			761
Nov	315630	721.3	12958	1691	2959			662
Dec 77	216212	636.8	10274	2252	3941			1119
TOTAL	4161142	887.2	164789	19696	34468			5434
Jan 78	212318	526.8	10168	1926	3370			974
Feb	208900	540.2	10076	2116	3703			912
Mar	316502	581.8	12982	2042	3574			603
Apr	341206	598.1	13649	1520	2660			261
May	402916	730.0	15315	1260	2205			0
June	400480	335.6	15249	1026	1796			0
July	482916	387.2	17474	1015	1776			0
Aug								
Sept								
Oct								
Nov								
Dec								

FIGURE 10-4
EXAMPLE OF COMPLETED UTILITY CONSUMPTION PROFILE

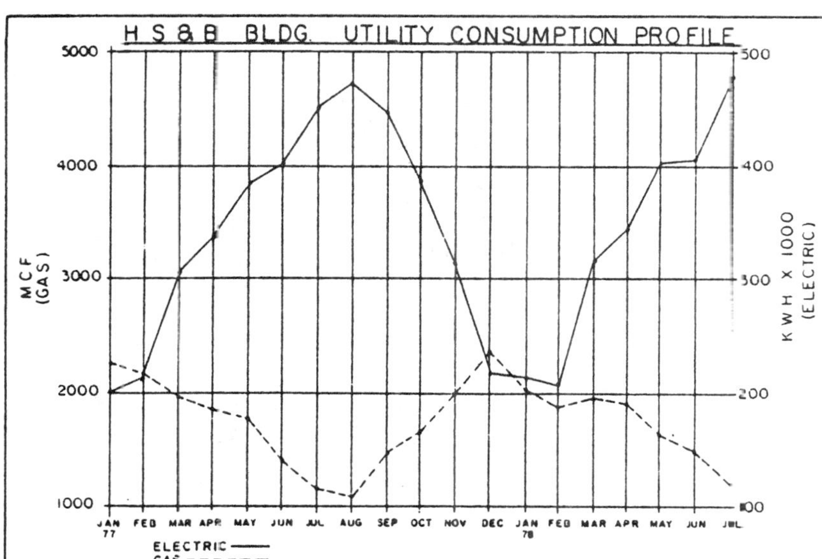

D. Building and Distribution Load Surveys

The following forms have been completed in accordance with procedures discussed in Chapter V (Step Five) and Chapter VIII, which provides detailed instructions regarding the use of these forms. The reader will note that the Building Survey Form provides space for data concerning a variety of individual "energy-users" corresponding with different functions and areas within the building.

FIGURE 10-5 EXAMPLE OF COMPLETED BUILDING LOAD SURVEY FORM FOR THE DRUG STORE SPACES

BUILDING LOAD SURVEY FORM

JOB: HB&S Building　AREA: Drug Store　WINTER: 75°-30%　SUMMER: 75°-50%　PECO NO: 0　DATE: 9-9-78

GLASS

Facing	SqFt	Shading Ext	Shading Interior	Type U= 1.13
S	200	0	Blinds S.F. 64	Single Fixed Tight Fit
E	880	0		
TOTAL	1080			

WALLS

Facing	SqFt	Shade	Type U=0.2
S	325	0	Light 4" Brick 4" Block Plas. Finish
E	1145	0	
TOTAL	1470		

ROOF OR FLOORS

SqFt	Type U=

LIGHTING

Area	Type	Occupied Kw	Occupied Fc	Occupied Hrs/Wk	Occupied Kwh/Yr	Custodial Kw	Custodial Fc	Custodial Hrs/Wk	Custodial Kwh/Yr
Main Store	96" Slimline	17.2	175	77	68869	17.2	175	12	10733
Store Room	96" Slimline	0.8	75	77	3203	0.8	75	12	499
	TOTAL	18.0		TOTAL	73072	18.0		TOTAL	11232

PEOPLE

Area	No	Hrs/Wk
Main Store	25	77

Infiltration Item	Crack Usage	Figure	Factor	Cfm
4 x 8 Door	150	8 - 15	250	250

DOMESTIC WATER

Use	Temp Hot	Temp Cold	Gal/Meals or Person	People/ Meals	Gallons Per Day	Btu/Yr E6	TOTAL %Eff. & Conversion	MCF /Yr
Hand Wash	110	60	3	25	75	8.1	60%	13.5

DISTRIBUTION LOADS

Item	Cfm	Gpm	Oa	Hp	Voltage	Amperage Rated	Amperage Opr	Kw	TOTAL Hrs/Wk	Kwh Per Year
A-H Unit No. 1	7500	60	750	7-1/2	460/3	11	10	6	168	52416

FIGURE 10-6 EXAMPLE OF COMPLETED BUILDING LOAD SURVEY FORM FOR THE OFFICE SPACES

BUILDING LOAD SURVEY FORM

JOB: HB&S Building　AREA: Office　WINTER: 75 - 30%　SUMMER: 75 - 50%　PECO NO: 0　DATE: 9-9-78

GLASS

Facing	SqFt	Shading Ext	Shading Interior	Type U= 1.13
N	1152	0	Blinds S.F.-.64	Single Metal Casem. Avg. Fit
S	3608	0		
E	4576	0		
W	5456	0		
TOTAL	14792			

WALLS

Facing	SqFt	Shade	Type U= 0.20
N	8368	0	Light 4" Brick 4" Block Plas. Finish
S	4587	0	
E	9464	0	
W	10609	0	
TOTAL	33028		

ROOF OR FLOORS

SqFt	Type U= 0.20
10300	6" Concrete Built-Up Roofing 2x2 Lay-in Ceiling
10300	

LIGHTING

Area	Type	Occupied Kw	Occupied Fc	Occupied Hrs/Wk	Occupied Kwh/Yr	Custodial Kw	Custodial Fc	Custodial Hrs/Wk	Custodial Kwh/Yr
Office - Exterior	2x4-200W Flourescent	142.4	150	40	296192	142.4	150	20	148096
Office - Interior	2x4-200W Flourescent	110.5	150	40	229840	110.5	150	20	114920
Halls	150W. Incandescent Spots	21.6	40	40	44928	21.6	40	20	22464
Restrooms	2x4-200W Flourescent	7.8	75	40	16224	7.8	75	20	8112
	TOTAL	282.3		TOTAL	587184	282.3		TOTAL	293592

PEOPLE

Area	No	Hrs/Wk
Office - Exterior	241	40
Office - Interior	181	40

Infiltration Item	Crack Usage	Figure	Factor	Cfm
Doors	250 Pas/Hr	8 - 15	400	400
Windows	14144 LF	8 - 14	.55	7780
			TOTAL	8180

DOMESTIC WATER

Use	Temp Hot	Temp Cold	Gal/Meals or Person	People/ Meals	Gallons Per Day	Btu/Yr E6	%Eff. & Conversion	MCF /Yr
Handwash	110	60	3	422	1266	136.4	60 x 1	227

DISTRIBUTION LOADS

Item	Cfm	Gpm	Oa	Hp	Voltage	Amperage Rated	Amperage Opr	Kw	TOTAL Hrs/Wk	Kwh Per Year
9-Exterior A-H Units	83500	456	8350	87.5	460/3	123	106	66.1	168	577450
8-Interior A-H Units	36000	216	3600	40.0	460/3	61	48	27.6	168	241100
2-Toilet Exhaust Fans	3200			1.0	230/1	8.5	7.5	0.8	168	6730

The Building Load Survey Form is also used to record related exterior building loads as appropriate.

FIGURE 10-7 EXAMPLE OF COMPLETED BUILDING LOAD SURVEY FORM FOR MISCELLANEOUS SPACES AND EXTERIOR AREAS

BUILDING LOAD SURVEY FORM

JOB: HB&S Building AREA: Build. & Parking WINTER: SUMMER: PECO NO 0 DATE: 9-9-78

GLASS					WALLS				ROOF OR FLOORS	
Facing	SqFt	Shading Ext	Shading Interior	Type U=	Facing	SqFt	Shade	Type U=	SqFt	Type U=
TOTAL					TOTAL					

LIGHTING Area	Type	OCCUPIED Kw	Fc	Hrs/Wk	Kwh/Yr	CUSTODIAL Kw	Fc	Hrs/Wk	Kwh/Yr
Building	250W Flood	4		84	17472				
Parking	96" 100W Fluorescent	9.5	5	84	41496				
Stair Wells	48" 50W Fluorescent	2	10	168	17472				
	TOTAL	15.5		TOTAL	76440			TOTAL	

PEOPLE Area	No	Hrs/Wk	Infiltration Item	Crack Usage	Figure	Factor	Cfm

DOMESTIC WATER Use	Temp Hot	Temp Cold	Gal/Meals or Person	People/Meals	Gallons Per Day	Btu/Yr E6	TOTAL Eff. & Conversion	/Yr

DISTRIBUTION LOADS Item	Cfm	Gpm	Oa	Hp	Voltage	Amperage Rated	Amperage Opr	Kw	TOTAL Hrs/Wk	Kwh Per Year

The Distribution Load Survey Form has been used to record the weekly and yearly energy consumption of specific items of energy-using equipment and (pumps, fans, chillers, etc.).

FIGURE 10-8 EXAMPLE OF COMPLETED DISTRIBUTION LOAD SURVEY FORM

DISTRIBUTION LOAD SURVEY FORM

JOB: HB&S Building AREA: Building PECO NO: 0 DATE: 9-9-78

Area Served	Item	Cfm	Gpm	O A	Hp	Voltage	Amperage Rated	Amperage Opr.	Kw	Hrs/Wk	Kwh/Yr
	CHW Pump 1		366		10	460/3	14	14	8.8	168	78877
	CHW Pump 2		366		10	460/3	14	14	8.8	168	78877
	Cond. Pump 1		450		15	460/3	21	21	13.2	168	115315
	Cond. Pump 2		450		15	460/3	21	21	13.2	168	115315
	Cool Twr. Fan				20	460/3	27	27	17.6	Auto	96096
	Boiler Fan 1				2	460/3	2.8	2.8	1.8	Auto	9828
	Boiler Fan 2				2	460/3	2.8	2.8	1.8	Auto	9828
	Elevator 1				40	460/3	56	Varies	Varies	40	16000
	Elevator 2				40	460/3	56	Varies	Varies	40	16000
	Elevator 3				40	460/3	56	Varies	Varies	40	16000
	Elevator 4				40	460/3	56	Varies	Varies	168	25000
	Dom Wtr Pump				2	460/3	2.8	2.8	1.8	168	15725
	Chiller - 1		366		150	460/3	210	-	132	-	-
	Chiller - 2		366		150	460/3	210	-	132	-	-

E. Annual Heating and Cooling Tabulation

The Climatic Zone Form below has been filled out using the general instructions in Chapter VI and VIII, and using the assumptions for the HB&S Building example. This data should be filled in prior to starting on the Annual Heating and Cooling Tabulation Form.

FIGURE 10-9 EXAMPLE OF COMPLETED CLIMATIC ZONE FORM

CLIMATIC ZONE FORM

Job: HB & S BUILDING   Date: 9-9-78

Design Temperature: Summer 98 Dbt 79 Wbt Winter -2 Dbt

Wind Velocity 7 Miles/Hr.

MAP	SOURCE	VALUE
Latitude	Fig 8-1	38°
Heating Degree Days	Fig 8-1	4,900
Solar Radiation in Langleys	Fig 8-2	375
Degree Hours Less Than 54F Wbt/68F Dbt	Fig 8-3	24,000
Degree Hours Greater Than 78F Dbt	Fig 8-4	6,500
Degree Hours Greater Than 66F Wbt	Fig 8-5	5,100
Degree Hours Greater Than 85F Dbt	Fig 8-6	1,800

REMARKS

The Annual Heating and Cooling Tabulation Forms are completed next. The examples below have been filled out using the general instructions. A form has been used for each of the areas covered by the Building Load Survey Forms. Note that the Annual Heating and Cooling Tabulation Form provides space for data concerning both the building envelope and the building's heating and cooling systems. Also note that most of the data on the Annual Heating and Cooling Form has been carried over from data on the Building Load Survey Form and the Distribution Load Survey Form.

FIGURE 10-10 EXAMPLE OF COMPLETED ANNUAL HEATING AND COOLING TABULATION FORM FOR THE DRUG STORE SPACES

ANNUAL HEATING AND COOLING TABULATION FORM

JOB: HB&S Building AREA Drug Store PECO NO: 0 DATE: 9-9-78

ITEM	UNITS	COOLING				HEATING		
	SqFt-Cfm	Fig No 8-	Factor	Shade Factor	Total Btu/Yr-E6	Fig No 8-	Factor	Total Btu/Yr-E6
North Glass	--	16 or 17	--		--	19 or 20	--	--
South Glass	200	16 or 17	135E3	.64	17.3	19 or 20	60E3	12
East Glass	880	16 or 17	135E3	.64	76.0	19 or 20	67E3	59
West Glass	--	16 or 17	--	--	--	19 or 20	--	--
Glass Conduction	1080	18	8E3		8.6	--	--	--
North Wall	--	22 or 23	--		--	25 or 26	--	--
South Wall	325	22 or 23	2.5E3		0.8	25 or 26	12E3	4
East Wall	1145	22 or 23	2.5E3		2.9	25 or 26	15E3	17
West Wall	--	22 or 23	--		--	25 or 26	--	--
Wall Conduction	1470	24	3.0E3		4.4	--	--	--
Roof Solar	--	27	--		--	--	--	--
Roof Conduction	--	24	--		--	28	--	--
Outside Air*	750	30	75.3E6		56.5	31	150E6	113
Humidification*	750	--	--		--	32	75E6	5.6
*Factor is Btu/1000Cfm		Sub-Total Cooling				Total Heating		261.E6

Item	Kwh/Yr	Conversion Factor	%Cooling Season/Yr	
Lighting	84304	3413 x 1.0	50	143.9
Fans	52416	3413 x 1.0	50	89.4

People	No	Hrs/Yr	Btu/Person	%Cooling Season/Yr	
	25	4004	450	50	22.5
				TOTAL COOLING	422.3

Summary	
Heating	
Conv. Factor	1E6
Efficiency	75%
Mcf/Yr	348
Cooling	
Conv. Factor	3413
COP	4.0
Kwh/Yr	30933

FIGURE 10-11 EXAMPLE OF COMPLETED ANNUAL HEATING AND COOLING TABULATION FORM FOR THE OFFICE AREA

ANNUAL HEATING AND COOLING TABULATION FORM

JOB: HB&S Building AREA Office PECO NO: 0 DATE: 9-9-78

ITEM	UNITS	COOLING				HEATING		
	SqFt-Cfm	Fig No 8-	Factor	Shade Factor	Total Btu/Yr-E6	Fig No 8-	Factor	Total Btu/Yr-E6
North Glass	1152	16 or 17	55E3	.64	40.6	19 or 20	85E3	97.9
South Glass	3608	16 or 17	135E3	.64	311.7	19 or 20	60E3	216.5
East Glass	4576	16 or 17	135E3	.64	395.4	19 or 20	67E3	306.6
West Glass	5456	16 or 17	135E3	.64	471.4	19 or 20	67E3	365.5
Glass Conduction	14792	18	8E3		118.3	--	--	--
North Wall	8368	22 or 23	1.5E3		12.6	25 or 26	17E3	142.3
South Wall	4587	22 or 23	2.5E3		11.5	25 or 26	12E3	55.0
East Wall	9464	22 or 23	2.5E3		23.7	25 or 26	15E3	142.0
West Wall	10609	22 or 23	2.5E3		26.5	25 or 26	15E3	159.1
Wall Conduction	33028	24	3.0E3		99.1	--	--	--
Roof Solar	10300	27	5.5E3		56.7	--	17E3	175.1
Roof Conduction	10300	24	1.5E3		15.5	28	--	--
Outside Air*	11380	30	75.3E6		856.9	31	150E6	1707.0
Humidification*	11380	--	--			32	75E6	853.5
*Factor is Btu/1000Cfm		Sub-Total Cooling			2439.9	Total Heating		4220.5

Item	Kwh/Yr	Conversion Factor	%Cooling Season/Yr	
Lighting - Exterior	444288	3413 x 1.0	50	758.2
- Interior	436488	3413 x 1.0	100	1489.7
Fans - Exterior	577450	3413 x 1.0	50	985.4
-Interior	241100	3413 x 1.0	100	822.9

People	No	Hrs/Yr	Btu/Person	%Cooling Season/Yr	
Office - Exterior	241	2080	450	50	112.8
- Interior	181	2080	450	100	169.4
				TOTAL COOLING	6778.3

Summary	
Heating	
Conv. Factor	1 x E6
Efficiency	75%
Mcf/Yr	5625
Cooling	
Conv. Factor	3413
COP	4.0
Kwh/Yr	496506

FIGURE 10-12 EXAMPLE OF COMPLETED ANNUAL HEATING AND COOLING TABULATION FORM FOR SYSTEMS EFFECTING THE ENTIRE BUILDING

ANNUAL HEATING AND COOLING TABULATION FORM

JOB: HB&S Building AREA Building PECO NO: 0 DATE: 9-9-78

ITEM	UNITS	COOLING				HEATING		
	SqFt-Cfm	Fig No 8-	Factor	Shade Factor	Total Btu/Yr-E6	Fig No 8-	Factor	Total Btu/Yr-E6
North Glass		16 or 17				19 or 20		
South Glass		16 or 17				19 or 20		
East Glass		16 or 17				19 or 20		
West Glass		16 or 17				19 or 20		
Glass Conduction		18				--		
North Wall		22 or 23				25 or 26		
South Wall		22 or 23				25 or 26		
East Wall		22 or 23				25 or 26		
West Wall		22 or 23				25 or 26		
Wall Conduction		24				--		
Roof Solar		27				--		
Roof Conduction		24				28		
Outside Air*		30				31		
Humidification*		--				32		
*Factor is Btu/1000Cfm		Sub-Total Cooling				Total Heating		

Item	Kwh/Yr	Conversion Factor	%Cooling Season/Yr	
Lighting				
Fans				
Chill Water Pumps	157754	3413 X .85	100	457.7

People	No	Hrs/Yr	Btu/Person	%Cooling Season/Yr	
				TOTAL COOLING	458E6

Summary	
Heating	
Conv. Factor	
Efficiency	75%
Mcf/Yr	
Cooling	
Conv. Factor	3413
COP	4.0
Kwh/Yr	33,548

F. Total Energy Tabulation

General instructions for using the Energy Tabulation Form are presented in Chapters VI and VIII. As here applied to the model office building, the form is used to record the total annual electrical demand, total annual electrical

FIGURE 10-13 EXAMPLE OF COMPLETED ENERGY TABULATION FORM FOR ELECTRICITY

ENERGY TABULATION FORM

JOB: HB&S Building FUEL Electrical UNITS Kwh/Yr PECO NO 0 DATE: 9-9-78

ITEM \ AREA	Drug Store	Office	Bldg. & Park.						TOTALS
Lighting	84304	880776	76440						1041520
A-H Units	52416	818550							870966
Cooling	30933	496506	33548						560987
CHW Pumps			157754						157754
Condenser Water Pumps			230630						230630
Cooling Tower			96096						96096
Elevators			73000						73000
Exhaust Fans		6730							6730
Boiler Fans			19656						19656
Domestic Water Pump			15725						15725
SUB-TOTAL	167653	2202562	702849						3073064
MISC.	8380	108670	32950						150000
REHEAT	0	938078	0						938078
TOTAL	176033	3249310	735799						4161142

Utility Bill Energy Use − (Base Year − Avg. Year) / (Avg. Year Degree Days) × Yearly Building Heating Use = Average Yearly Energy Use   4161142

consumption, and total annual gas consumption of the building under analysis. These totals have been drawn from data recorded on all the other survey and tabulation forms. (The reader will note that space has been provided for tabulating "reheat" energy consumption, as discussed in Chapter VI.) The totals on this form must correspond with the requisite data on the Utility Cost/Consumption Forms. Otherwise, an error has been made in the tabulation of the building's energy use, and the consultant must review his calculation.

FIGURE 10-14 EXAMPLE OF COMPLETED DEMAND ENERGY TABULATION FORM FOR ELECTRICITY

ENERGY TABULATION FORM

JOB: HB&S BUILDING FUEL Demand UNITS Kw/Yr PECO NO 0 DATE: 9-9-78

ITEM \ AREA	Drug Store	Office	Bldg. & Park.					TOTALS
Lighting	18.0	282.3	15.5					315.8
A-H Units	6.0	93.7	-					99.7
Cooling			264.0					264.0
CHW Pumps			17.6					17.6
Condenser Water			26.4					26.4
Cooling Tower			17.6					17.6
Elevators			140.3					140.3
Exhaust Fans		0.8						0.8
Boiler Fans			3.2					3.2
Domestic Water Pump			1.8					1.8
SUB-TOTAL	24.0	376.8	486.4					887.2
MISC.								
REHEAT								
TOTAL								

Utility Bill Energy Use - (Base Year - Avg. Year) / (Avg. Year Degree Days) x Yearly Building Heating Use = Average Yearly Energy Use — 887.2

FIGURE 10-15 EXAMPLE OF COMPLETED ENERGY TABULATION FORM FOR GAS

ENERGY TABULATION FORM

JOB: HB&S Building FUEL Gas UNITS mcf/Yr PECO NO 0 DATE: 9-9-78

ITEM \ AREA	Drug Store	Office						TOTALS
Domestic Water	13.5	227.0						240.5
Heating	348.0	5627.0						5975.0
SUB-TOTAL	361.5	5854.0						
MISC.	10.0	150.0						160.0
REHEAT	-0-	12669.0						12669.0
TOTAL	371.5	18673.0						19044.5

19,696 - (5434 - 4900) / 4900 x 5975 = 19044.5

Utility Bill Energy Use - (Base Year - Avg. Year) / (Avg. Year Degree Days) x Yearly Building Heating Use = Average Yearly Energy Use — 19044.5

G. Project Energy Conservation Opportunity Analysis

The next step in the process is to determine which potential ECO's or Project ECO's (PECO's) are appropriate to carry on into the implementation phase. The books would have been balanced between the survey and the actual energy use. A number of potential ECO's would have been identified. They must now be analyzed.

As mentioned in the Introduction to this chapter, the seven PECO's here applied to the model office building are representative of as many as 30 PECO's which might be applied in an actual building study. Each of the seven PECO's has been selected to represent and illustrate each of the seven ECO-groups discussed in Chapter IX. Furthermore, in keeping with procedures discussed in earlier chapters, the following PECO's have been analyzed in the order in which would ordinarily be implemented:

PECO #1 (Text ECO #1):	Reduce Operating Time
PECO #2 (Text ECO #8):	Reduce Illumination Levels
PECO #3 (Text ECO #2 & 20)	Reduce Outside Air & Infiltration
PECO #4 (Text ECO #21)	Reduce Air Flow
PECO #5 (Text ECO #31)	Reduce Reheat
PECO #6 (Text ECO #44)	Insulate Storage Tank
PECO #7 (Text ECO #50)	Replace Oversize Motors

In addition, the PECO Tabulation Forms accompanying each PECO reflect the various energy reductions of PECO's coming before (see "Accounting for the Cumulative Principle" in Chapter VII). The PECO Tabulation Forms themselves --comparing the building's present energy use with its future energy use (assuming implementation of the PECO) are the basis for the cost-benefit information (including "payback") supplied at the conclusion of each PECO.

1. The analysis of PECO#1 for the HB&S model building is shown below. The "Theory" is repeated for ease of understanding.

PECO #1 (Text ECO#1): Reduce Operating Time

THEORY

Reduce the operating time of all possible equipment during unoccupied hours. This saves energy in all the following ways:

(A) Reduces operating hours
(B) Reduces cooling loads where fan, pump and lighting energy is absorbed by the cooling system.
(C) Reduces outside air load by reducing the number of hours the outside air is conditioned.
(D) Reduces reheat during "OFF" hours.

PRODEDURE

(A) Reduce operating time of the fan in the drug store to 22 hours and the office area fans to 10 hours above the occupied hours.

(B) Reduce the "ON" time of office area lights to 10 hours during custodial hours. Reschedule cleaning as required/

(C) Cut off domestic water pump and exhaust fans except for occupied hours.

(D) Cut off chilled and concenser water pumps and cooling tower fans when drug store is secured.

All of these items can be accomplished manually if operating personnel are available. It is recommended, however, that time clocks be provided so that the schedules operate automatically.

ANALYSIS

Substitute the new operating hours on the present energy use"PECO No. 0", forms, after a copy has first been made for the project records. In this example note that the original 168 hours per week to 99 hours for the Drug Store "A-H Unit No. 1" (Distribution Loads Building Load Survey Form). The resulting kwh/year is thus reduced to 30,888 kwh/year. (See Figure 10-16 below.)

FIGURE 10-16 DRUG STORE BUILDING LOAD SURVEY FORM MODIFIED FOR PECO #1

BUILDING LOAD SURVEY FORM

JOB: HB&S Building — AREA: Drug Store — WINTER: 75°-30% — SUMMER: 75°-50% — PECO NO: (1) — DATE: 9-9-78

GLASS

Facing	SqFt	Shading Ext	Shading Interior	Type U= 1.13
S	200	0	Blinds	Single
E	880	0	S.F. 64	Fixed
				Tight Fit
TOTAL	1080			

WALLS

Facing	SqFt	Shade	Type U= 0.2
S	325	0	Light
E	1145	0	4" Brick
			4" Block
			Plas. Finish
TOTAL	1470		

ROOF OR FLOORS

SqFt	Type U=

LIGHTING

Area	Type	Occupied Kw	Occupied Fc	Occupied Hrs/Wk	Occupied Kwh/Yr	Custodial Kw	Custodial Fc	Custodial Hrs/Wk	Custodial Kwh/Yr
Main Store	96" Slimline	17.2	175	77	68869	17.2	175	12	10733
Store Room	96" Slimline	0.8	75	77	3203	0.8	75	12	499
	TOTAL	18.0		TOTAL	73072	18.0		TOTAL	11232

PEOPLE

Area	No	Hrs/Wk	Infiltration Item	Crack Usage	Figure	Factor	Cfm
Main Store	25	77	4 x 8 Door	150	8 - 15	250	250

DOMESTIC WATER

Use	Temp Hot	Temp Cold	Gal/Meals or Person	People/ Meals	Gallons Per Day	Btu/Yr E6	TOTAL %Eff. & Conversion	MCF /Yr
Hand Wash	110	60	3	25	75	8.1	60%	13.5

DISTRIBUTION LOADS

Item	Cfm	Gpm	Oa	Hp	Voltage	Amperage Rated	Amperage Opr	Kw	TOTAL Hrs/Wk	Kwh Per Year
A-H Unit No. 1	7500	60	750	7-1/2	460/3	11	10	6	99	30888

This is carried over to the Drug Store Annual Heating and Cooling Tabulation Form as shown in Figure 10-17. The numbers for the outside air and humidification are also modified as indicated. These changes in turn modify the totals.

FIGURE 10-17 DRUG STORE ANNUAL HEATING AND COOLING TABULATION FORM MODIFIED FOR PECO #1

ANNUAL HEATING AND COOLING TABULATION FORM

JOB: HB&S Building AREA Drug Store PECO NO: (1) DATE: 9-9-78

ITEM	UNITS SqFt-Cfm	COOLING Fig No 8-	COOLING Factor	COOLING Shade Factor	COOLING Total Btu/Yr-E6	HEATING Fig No 8-	HEATING Factor	HEATING Total Btu/Yr-E6
North Glass	--	16 or 17	--		--	19 or 20	--	--
South Glass	200	16 or 17	135E3	.64	17.3	19 or 20	60E3	12
East Glass	880	16 or 17	135E3	.64	76.0	19 or 20	67E3	59
West Glass	--	16 or 17	--	--	--	19 or 20	--	--
Glass Conduction	1080	18	8E3		8.6	--	--	--
North Wall	--	22 or 23	--		--	25 or 26	--	--
South Wall	325	22 or 23	2.5E3		0.8	25 or 26	12E3	4
East Wall	1145	22 or 23	2.5E3		2.9	25 or 26	15E3	17
West Wall	--	22 or 23	--		--	25 or 26	--	--
Wall Conduction	1470	24	3.0E3		4.4	--	--	--
Roof Solar	--	27	--		--	--	--	--
Roof Conduction	--	24	--		--	28	--	--
Outside Air*	750	30	53.1E6		39.8	31	87E6	65.3
Humidification*	750	--	--		--	32	42E6	31.5
*Factor is Btu/1000Cfm		Sub-Total Cooling			149.8	Total Heating		188.8

Item	Kwh/Yr	Conversion Factor	%Cooling Season/Yr	
Lighting	84304	3413 x 1.0	50	143.9
Fans	30888	3413 x 1.0	50	52.7

People	No	Hrs/Yr	Btu/ Person	%Cooling Season/Yr	
	25	4004	450	50	22.5
				TOTAL COOLING	368.9

Summary	
Heating	
Conv. Factor	1E6
Efficiency	75%
Mcf/Yr	251.7
Cooling	
Conv. Factor	3413
COP	4.0
Kwh/Yr	27021

The forms for the office area are similarly modified as shown in figures number 10-18 and 10-19. Note the change in custodial lighting hours from 20 to 10 hours.

FIGURE 10-18 OFFICE BUILDING LOAD SURVEY FORM MODIFIED FOR PECO #1

BUILDING LOAD SURVEY FORM

JOB: HB&S Building AREA: Office WINTER: 75 - 30% SUMMER: 75 - 50% PECO NO: (1) DATE: 9-9-78

GLASS

Facing	SqFt	Shading Ext	Shading Interior	Type U= 1.13
N	1152	0	Blinds	Single
S	3608	0	S.F.-.64	Metal
E	4576	0		Casem.
W	5456	0		Avg. Fit
TOTAL	14792			

WALLS

Facing	SqFt	Shade	Type U= 0.20
N	8368	0	Light
S	4587	0	4" Brick
E	9464	0	4" Block
W	10609	0	Plas. Finish
TOTAL	33028		

ROOF OR FLOORS

SqFt	Type U= 0.20
10300	6" Concrete
	Built-Up
	Roofing
	2x2 Lay-in Ceiling
10300	

LIGHTING

Area	Type	OCCUPIED Kw	OCCUPIED Fc	OCCUPIED Hrs/Wk	OCCUPIED Kwh/Yr	CUSTODIAL Kw	CUSTODIAL Fc	CUSTODIAL Hrs/Wk	CUSTODIAL Kwh/Yr
Office - Exterior	2x4-200W Flourescent	142.4	150	40	296192	142.4	150	10	74048
Office - Interior	2x4-200W Flourescent	110.5	150	40	229840	110.5	150	10	57460
Halls	150W. Incandescent Spots	21.6	40	40	44928	21.6	40	10	11232
Restrooms	2x4-200W Flourescent	7.8	75	40	16224	7.8	75	10	4060
	TOTAL	282.3		TOTAL	587184	282.3		TOTAL	146796

PEOPLE

Area	No	Hrs/Wk
Office - Exterior	241	40
Office - Interior	181	40

Infiltration Item	Crack Usage	Figure	Factor	Cfm
Doors	250 Pas/Hr	8 - 15	400	400
Windows	14144 LF	8 - 14	.55	7780
			TOTAL	8180

DOMESTIC WATER

Use	Temp Hot	Temp Cold	Gal/Meals or Person	People/ Meals	Gallons Per Day	Btu/Yr E6	%Eff. & Conversion	MCF /Yr
Handwash	110	60	3	422	1266	136.4	60 x 1	227

DISTRIBUTION LOADS

Item	Cfm	Gpm	Qa	Hp	Voltage	Amperage Rated	Amperage Opr	Kw	TOTAL Hrs/Wk	Kwh Per Year
9-Exterior A-H Units	83500	456	8350	87.5	460/3	123	106	66.1	50	171860
8-Interior A-H Units	36000	216	3600	40.0	460/3	61	48	27.6	50	71760
2-Toilet Exhaust Fans	3200			1.0	230/1	8.5	7.5	0.8	40	1602

FIGURE 10-19 OFFICE HEATING AND COOLING TABULATION FORM MODIFIED FOR PECO#1

ANNUAL HEATING AND COOLING TABULATION FORM

JOB: HB&S Building AREA: Office PECO NO: (1) DATE: 9-9-78

ITEM	UNITS	COOLING				HEATING		
	SqFt-Cfm	Fig No 8-	Factor	Shade Factor	Total Btu/Yr-E6	Fig No 8-	Factor	Total Btu/Yr-E6
North Glass	1152	16 or 17	55E3	.64	40.6	19 or 20	85E3	97.[illegible]
South Glass	3608	16 or 17	135E3	.64	311.7	19 or 20	60E3	216.5
East Glass	4576	16 or 17	135E3	.64	395.4	19 or 20	67E3	306.[illegible]
West Glass	5456	16 or 17	135E3	.64	471.4	19 or 20	67E3	365.5
Glass Conduction	14792	18	8E3		118.3	--	--	--
North Wall	8368	22 or 23	1.5E3		12.6	25 or 26	17E3	142.[illegible]
South Wall	4587	22 or 23	2.5E3		11.5	25 or 26	12E3	55.[illegible]
East Wall	9464	22 or 23	2.5E3		23.7	25 or 26	15E3	142.[illegible]
West Wall	10609	22 or 23	2.5E3		26.5	25 or 26	15E3	159.[illegible]
Wall Conduction	33028	24	3.0E3		99.1	--	--	--
Roof Solar	10300	27	5.5E3		56.7	--	17E3	175.1
Roof Conduction	10300	24	1.5E3		15.5	28	--	--
Outside Air*	11380	30	32.2E6		366.4	31	47E6	534.[illegible]
Humidification*	11380	--	--			32	23E6	261.[illegible]
*Factor is Btu/1000Cfm		Sub-Total Cooling			1949.4	Total Heating		2456.6

Item	Kwh/Yr	Conversion Factor	%Cooling Season/Yr	
Lighting - Exterior	370240	3413 x 1.0	50	631.8
- Interior	363736	3413 x 1.0	100	1241.4
Fans - Exterior	171860	3413 x 1.0	50	293.3
- Interior	71760	3413 x 1.0	100	244.9

People	No	Hrs/Yr	Btu/Person	%Cooling Season/Yr	
Office - Exterior	241	2080	450	50	112.8
- Interior	181	2080	450	100	169.4
				TOTAL COOLING	4643.0

Summary	
Heating	
Conv. Factor	1 x E6
Efficiency	75%
Mcf/Yr	3275
Cooling	
Conv. Factor	3413
COP	4.0
Kwh/Yr	3[illegible]0096

Note that only those items affected are changed along with their related totals. In this PECO, the Building and Parking form remains unchanged except to reflect the new PECO number, "PECO No. 1". The hours of operation of pumps will also change so that the Distribution Load Survey Form and related tabulation form will change as shown in Figures 10-20 and 10-21.

FIGURE 10-20 BUILDING DISTRIBUTION LOAD SURVEY FORM MODIFIED FOR PECO # 1

DISTRIBUTION LOAD SURVEY FORM

JOB: HB&S Building AREA: Building PECO NO: (1) DATE: 9-9-78

Area Served	Item	Cfm	Gpm	O A	Hp	Voltage	Amperage Rated	Amperage Opr.	Kw	Hrs/Wk	Kwh/Yr
	CHW Pump 1		366		10	460/3	14	14	8.8	99	4648
	CHW Pump 2		366		10	460/3	14	14	8.8	99	4648
	Cond. Pump 1		450		15	460/3	21	21	13.2	99	6795[illegible]
	Cond. Pump 2		450		15	460/3	21	21	13.2	99	6795[illegible]
	Cool Twr. Fan				20	460/3	27	27	17.6	Auto	7207[illegible]
	Boiler Fan 1				2	460/3	2.8	2.8	1.8	Auto	737[illegible]
	Boiler Fan 2				2	460/3	2.8	2.8	1.8	Auto	737[illegible]
	Elevator 1				40	460/3	56	Varies	Varies	40	1600[illegible]
	Elevator 2				40	460/3	56	Varies	Varies	40	1600[illegible]
	Elevator 3				40	460/3	56	Varies	Varies	40	1600[illegible]
	Elevator 4				40	460/3	56	Varies	Varies	168	2500[illegible]
	Dom Wtr Pump				2	460/3	2.8	2.8	1.8	77	720[illegible]
	Chiller - 1		366		150	460/3	210	-	132	-	-
	Chiller - 2		366		150	460/3	210	-	132	-	-

FIGURE 10-21 BUILDING ANNUAL HEATING AND COOLING TABULATION FORM MODIFIED FOR PECO #1

ANNUAL HEATING AND COOLING TABULATION FORM

JOB: HB&S Building AREA Building PECO NO: (1) DATE: 9-9-78

ITEM	UNITS Sqft-Cfm	COOLING Fig No 8-	COOLING Factor	COOLING Shade Factor	COOLING Total Btu/Yr-E6	HEATING Fig No 8-	HEATING Factor	HEATING Total Btu/Yr-E6
North Glass		16 or 17				19 or 20		
South Glass		16 or 17				19 or 20		
East Glass		16 or 17				19 or 20		
West Glass		16 or 17				19 or 20		
Glass Conduction		18				--		
North Wall		22 or 23				25 or 26		
South Wall		22 or 23				25 or 26		
East Wall		22 or 23				25 or 26		
West Wall		22 or 23				25 or 26		
Wall Conduction		24				--		
Roof Solar		27				--		
Roof Conduction		24				28		
Outside Air*		30				31		
Humidification*		--				32		
*Factor is Btu/1000Cfm		Sub-Total Cooling				Total Heating		

Item	Kwh/Yr	Conversion Factor	%Cooling Season/Yr	
Lighting				
Fans				
Chill Water Pumps	92962	3413 X .85	100	269.7

People	No	Hrs/Yr	Btu/Person	%Cooling Season/Yr	
				TOTAL COOLING	269.7

Summary	
Heating	
Conv. Factor	
Efficiency	75%
Mcf/Yr	
Cooling	
Conv. Factor	3413
COP	4.0
Kwh/Yr	19755

The Energy Tabulation Forms can now be modified using the revised figures calculated above. The reduction of reheat resulting from this PECO is reflected. This is judgemental on the part of the consultant and in this case a reduction of 65% was used for both electricity and gas. These revisions are shown below in figures 10-22 and 10-23. As there is no change in the electrical demand in implementing this PECO, just change the PECO number on that form from zero to one.

FIGURE 10-22 ELECTRICAL ENERGY TABULATION FORM MODIFIED FOR PECO # 1

ENERGY TABULATION FORM

JOB: HB&S Building FUEL Electrical UNITS Kwh/Yr PECO NO (1) DATE: 9-9-78

ITEM / AREA	Drug Store	Office	Bldg. & Park.					TOTALS
Lighting	84304	733980	76440					894724
A-H Units	30888	243620						274508
Cooling	27021	340096	19755					386872
CHW Pumps			92962					92962
Condenser Water Pumps			135906					135906
Cooling Tower			72072					72072
Elevators			73000					73000
Exhaust Fans		1602						1602
Boiler Fans			14742					14742
Domestic Water Pump			7207					7207
SUB-TOTAL	142213	1319298	492084					1953595
MISC.	8380	108670	32950					150000
REHEAT	0	328327	0					328327
TOTAL	150593	1756295	525034					2431922

Utility Bill Energy Use - (Base Year - Avg. Year) / (Avg. Year Degree Days) x Yearly Building Heating Use = Average Yearly Energy Use

FIGURE 10-23 GAS ENERGY TABULATION FORM MODIFIED FOR PECO # 1

ENERGY TABULATION FORM

JOB: HB&S Building   FUEL Gas   UNITS Mcf/Yr   PECO NO 1   DATE: 9-9-78

ITEM \ AREA	Drug Store	Office	TOTALS
Domestic Water	13.5	227.0	240.5
Heating	(251.7)	(3275.0)	(3526.7)
SUB-TOTAL	(265.2)	(3502.0)	(3767.2)
MISC.	10.0	150.0	160.0
REHEAT	-0-	(4434.0)	(4434.0)
TOTAL	(275.2)	(8086.0)	(8361.2)

19,696 $-$ $\frac{5434 - 4900}{4900}$ x 5975 = 19044.5

Utility Bill Energy Use $-$ $\frac{\text{Base Year - Avg. Year}}{\text{Avg. Year Degree Days}}$ x Yearly Building Heating Use = Average Yearly Energy Use   19044.5

A readily comparable set of results is now available from the above two tabulation forms. The lower right hand box on each shows the current energy use for the fuel while the total just above it reflects the revised projected use should the PECO be implemented. These are then entered on the Energy Tabulation Form Figure 10-24.

FIGURE 10-24 ENERGY TABULATION FORM FOR PECO # 1

PECO ENERGY TABULATION FORM

JOB: HB&S Building   PECO NO: 1   DATE 9-9-78

FUEL	ITEM \ AREA	Drug Store	Office	Bldg. & Park	TOTALS
ELECTRICAL	Before ECO Energy Use	176033	3249310	735799	4161142
	After ECO Energy Use	150593	1756295	525034	2431922
	Kwh Savings	25540	1493015	210765	1729220
	Before ECO Demand Use	24.0	376.8	486.4	387.2
	After ECO Demand Use	24.0	376.8	486.4	387.2
	Kw Savings	0	0	0	0
GAS	Before ECO Energy Use	371.5	18673		19044.5
	After ECO Energy Use	275.2	8086		8361.2
	Mcf Savings	96.3	10587		10683.3

Energy costs are then applied to these results and compared to the estimated cost of implementation (in this example $3,500.00) to arrive at the simple payback. They may also be used in a more comprehensive life cycle cost analysis if warrented.

Energy Savings

Electrical	1729220 kwh @ 2.7¢	=$47,789.00
Gas	10683.3 mcf @ 1.75	=$18,696.00
TOTAL SAVINGS		$65,385.00

Estimated Cost $3,500.00
Payback - 2.8 weeks

A copy of the PECO #1 set of forms is then made prior to starting the analysis of PECO #2.

2. The analysis of PECO #2 represents those PECO's that reduce the building load. In this example a reduction of illumination levels in the HB&S Building is shown.

PECO # 2 (Text ECO #8): Reduce Illumination Levels

Theory

Lighting levels in all areas exceed required levels. Reducing lighting saves energy in the following ways:

(A) Less kw used while lighting is "ON".
(B) Less cooling required because of less heat produced.

PROCEDURE

Reduce all lighting by 50 percent by removing lamps and disconnecting ballasts or substituting smaller wattage in the case of incandescent.

Analysis

The new wattage resulting from the reduction for both "Occupied" and "Custodial" are entered on the Building Load Survey Form for the Drug Store and Office Building Figures 10-25 and 10-26. No reduction is available in the Building and Parking area and the Building Distribution is not effected so these forms are only changes to read PECO #2.

FIGURE 10-25 DRUG STORE BUILDING LOAD SURVEY FORM MODIFIED FOR PECO # 2

BUILDING LOAD SURVEY FORM

JOB: HB&S Building AREA: Drug Store WINTER: 75°-30% SUMMER: 75°-50% PECO NO: (2) DATE: 9-9-78

GLASS					WALLS				ROOF OR FLOORS	
Facing	SqFt	Shading Ext	Shading Interior	Type U= 1.13	Facing	SqFt	Shade	Type U= 0.2	SqFt	Type U=
S	200	0	Blinds	Single	S	325	0	Light		
E	880	0	S.F. 64	Fixed	E	1145	0	4" Brick		
				Tight Fit				4" Block		
								Plas. Finish		
TOTAL	1080				TOTAL	1470				

LIGHTING Area	Type	OCCUPIED Kw	Fc	Hrs/Wk	Kwh/Yr	CUSTODIAL Kw	Fc	Hrs/Wk	Kwh/Yr
Main Store	96" Slimline	8.6	175	77	34434	8.6	175	12	5366
Store Room	96" Slimline	0.4	75	77	1601	0.4	75	12	250
	TOTAL	9.0		TOTAL	36535	9.0		TOTAL	5616

PEOPLE Area	No	Hrs/Wk	Infiltration Item	Crack Usage	Figure	Factor	Cfm
Main Store	25	77	4 x 8 Door	150	8 - 15	250	250

DOMESTIC WATER Use	Temp Hot	Temp Cold	Gal/Meals or Person	People/ Meals	Gallons Per Day	Btu/Yr E6	TOTAL %Eff. & Conversion	MCF /Yr
Hand Wash	110	60	3	25	75	8.1	60%	3.5

DISTRIBUTION LOADS Item	Cfm	Gpm	Oa	Hp	Voltage	Amperage Rated	Amperage Opr	Kw	TOTAL Hrs/Wk	Kwh Per Year
A-H Unit No. 1	7500	60	750	7-1/2	460/3	11	10	6	99	3088

FIGURE 10-26 OFFICE BUILDING LOAD SURVEY FORM MODIFIED FOR PECO # 2

BUILDING LOAD SURVEY FORM

JOB: HB&S Building AREA: Office WINTER: 75 - 30% SUMMER: 75 - 50% PECO NO: 2 DATE: 9-9-78

GLASS

Facing	SqFt	Shading Ext	Shading Interior	Type U= 1.13
N	1152	0	Blinds	Single
S	3608	0	S.F.-.64	Metal
E	4576	0		Casem.
W	5456	0		Avg. Fit
TOTAL	14792			

WALLS

Facing	SqFt	Shade	Type U= 0.20
N	8368	0	Light
S	4587	0	4" Brick
E	9464	0	4" Block
W	10609	0	Plas. Finish
TOTAL	33028		

ROOF OR FLOORS

SqFt	Type U= 0.20
10300	6" Concrete
	Built-Up
	Roofing
	2x2 Lay-In Ceiling
10300	

LIGHTING

Area	Type	Occupied Kw	Occupied Fc	Occupied Hrs/Wk	Occupied Kwh/Yr	Custodial Kw	Custodial Fc	Custodial Hrs/Wk	Custodial Kwh/Yr
Office - Exterior	2x4-200W Flourescent	71.2	75	40	148096	71.2	75	10	37024
Office - Interior	2x4-200W Flourescent	52.3	75	40	114920	52.3	75	10	28730
Halls	150W. Incandescent Spots	10.8	20	40	22464	10.8	20	10	5616
Restrooms	2x4-200W Flourescent	3.9	40	40	8112	3.9	40	10	2028
	TOTAL	141.2		TOTAL	293592	282.3		TOTAL	73398

PEOPLE

Area	No	Hrs/Wk
Office - Exterior	241	40
Office - Interior	181	40

Infiltration Item	Crack Usage	Figure	Factor	Cfm
Doors	250 Pas/Hr	8 - 15	400	400
Windows	14144 LF	8 - 14	.55	7780
			TOTAL	8180

DOMESTIC WATER

Use	Temp Hot	Temp Cold	Gal/Meals or Person	People/ Meals	Gallons Per Day	Btu/Yr E6	%Eff. & Conversion	MCF /Yr
Handwash	110	60	3	422	1266	136.4	60 x 1	227

DISTRIBUTION LOADS

Item	Cfm	Gpm	Oa	Hp	Voltage	Amperage Rated	Amperage Opr	Kw	TOTAL Hrs/Wk	Kwh Per Year
9-Exterior A-H Units	83500	456	8350	87.5	460/3	123	106	66.1	50	171860
8-Interior A-H Units	36000	216	3600	40.0	460/3	61	48	27.6	50	71760
2-Toilet Exhaust Fans	3200			1.0	230/1	8.5	7.5	0.8	40	1602

The Annual Heating and Cooling Forms for the Drug Store and Office areas are changed to reflect the reduced lighting and a lower cooling energy as calculated in Figures 10-27 and 10-28. The Building and Parking area form is unaffected except the change to PECO #2.

FIGURE 10-27 DRUG STORE ANNUAL HEATING AND COOLING TABULATION FORM MODIFIED FOR PECO #2

ANNUAL HEATING AND COOLING TABULATION FORM

JOB: HB&S Building AREA Drug Store PECO NO: 2 DATE: 9-9-78

ITEM	UNITS SqFt-Cfm	COOLING Fig No 8-	COOLING Factor	COOLING Shade Factor	COOLING Total Btu/Yr-E6	HEATING Fig No 8-	HEATING Factor	HEATING Total Btu/Yr-E6
North Glass	--	16 or 17	--		--	19 or 20	--	--
South Glass	200	16 or 17	135E3	.64	17.3	19 or 20	60E3	12
East Glass	880	16 or 17	135E3	.64	76.0	19 or 20	67E3	59
West Glass	--	16 or 17	--	--	--	19 or 20	--	--
Glass Conduction	1080	18	8E3		8.6	--	--	--
North Wall	--	22 or 23	--		--	25 or 26	--	--
South Wall	325	22 or 23	2.5E3		0.8	25 or 26	12E3	4
East Wall	1145	22 or 23	2.5E3		2.9	25 or 26	15E3	17
West Wall	--	22 or 23	--		--	25 or 26	--	--
Wall Conduction	1470	24	3.0E3		4.4	--	--	--
Roof Solar	--	27	--		--	--	--	--
Roof Conduction	--	24	--		--	28	--	--
Outside Air*	750	30	53.1E6		39.8	31	87E6	65.3
Humidification*	750	--	--		--	32	42E6	31.5
*Factor is Btu/1000Cfm		Sub-Total Cooling			149.8	Total Heating		188.8

Item	Kwh/Yr	Conversion Factor	%Cooling Season/Yr	
Lighting	42152	3413 x 1.0	50	72.0
Fans	30808	3413 x 1.0	50	52.7

People	No	Hrs/Yr	Btu/ Person	%Cooling Season/Yr	
	25	4004	450	50	22.5
			TOTAL COOLING		297.0

Summary	
Heating	
Conv. Factor	1E6
Efficiency	75%
Mcf/Yr	251.7
Cooling	
Conv. Factor	3413
COP	4.0
Kwh/Yr	21755

FIGURE 10-28 OFFICE HEATING AND COOLING TABULATION FORM MODIFIED FOR PECO #2

ANNUAL HEATING AND COOLING TABULATION FORM

JOB: HB&S Building AREA Office PECO NO. (2) DATE: 9-9-78

ITEM	UNITS	COOLING				HEATING		
	SqFt-Cfm	Fig No 8-	Factor	Shade Factor	Total Btu/Yr-E6	Fig No 8-	Factor	Total Btu/Yr-E6
North Glass	1152	16 or 17	55E3	.64	40.6	19 or 20	85E3	97.9
South Glass	3608	16 or 17	135E3	.64	311.7	19 or 20	60E3	216.5
East Glass	4576	16 or 17	135E3	.64	395.4	19 or 20	67E3	306.6
West Glass	5456	16 or 17	135E3	.64	471.4	19 or 20	67E3	365.5
Glass Conduction	14792	18	8E3		118.3	--	--	--
North Wall	8368	22 or 23	1.5E3		12.6	25 or 26	17E3	142.3
South Wall	4587	22 or 23	2.5E3		11.5	25 or 26	12E3	55.0
East Wall	9464	22 or 23	2.5E3		23.7	25 or 26	15E3	142.0
West Wall	10609	22 or 23	2.5E3		26.5	25 or 26	15E3	159.1
Wall Conduction	33028	24	3.0E3		99.1	--	--	--
Roof Solar	10300	27	5.5E3		56.7	--	17E3	175.1
Roof Conduction	10300	24	1.5E3		15.5	28	--	--
Outside Air*	11380	30	32.2E6		366.4	31	47E6	534.9
Humidification*	11380	--	--			32	23E6	261.7
*Factor is Btu/1000Cfm		Sub-Total Cooling			1949.4	Total Heating		2456.6

Item	Kwh/Yr	Conversion Factor	%Cooling Season/Yr	
Lighting - Exterior	185120	3413 x 1.0	50	315.9
- Interior	181870	3413 x 1.0	100	620.7
Fans - Exterior	171860	3413 x 1.0	50	293.3
- Interior	71760	3413 x 1.0	100	244.9

People	No	Hrs/Yr	Btu/Person	%Cooling Season/Yr	
Office - Exterior	241	2080	450	50	112.8
- Interior	181	2080	450	100	169.4
TOTAL COOLING					3706.4

Summary	
Heating	
Conv. Factor	x E6
Efficiency	75%
Mcf/Yr	3275
Cooling	
Conv. Factor	3413
COP	4.0
Kwh/Yr	27[illegible]491

The resultant energy consumption is ready now to be entered on the Energy Tabulation Forms. Only electrical consumption and electrical demand are affected. These are shown on Figures 10-29 and 10-30.

FIGURE 10-29 ELECTRICAL ENERGY TABULATION FORM MODIFIED FOR PECO # 2

ENERGY TABULATION FORM

JOB: HB&S Building FUEL Electrical UNITS Kwh/Yr PECO NO (2) DATE: 9-9-78

ITEM \ AREA	Drug Store	Office	Bldg. & Park.					TOTALS
Lighting	42152	366990	76440					48559[illegible]
A-H Units	30888	243620						27450[illegible]
Cooling	21755	271491	19755					31300[illegible]
CHW Pumps			[illegible]962					9296[illegible]
Condenser Water Pumps			135906					13590[illegible]
Cooling Tower			72072					7207[illegible]
Elevators			73000					7300[illegible]
Exhaust Fans		1602						160[illegible]
Boiler Fans			14742					1474[illegible]
Domestic Water Pump			7207					720[illegible]
SUB-TOTAL	94795	883703	492084					147058[illegible]
MISC.	8380	108670	32950					15000[illegible]
REHEAT	0	328327	0					32832[illegible]
TOTAL	103175	1320700	525034					194890[illegible]

$$\frac{\text{Utility Bill}}{\text{Energy Use}} - \frac{\text{Base Year - Avg. Year}}{\text{Avg. Year Degree Days}} \times \frac{\text{Yearly Building}}{\text{Heating Use}} = \frac{\text{Average Yearly}}{\text{Energy Use}}$$

FIGURE 10-30 ELECTRICAL ENERGY DEMAND TABULATION FORM MODIFIED FOR PECO #2

ENERGY TABULATION FORM

JOB: HB&S BUILDING FUEL Demand UNITS Kw/Yr PECO NO (2) DATE: 9-9-78

ITEM \ AREA	Drug Store	Office	Bldg. & Park.					TOTALS
Lighting	(9.0)	(141.2)	15.5					(165.7)
A-H Units	6.0	93.7						99.7
Cooling			264.0					264.0
CHW Pumps			17.6					17.6
Condenser Water			26.4					26.4
Cooling Tower			17.6					17.6
Elevators			140.3					140.3
Exhaust Fans		0.8						0.8
Boiler Fans			3.2					3.2
Domestic Water Pump			1.8					1.8
SUB-TOTAL	(15.0)	(235.7)	486.4					(737.1)
MISC.								
REHEAT								
TOTAL								

Utility Bill Energy Use − $\frac{\text{Base Year - Avg. Year}}{\text{Avg. Year Degree Days}}$ x Yearly Building Heating Use = Average Yearly Energy Use

The PECO #2 Energy Tabulation form is now used to calculate the energy savings as shown in Figure 10-31. The following information is provided:

Energy Savings

Electrical 4,830,013 @ 2.7¢ = $13,041.00   Estimated Cost $2,500.00

Demand 150.1 x 5.00 x 1 = 9,006.00

TOTAL SAVINGS $22,047.00   Payback - 5.9 weeks

FIGURE 10-31 ENERGY TABULATION FORM FOR PECO #2

PECO ENERGY TABULATION FORM

JOB: HB&S Building PECO NO: 2 DATE: 9-9-78

FUEL	ITEM \ AREA	Drug Store	Office	Bldg. & Park					TOTALS
ELECTRICAL	Before ECO Energy Use	150593	1756295	525034					2431922
	After ECO Energy Use	103175	1320700	525034					1948909
	Kwh Savings	47418	435595	0					483013
	Before ECO Demand Use	24.0	376.8	486.4					887.2
	After ECO Demand Use	15.0	235.7	486.4					737.1
	Kw Savings	9.0	141.1	0					150.1
GAS	Before ECO Energy Use	275.2	8086						8361.2
	After ECO Energy Use	275.2	8086						8361.2
	Mcf Savings	0	0						0

Copies of PECO # 2 Forms are made for use in analyzing PECO #3.

3. The analysis of PECO # 3 deals with the reduction of the Building Envelope load namely infiltration. In this case an operational ECO, reduction of outside air, cannot be accomplished without reducing infiltration.

PECO # 3 (Text ECO's #2 and #20): Reduce Outside Air and Infiltration

THEPRY

The Drug Store ventilation requirement at 5 cfm per person is 125 cfm and the office is 2110 cfm. Infiltration into the Drug Store is 250 cfm and into the office building 8180 cfm plus 3200 cfm exhaust. Outside air is 750 cfm for the drug store and 11950 cfm for the office area. In order to reduce ventilation, infiltration must be reduced.

PROCEDURE

Reducing ventilation air saves energy by reducing heating and cooling energy required to condition the outside air.

The cost of reducing the door infiltration to 125 cfm is not practical, so reduce drug store ventilation to 250 cfm. The 3200 cfm exhaust is required plus 400 cfm of infiltration through the door. Thus, reduce ventilation in the office areas to 3600 cfm by weatherstripping and sealing the windows.

ANALYSIS

The "Factor" for window infiltration is changed on the Office Building Load Survey Form to reflect the sealed windows and the resulting infiltration calculated. The outside air of both the Drug Store and Office area is reduced to the amount required to effect the remaining infiltration plus the exhaust requirements. (Distribution Loads-Building Load Survey Forms Figure 10-32 and 10-33)

FIGURE 10-32 DRUG STORE BUILDING LOAD SURVEY FORM MODIFIED FOR PECO #3

BUILDING LOAD SURVEY FORM

JOB: HB&S Building AREA: Drug Store WINTER: 75°-30% SUMMER: 75°-50% PECO NO: 3 DATE: 9-9-78

GLASS

Facing	SqFt	Shading Ext	Shading Interior	Type U= 1.13
S	200	0	Blinds S.F. 64	Single Fixed Tight Fit
E	880	0		
TOTAL	1080			

WALLS

Facing	SqFt	Shade	Type U=C.2
S	325	0	Light 4" Brick 4" Block Plas. Finish
E	1145	0	
TOTAL	1470		

ROOF OR FLOORS

SqFt	Type U=

LIGHTING

Area	Type	Occupied Kw	Occupied Fc	Occupied Hrs/Wk	Occupied Kwh/Yr	Custodial Kw	Custodial Fc	Custodial Hrs/Wk	Custodial Kwh/Yr
Main Store	96" Slimline	8.6	175	77	34434	8.6	175	2	5366
Store Room	96" Slimline	0.4	75	77	1601	0.4	75	2	250
TOTAL		9.0		TOTAL	36535	9.0		TOTAL	5616

PEOPLE

Area	No	Hrs/Wk
Main Store	25	77

Infiltration Item	Crack Usage	Figure	Factor	Cfm
4 x 8 Door	150	8 - 15	250	250

DOMESTIC WATER

Use	Temp Hot	Temp Cold	Gal/Meals or Person	People/ Meals	Gallons Per Day	Btu/Yr E6	TOTAL %Eff. & Conversion	MCF /Yr
Hand Wash	110	60	3	25	75	8.1	60%	13.5

DISTRIBUTION LOADS

Item	Cfm	Gpm	Oa	Hp	Voltage	Amperage Rated	Amperage Opr	Kw	TOTAL Hrs/Wk	Kwh Per Year
A-H Unit No. 1	7500	60	250	7-1/2	460/3	11	10	6	99	30888

FIGURE 10-33 OFFICE BUILDING LOAD SURVEY FORM MODIFIED FOR PECO # 3

BUILDING LOAD SURVEY FORM

JOB: HB&S Building AREA: Office WINTER: 75 - 30% SUMMER: 75 - 50% PECO NO: (3) DATE: 9-9-78

GLASS Facing	SqFt	Shading Ext	Shading Interior	Type U= 1.13	WALLS Facing	SqFt	Shade	Type U= 0.20	ROOF OR FLOORS SqFt	Type U= 0.20
N	1152	0	Blinds	Single	N	8368	0	Light	10300	6" Concrete
S	3608	0	S.F.-.64	Metal	S	4587	0	4" Brick		Built-Up
E	4576	0		Casem.	E	9464	0	4" Block		Roofing
W	5456	0		Avg. Fit	W	10609	0	Plas. Finish		2x2 Lay-in Ceiling
TOTAL	14792				TOTAL	33028			10300	

LIGHTING Area	Type	Occupied Kw	Occupied Fc	Occupied Hrs/Wk	Occupied Kwh/Yr	Custodial Kw	Custodial Fc	Custodial Hrs/Wk	Custodial Kwh/Yr
Office - Exterior	2x4-200W Fluorescent	71.2	75	40	148096	71.2	75	10	37024
Office - Interior	2x4-200W Fluorescent	52.3	75	40	114920	52.3	75	10	28730
Halls	150W. Incandescent Spots	10.8	20	40	22464	10.8	20	10	5616
Restrooms	2x4-200W Fluorescent	3.9	40	40	8112	3.9	40	10	2028
	TOTAL	141.2		TOTAL	293592	282.3		TOTAL	73398

PEOPLE Area	No	Hrs/Wk	Infiltration Item	Crack Usage	Figure	Factor	Cfm
Office - Exterior	241	40	Doors	250 Pas/Hr	8 - 15	400	400
Office - Interior	181	40	Windows	14144 LF	8 - 14	(0)	(0)
						TOTAL	(400)

DOMESTIC WATER Use	Temp Hot	Temp Cold	Gal/Meals or Person	People/ Meals	Gallons Per Day	Btu/Yr E6	%Eff. & Conversion	MCF /Yr
Handwash	110	60	3	422	1266	136.4	60 x 1	227

DISTRIBUTION LOADS Item	Cfm	Gpm	Oa	Hp	Voltage	Amperage Rated	Amperage Opr	Kw	TOTAL Hrs/Wk	Kwh Per Year
9-Exterior A-H Units	83500	456	(1900)	87.5	460/3	123	106	66.1	50	171860
8-Interior A-H Units	36000	216	(1700)	40.0	460/3	61	48	27.6	50	71760
2-Toilet Exhaust Fans	3200			1.0	230/1	8.5	7.5	0.8	40	1602

The new outside air requirements are entered on the Annual Heating and Cooling Tabulation Form and resulting heating and cooling energy calculated, as shown in Figures 10-34 and 10-35. The Building and Parking areas as well as Building distribution are not effected so the forms are only changed to reflect PECO #3.

FIGURE 10-34 DRUG STORE ANNUAL HEATING AND COOLING TABULATION FORM MODIFIED FOR PECO #3

ANNUAL HEATING AND COOLING TABULATION FORM

JOB: HC&S Building AREA Drug Store PECO NO: (3) DATE: 9-9-78

ITEM	UNITS SqFt-Cfm	COOLING Fig No 8-	COOLING Factor	COOLING Shade Factor	COOLING Total Btu/Yr-E6	HEATING Fig No 8-	HEATING Factor	HEATING Total Btu/Yr-E6
North Glass	--	16 or 17	--		--	19 or 20	--	--
South Glass	200	16 or 17	135E3	.64	17.3	19 or 20	60E3	12
East Glass	880	16 or 17	135E3	.64	76.0	19 or 20	67E3	59
West Glass	--	16 or 17	--	--	--	19 or 20	--	--
Glass Conduction	1080	18	8E3		8.6	--	--	--
North Wall	--	22 or 23	--		--	25 or 26	--	--
South Wall	325	22 or 23	2.5E3		0.8	25 or 26	12E3	4
East Wall	1145	22 or 23	2.5E3		2.9	25 or 26	15E3	17
West Wall	--	22 or 23	--		--	25 or 26	--	--
Wall Conduction	1470	24	3.0E3		4.4	--	--	--
Roof Solar	--	27	--		--	--	--	--
Roof Conduction	--	24	--			28	--	
Outside Air*	(250)	30	53.1E6		(13.3)	31	87E6	(21.8)
Humidification*	(250)	--	--		--	32	42E6	(10.5)
*Factor is Btu/1000Cfm		Sub-Total Cooling			(123.3)	Total Heating		(124.3)

Item	Kwh/Yr	Conversion Factor	%Cooling Season/Yr	
Lighting	42152	3413 x 1.0	50	72.0
Fans	30888	3413 x 1.0	50	52.7

People	No	Hrs/Yr	Btu/ Person	%Cooling Season/Yr	
	25	4004	450	50	22.5
TOTAL COOLING					(270.5)

Summary	
Heating	
Conv. Factor	1E6
Efficiency	75%
Mcf/Yr	(165.7)
Cooling	
Conv. Factor	3413
COP	4.0
Kwh/Yr	(19814)

FIGURE 10-35 OFFICE HEATING AND COOLING TABULATION FORM MODIFIED FOR PECO #3

ANNUAL HEATING AND COOLING TABULATION FORM

JOB: HB&S Building AREA Office PECO NO: (3) DATE: 9-9-78

ITEM	UNITS	COOLING				HEATING		
	SqFt-Cfm	Fig No 8-	Factor	Shade Factor	Total Btu/Yr-E6	Fig No 8-	Factor	Total Btu/Yr-E6
North Glass	1152	16 or 17	55E3	.64	40.6	19 or 20	85E3	97.9
South Glass	3608	16 or 17	135E3	.64	311.7	19 or 20	60E3	216.5
East Glass	4576	16 or 17	135E3	.64	395.4	19 or 20	67E3	306.6
West Glass	5456	16 or 17	135E3	.64	471.4	19 or 20	67E3	365.5
Glass Conduction	14792	18	8E3		118.3	--	--	--
North Wall	8368	22 or 23	1.5E3		12.6	25 or 26	17E3	142.3
South Wall	4587	22 or 23	2.5E3		11.5	25 or 26	12E3	55.0
East Wall	9464	22 or 23	2.5E3		23.7	25 or 26	15E3	142.0
West Wall	10609	22 or 23	2.5E3		26.5	25 or 26	15E3	159.1
Wall Conduction	33028	24	3.0E3		99.1	--	--	--
Roof Solar	10300	27	5.5E3		56.7	--	17E3	175.1
Roof Conduction	10300	24	1.5E3		15.5	28	--	
Outside Air*	(3600)	30	32.2E6		(115.9)	31	47E6	(169.2)
Humidification*	(3600)	--	--			32	23E6	(82.8)
*Factor is Btu/1000Cfm		Sub-Total Cooling			(1698.9)	Total Heating		(912.0)

Item	Kwh/Yr	Conversion Factor	%Cooling Season/Yr	
Lighting - Exterior	185120	3413 x 1.0	50	315.9
- Interior	181870	3413 x 1.0	100	620.7
Fans - Exterior	171860	3413 x 1.0	50	293.3
- Interior	71760	3413 x 1.0	100	244.9

People	No	Hrs/Yr	Btu/Person	%Cooling Season/Yr	
Office - Exterior	241	2080	450	50	112.8
- Interior	181	2080	450	100	169.4
TOTAL COOLING					(3455.9)

Summary	
Heating	
Conv. Factor	1 x E6
Efficiency	75%
Mcf/Yr	(2[illegible]49.3)
Cooling	
Conv. Factor	3413
COP	[illegible].0
Kwh/Yr	(253.42)

The electrical and gas consumption are reduced but the electrical demand is unchanged. These changes are entered on Figures 10-36 and 10-37.

FIGURE 10-36 ELECTRICAL ENERGY TABULATION FORM MODIFICATION FOR PECO # 3

ENERGY TABULATION FORM

JOB: HB&S Building FUEL Electrical UNITS Kwh/Yr PECO NO (3) DATE: 9-9-78

ITEM \ AREA	Drug Store	Office	Bldg. & Park.					TOTALS
Lighting	42152	366990	76440					[illegible]85592
A-H Units	30888	243620						[illegible]74508
Cooling	(19814)	(253142)	19755					([illegible]92711)
CHW Pumps			92962					92962
Condenser Water Pumps			135906					35906
Cooling Tower			72072					72072
Elevators			73000					73000
Exhaust Fans		1602						1602
Boiler Fans			14742					14742
Domestic Water Pump			7207					7207
SUB-TOTAL	(92854)	(865354)	492084					(1[illegible]50292)
MISC.	8380	108670	32950					50000
REHEAT	0	328327	0					[illegible]28327
TOTAL	(101234)	(1302351)	525034					(1[illegible]28619)

Utility Bill Energy Use - (Base Year - Avg. Year) / (Avg. Year Degree Days) x Yearly Building Heating Use = Average Yearly Energy Use

FIGURE 10-37 GAS ENERGY TABULATION FORM MODIFIED FOR PECO #3

ENERGY TABULATION FORM

JOB: HB&S Building FUEL Gas UNITS Mcf/Yr PECO NO 3 DATE: 9-9-78

ITEM \ AREA	Drug Store	Office						TOTALS
Domestic Water	13.5	227.0						240.5
Heating	165.7	2549.3						2715.0
SUB-TOTAL	179.2	2776.3						2955.5
MISC.	10.0	150.0						160.0
REHEAT	-0-	4434.0						4434.0
TOTAL	189.2	7360.3						7549.5

19.696 (Utility Bill Energy Use) − (5434 - 4900) / 4900 (Base Year - Avg. Year / Avg. Year Degree Days) x 5975 (Yearly Building Heating Use) = 19044.5 (Average Yearly Energy Use) 19044.5

PECO #3 Energy Tabulation Form Figure 10-38 is now filled out and provides the following information:

Energy Savings

Electrical - 20290 kwh @ 2.7¢	=	$ 548.00	Estimated Cost	$15,000.00
Gas 811.7 mcf @1.75	=	1421.00	Payback	7.6 years
TOTAL SAVINGS		$ 1969.00		

FIGURE 10-38 ENERGY TABULATION FORM FOR PECO #3

PECO ENERGY TABULATION FORM

JOB: HB&S Building PECO NO: 3 DATE: 9-9-78

	ITEM \ AREA	Drug Store	Office	Bldg. & Park					TOTALS
FUEL									
ELECTRICAL	Before ECO Energy Use	103175	1320700	525034					1948909
	After ECO Energy Use	101234	1302351	525034					1928619
	Kwh Savings	1941	18349	0					20290
	Before ECO Demand Use	15.0	235.7	486.4					737.1
	After ECO Demand Use	15.0	235.7	486.4					737.1
	Kw Savings	0	0	0					0
GAS	Before ECO Energy Use	275.2	8086						8361.2
	After ECO Energy Use	189.2	7360.3						7549.5
	Mcf Savings	86.0	725.7						811.7

Copies of PECO #3 Forms are made for use in analyzing PECO #4.

4, The analysis of PECO #4 deals with energy reduction in distribution systems. In this case, reduction of air in the heating and air conditioning system.

PECO # 4 (Text ECO #21): Reduce Energy Consumption for Fans by Reducing Air Flow Rates.

THEORY

A reduction in the instantaneous cooling load makes possible a reduction in air flow. Air flow is determined by this formula:

$$cfm = \frac{Btuh}{T\ D \quad x\ 1.08} \quad \text{(ASHRAE Guide)}$$

The power requirements vary with the cube of the cfm, provided cfm reduction is achieved by speed reduction.

Air flow reduction is made possible on the project by the following PECO's:

PECO #2 - Reduce Lighting Levels
PECO #3 - Reduce Outside Air and Infiltration (Infiltration is a room load while outside air is a coil load. In this project, infiltration is offset by outside air, and no air flow reduction can be made.)

PRODEDURE

Determine the potential flow reduction by the above formula. Air flow varies directly with the speed, so reduce the rpm of the fan by the same percentage as the desired flow reduction. This speed reduction is accomplished by adjusting the drive pulleys if possible or by changing to new pulleys and drives if necessary.

ANALYSIS

Possible kw reduction in lighting (PECO #2);

(A) Drug Store 9kw
(B) Office 141.1 kw

Drug Store

$$cfm = \frac{9\ kw\ x\ 3414}{20\ x\ 1.08} = 1422\ cfm$$

Original cfm = 7500
1422 cfm reduction = 19%
19% cfm reduction = 47% fan power reduction
Fan motor efficiency will decrease, so assume a 40% reduction in fan kw.

Office

$$cfm = \frac{141.1 \text{ kw} \times 3413}{20 \times 108} = 22{,}295 \text{ cfm}$$

Original cfm = 119.500
22,295 cfm reduction = 18.6%
18.6% cfm reduction = 46% fan power reduction
Fan motor efficiency will decrease, so assume a 40% reduction in fan kw.

Enter the new cfm and kw on the Building Load Survey Form of both the Drug Store and Office areas under "Distribution Loads" Figures 10-39 and 10-40. Calculate the reduced energy consumption.

FIGURE 10-39 DRUG STORE BUILDING LOAD SURVEY FORM MODIFIED FOR PECO #4

BUILDING LOAD SURVEY FORM

JOB: HB&S Building AREA: Drug Store WINTER: 75°-30% SUMMER: 75°-50% PECO NO: (4) DATE: 9-9-78

GLASS

Facing	SqFt	Shading Ext	Shading Interior	Type U= 1.13
S	200	0	Blinds S.F. 64	Single Fixed Tight Fit
E	880	0		
TOTAL	1080			

WALLS

Facing	SqFt	Shade	Type U=0.2
S	325	0	Light 4" Brick 4" Block Plas. Finish
E	1145	0	
TOTAL	1470		

ROOF OR FLOORS

SqFt	Type U=

LIGHTING

Area	Type	Occupied Kw	Occupied Fc	Occupied Hrs/Wk	Occupied Kwh/Yr	Custodial Kw	Custodial Fc	Custodial Hrs/Wk	Custodial Kwh/Yr
Main Store	96" Slimline	8.6	175	77	34434	8.6	175	12	5366
Store Room	96" Slimline	0.4	75	77	1601	0.4	75	12	250
	TOTAL	9.0		TOTAL	36535	9.0		TOTAL	5616

PEOPLE

Area	No	Hrs/Wk
Main Store	25	77

Infiltration Item	Crack Usage	Figure	Factor	Cfm
4 x 8 Door	150	8 - 15	250	250

DOMESTIC WATER

Use	Temp Hot	Temp Cold	Gal/Meals or Person	People/ Meals	Gallons Per Day	Btu/Yr E6	TOTAL %Eff. & Conversion	MCF /Yr
Hand Wash	110	60	3	25	75	8.1	60%	13.5

DISTRIBUTION LOADS

Item	Cfm	Gpm	Oa	Hp	Voltage	Amperage Rated	Amperage Opr	Kw	TOTAL Hrs/Wk	Kwh Per Year
A-H Unit No. 1	6078	60	250	7-1/2	460/3	11	10	3.6	99	18533

FIGURE 10-40 OFFICE BUILDING LOAD SURVEY FORM MODIFIED FOR PECO #4

BUILDING LOAD SURVEY FORM

JOB: HB&S Building AREA: Office WINTER: 75 - 30% SUMMER: 75 - 50% PECO NO: (4) DATE: 9-9-78

GLASS

Facing	SqFt	Shading Ext	Shading Interior	Type U= 1.13
N	1152	0	Blinds S.F.-.64	Single Metal Casem. Avg. Fit
S	3608	0		
E	4576	0		
W	5456	0		
TOTAL	14792			

WALLS

Facing	SqFt	Shade	Type U= 0.20
N	8368	0	Light 4" Brick 4" Block Plas. Finish
S	4587	0	
E	9464	0	
W	10609	0	
TOTAL	33028		

ROOF OR FLOORS

SqFt	Type U= 0.20
10300	6" Concrete Built-Up Roofing 2x2 Lay-in Ceiling
10300	

LIGHTING

Area	Type	Occupied Kw	Occupied Fc	Occupied Hrs/Wk	Occupied Kwh/Yr	Custodial Kw	Custodial Fc	Custodial Hrs/Wk	Custodial Kwh/Yr
Office - Exterior	2x4-200W Fluorescent	71.2	75	40	140096	71.2	75	10	37024
Office - Interior	2x4-200W Fluorescent	52.3	75	40	114920	52.3	75	10	28730
Halls	150W. Incandescent Spots	10.8	20	40	22464	10.8	20	10	5616
Restrooms	2x4-200W Fluorescent	3.9	40	40	8112	3.9	40	10	2028
	TOTAL	141.2		TOTAL	293592	282.3		TOTAL	73398

PEOPLE

Area	No	Hrs/Wk
Office - Exterior	241	40
Office - Interior	181	40

Infiltration Item	Crack Usage	Figure	Factor	Cfm
Doors	250 Pas/Hr	8 - 15	400	400
Windows	14144 LF	8 - 14	0	0
			TOTAL	400

DOMESTIC WATER

Use	Temp Hot	Temp Cold	Gal/Meals or Person	People/ Meals	Gallons Per Day	Btu/Yr E6	%Eff. & Conversion	MCF /Yr
Handwash	110	60	3	422	1266	136.4	60 x 1	227

DISTRIBUTION LOADS

Item	Cfm	Gpm	Oa	Hp	Voltage	Amperage Rated	Amperage Opr	Kw	TOTAL Hrs/Wk	Kwh Per Year
9-Exterior A-H Units	67900	456	1900	87.5	460/3	123	106	39.8	50	103480
8-Interior A-H Units	29300	216	1700	40.0	460/3	61	48	16.6	50	43160
2-Toilet Exhaust Fans	3200			1.0	230/1	8.5	7.5	0.8	40	1602

This fan reduction causes a reduction in the cooling energy so enter the new fan energy on the Annual Heating and Cooling Tabulation Form and calculate the required cooling energy as shown in Figures 10-41 and 10-42.

FIGURE 10-41 DRUG STORE ANNUAL HEATING AND COOLING TABULATION FORM MODIFIED FOR PECO # 4

ANNUAL HEATING AND COOLING TABULATION FORM

JOB: HB&S Building AREA Drug Store PECO NO: 4 DATE: 9-9-78

ITEM	UNITS SqFt-Cfm	COOLING Fig No 8-	Factor	Shade Factor	Total Btu/Yr-E6	HEATING Fig No 8-	Factor	Total Btu/Yr-E6
North Glass	--	16 or 17	--		--	19 or 20	--	--
South Glass	200	16 or 17	135E3	.64	17.3	19 or 20	60E3	1[illegible]
East Glass	880	16 or 17	135E3	.64	76.0	19 or 20	67E3	5[illegible]
West Glass	--	16 or 17	--	--	--	19 or 20	--	--
Glass Conduction	1080	18	8E3		8.6	--	--	--
North Wall	--	22 or 23	--		--	25 or 26	--	--
South Wall	325	22 or 23	2.5E3		0.8	25 or 26	12E3	[illegible]
East Wall	1145	22 or 23	2.5E3		2.9	25 or 26	15E3	1[illegible]
West Wall	--	22 or 23	--		--	25 or 26	--	--
Wall Conduction	1470	24	3.0E3		4.4	--	--	--
Roof Solar	--	27	--		--	--	--	--
Roof Conduction	--	24	--		--	28	--	--
Outside Air*	250	30	53.1E6		13.3	31	87E6	21.[illegible]
Humidification*	250	--	--		--	32	42E6	10.[illegible]
*Factor is Btu/1000Cfm		Sub-Total Cooling			123.3	Total Heating		124.[illegible]

Item	Kwh/Yr	Conversion Factor	%Cooling Season/Yr	
Lighting	42152	3413 x 1.0	50	72.[illegible]
Fans	18533	3413 x 1.0	50	31.[illegible]

People	No	Hrs/Yr	Btu/Person	%Cooling Season/Yr	
	25	4004	450	50	22.5
TOTAL COOLING					249.[illegible]

Summary	
Heating	
Conv. Factor	1E6
Efficiency	75%
Mcf/Yr	165 7
Cooling	
Conv. Factor	3413
COP	4.0
Kwh/Yr	18268

FIGURE 10-42 OFFICE HEATING AND COOLING TABULATION FORM MODIFIED FOR PECO #4

ANNUAL HEATING AND COOLING TABULATION FORM

JOB: HB&S Building AREA Office PECO NO: 4 DATE: 9-9-73

ITEM	UNITS SqFt-Cfm	COOLING Fig No 8-	Factor	Shade Factor	Total Btu/Yr-E6	HEATING Fig No 8-	Factor	Total Btu/Yr-E6
North Glass	1152	16 or 17	55E3	.64	4[illegible].5	19 or 20	85E3	[illegible]7.9
South Glass	3608	16 or 17	135E3	.64	31[illegible].7	19 or 20	60E3	216.5
East Glass	4576	16 or 17	135E3	.64	39[illegible].[illegible]	19 or 20	67E3	306.6
West Glass	5456	16 or 17	135E3	.64	47[illegible].[illegible]	19 or 20	67E3	365.5
Glass Conduction	14792	18	8E3		11[illegible].3	--	--	--
North Wall	8368	22 or 23	1.5E3		1[illegible].6	25 or 26	17E3	142.3
South Wall	4587	22 or 23	2.5E3		1[illegible].5	25 or 26	12E3	55.0
East Wall	9464	22 or 23	2.5E3		2[illegible].7	25 or 26	15E3	142.0
West Wall	10609	22 or 23	2.5E3		25.5	25 or 26	15E3	159.1
Wall Conduction	33028	24	3.0E3		99.1	--	--	--
Roof Solar	10300	27	5.5E3		56.7	--	17E3	175.1
Roof Conduction	10300	24	1.5E3		15.5	28	--	--
Outside Air*	3600	30	32.2E6		115.9	31	47E6	169.2
Humidification*	3600	--	--			32	23E6	82.8
*Factor is Btu/1000Cfm		Sub-Total Cooling			16[illegible]8 9	Total Heating		1[illegible]12.0

Item	Kwh/Yr	Conversion Factor	%Cooling Season/Yr	
Lighting - Exterior	185120	3413 x 1.0	50	3[illegible]5.9
- Interior	181870	3413 x 1.0	100	6[illegible]0.7
Fans - Exterior	103480	3413 x 1.0	50	1[illegible]6.5
- Interior	43160	3413 x 1.0	100	1[illegible]7.3

People	No	Hrs/Yr	Btu/Person	%Cooling Season/Yr	
Office - Exterior	241	2080	450	50	112.8
- Interior	181	2080	450	100	1[illegible]5.4
TOTAL COOLING					3292.5

Summary	
Heating	
Conv. Factor	[illegible] x E6
Efficiency	75%
Mcf/Yr	[illegible]549.3
Cooling	
Conv. Factor	3413
COP	4.0
Kwh/Yr	23[illegible]778

The Building and Parking area as well as the Building distribution equipment are not affected by this change so these forms will only reflect the PECO number chang to #4.

The electrical consumption and demand are affected by this PECO, so the reduced energy for fans and cooling energy is recorded on the Energy Tabulation Forms shown in Figures 10-43 and 10-44.

FIGURE 10-43 ELECTRICAL ENERGY TABULATION FORM MODIFIED FOR PECO #4

ENERGY TABULATION FORM

JOB: HB&S Building FUEL Electrical UNITS Kwh/Yr PECO NO 4 DATE: 9-9-78

ITEM \ AREA	Drug Store	Office	Bldg. & Park.					TOTALS
Lighting	42152	366990	76440					485592
A-H Units	18533	146640						165173
Cooling	18268	236778	19755					274801
CHW Pumps			92962					92962
Condenser Water Pumps			135906					135906
Cooling Tower			72072					72072
Elevators			73000					73000
Exhaust Fans		1602						1602
Boiler Fans			14742					14742
Domestic Water Pump			7207					7207
SUB-TOTAL	78953	752010	492084					1323047
MISC.	8380	108670	32950					150000
REHEAT	0	328327	0					328327
TOTAL	87333	1189007	525034					1801374

Utility Bill Energy Use - Base Year - Avg. Year / Avg. Year Degree Days x Yearly Building Heating Use = Average Yearly Energy Use

FIGURE 10-44 ELECTRICAL ENERGY DEMAND TABULATION FORM MODIFIED FOR PECO #4

ENERGY TABULATION FORM

JOB: HB&S BUILDING FUEL Demand UNITS Kw/Yr PECO NO 4 DATE: 9-9-78

ITEM \ AREA	Drug Store	Office	Bldg. & Park.					TOTALS
Lighting	9.0	141.2	15.5					165.7
A-H Units	3.6	56.4						60.0
Cooling			264.0					264.0
CHW Pumps			17.6					17.6
Condenser Water			26.4					26.4
Cooling Tower			17.6					17.6
Elevators			140.3					140.3
Exhaust Fans		0.8						0.8
Boiler Fans			3.2					3.2
Domestic Water Pump			1.8					1.8
SUB-TOTAL	12.6	198.4	486.4					697.4
MISC.								
REHEAT								
TOTAL								

Utility Bill Energy Use - Base Year - Avg. Year / Avg. Year Degree Days x Yearly Building Heating Use = Average Yearly Energy Use

The new energy consumption and demand is then recorded and compared with the previous use as shown in Figure 10-45. The following information is obtained:

Electrical Energy Savings:	127145 kwh @ .027	=$3436.00
Electrical Demand Savings:	39.7 kw x 12 mo. x $5.00	= 2382.00
TOTAL SAVINGS		$5818.00

Estimated Cost = $2,700.00

Payback- 5.6 months

FIGURE 10-45 ENERGY TABULATION FORM FOR PECO # 4

PECO ENERGY TABULATION FORM

JOB: HB&S Building

PECO NO: 4
DATE: 9-9-78

FUEL	ITEM \ AREA	Drug Store	Office	Bldg. & Park					TOTALS
ELECTRICAL	Before ECO Energy Use	101234	1302351	525034					1928619
	After ECO Energy Use	87333	1189007	525034					1[illegible]01374
	Kwh Savings	13901	113344	0					127245
	Before ECO Demand Use	15.0	235.7	486.4					737.1
	After ECO Demand Use	12.6	198.4	486.4					697.4
	Kw Savings	2.4	37.3	0					39.7
GAS	Before ECO Energy Use	189.2	7360.3						7549.5
	After ECO Energy Use	189.2	7360.3						7549.5
	Mcf Savings	0	0						0

A copy of PECO #4 Forms is then made for use in analyzing PECO #5.

(5) The analysis of PECO #5 deals with energy reductions in HVAC Equipment. In this case, "reheat energy", which is provided by both the boilers and the chillers.

PECO #5 (Text ECO#31) - Reduce Reheat

THEORY

Reheat is required on this project to maintain zone temperatures, and the present system maintains a 55° discharge temperature.

PROCEDURE

Connect all of the pneumatic thermostats served by each air handling unit together through a load analyzer relay so that the one calling for the most cooling controls the discharge air temperature. Install volume dampers behind each reheat coil and reduce the air 50 percent before introducing any reheat. This will be controlled by the room thermostat. Turn off heating when outside temperature is 65°.

ANALYSIS

The consultant must make a judgement as to the amount of reheat that will still be required with this system. If the upper limit of humidity is critical, then some reheat will be necessary for humidity control. Since humidity is not critical in this building, assume reheat will be required only during mild weather and base projected savings on a 75% reduction in reheat.

The only place on the form that reheat energy is recorded is on the Energy Tabulation Forms pertaining to the types of fuel involved as explained in Chapter VI - E - page 39 "Reheat Use Tabulation". Electrical and Gas consumption are affected in this case and the amount of reheat shown after PECO #4 will be reduced by 75% as shown in Figures 10-46 and 10-47.

FIGURE 10-46 EXAMPLE OF COMPLETED ENERGY TABULATION FORM FOR ELECTRICITY MODIFIED FOR PECO # 5

ENERGY TABULATION FORM

JOB: HB&S Building FUEL Electrical UNITS Kwh/Yr PECO NO (5) DATE: 9-9-78

ITEM \ AREA	Drug Store	Office	Bldg. & Park.					TOTALS
Lighting	42152	366990	76440					485592
A-H Units	18533	146640						165173
Cooling	18268	236778	19755					274801
CHW Pumps			92962					92962
Condenser Water Pumps			135906					135906
Cooling Tower			72072					72072
Elevators			73000					73000
Exhaust Fans		1602						1602
Boiler Fans			14742					14742
Domestic Water Pump			7207					7207
SUB-TOTAL	78953	752010	492084					1323047
MISC.	8380	108670	32950					150000
REHEAT	0	(82082)	0					(82082)
TOTAL	87333	(942762)	525034					(1555129)

Utility Bill Energy Use - (Base Year - Avg. Year) / (Avg. Year Degree Days) x Yearly Building Heating Use = Average Yearly Energy Use

All other forms remain the same except to change the PECO number to #5.

FIGURE 10-47 EXAMPLE OF COMPLETED ENERGY TABULATION FORM FOR GAS MODIFIED FOR PECO # 5

ENERGY TABULATION FORM

JOB: HB&S Building FUEL Gas UNITS Mcf/Yr PECO NO [illegible] DATE: 9-9-78

ITEM \ AREA	Drug Store	Office	TOTALS
Domestic Water	13.5	227.0	240 5
Heating	165.7	2549.3	2715 0
SUB-TOTAL	179.2	2776.3	2955.5
MISC.	10.0	150.0	160.0
REHEAT	-0-	(1108.5)	(1108.5)
TOTAL	189.2	(4034.8)	(4244.0)

$$\underset{\text{Utility Bill Energy Use}}{19{,}696} - \underset{\text{Base Year - Avg. Year / Avg. Year Degree Days}}{\frac{5434 - 4900}{4900}} \times \underset{\text{Yearly Building Heating Use}}{5975} = \underset{\text{Average Yearly Energy Use}}{19044.5}$$

TOTALS: 19044.5

The PECO #5 Energy Tabulation Form is filled out using these new energy consumptions as shown in Figure 10-48. The following information is obtained:

Energy Savings

Electrical	246245 kwh @ .027	=	$6,649.00	Cost Estimate	= $19,600
Gas	3325.5 mcf @ 1.57	=	5,820.00	Payback	= 1.6 Years
TOTAL SAVINGS			$12,469.00		

FIGURE 10-48 ENERGY TABULATION FORM FOR PECO # 5

PECO ENERGY TABULATION FORM

JOB: HB&S Building PECO NO: 5 DATE: 9-9-78

FUEL	ITEM \ AREA	Drug Store	Office	Bldg. & Park	TOTALS
ELECTRICAL	Before ECO Energy Use	87333	1189007	525034	18[illegible]374
	After ECO Energy Use	87333	942762	525034	15[illegible]129
	Kwh Savings	0	246245	0	2[illegible]245
	Before ECO Demand Use	12.6	198.4	486.4	6[illegible]7.4
	After ECO Demand Use	12.6	198.4	486.4	6[illegible]7.4
	Kw Savings	0	0	0	0
GAS	Before ECO Energy Use	189.2	7360.3		75[illegible]9.5
	After ECO Energy Use	189.2	4034.8		42[illegible]4.0
	Mcf Savings	0	3325.5		33[illegible]3.5

A copy of PECO #5 forms is made for analyzing PECO #6.

(6) The analysis of PECO #6 deals with energy reduction in domestic water systems. This example insulates a previously uninsulated water storage tank.

PECO #6 (Text ECO #44): Insulate Hot/Water storage/Tanks and Piping

THEORY

The 250 gallon hot water storage tank is uninsulated, but the piping is well insulated. The tank is heated 77 hrs/week (see PECO #1).

PROCEDURE

Insulate the storage tank with 3 inches of insulation having a "k" of 0.3

ANALYSIS

Present Btu/hr @ 110° - 5511 (Figure 9-30)

Future Btu/hr @ 110° - 283 (Figure 9-30)

Hourly Savings 5228 Btuh

Yearly Savings = 5228 x 4004 = 20.9 E6 Btu/yr

Gas Savings = 20.9 divided by 75% = 28.0 Mcf (use only heater efficiency for savings calculation)

Yearly Savings = 28.0 x 1.75 = $49.00

Estimated Cost - $240.00

Payback = 4.9 years

To calculate the new system efficiency divide the total of the Drug Store and Office area domestic water heating requirement (Building Load Survey Form for PECO #5) by the total domestic water gas energy use as reduced by the above.

EXAMPLE: $\dfrac{8.1 + 136.4 \text{ Btu/yr E6}}{212.5 \text{ Mcf/yr}} = \dfrac{144.5}{212.5} = 68\%$

The forms in the particular case are used only to record the results of the above calculations. Enter the new efficiency on the Building Load Survey Form for the Drug Store and Office areas and record the decreased gas energy consumption as shown in Figures 10-49 and 10-50.

FIGURE 10-49 DRUG STORE BUILDING LOAD SURVEY FORM MODIFIED FOR PECO #6

BUILDING LOAD SURVEY FORM

JOB: HB&S Building | AREA: Drug Store | WINTER: 75°-30% | SUMMER: 75°-50% | PECO NO: 6 | DATE: 9-9-78

GLASS

Facing	SqFt	Shading Ext	Shading Interior	Type U= 1.13
S	200	0	Blinds S.F. 64	Single Fixed Tight Fit
E	880	0		
TOTAL	1080			

WALLS

Facing	SqFt	Shade	Type U= 0.2
S	325	0	Light 4" Brick 4" Block Plas. Finish
E	1145	0	
TOTAL	1470		

ROOF OF FLOORS

SqFt	Type U=

LIGHTING

Area	Type	Occupied Kw	Occupied Fc	Occupied Hrs/Wk	Occupied Kwh/Yr	Custodial Kw	Custodial Fc	Custodial Hrs/Wk	Custodial Kwh/Yr
Main Store	96" Slimline	8.6	175	77	34434	8.6	175	12	5366
Store Room	96" Slimline	0.4	75	77	1601	0.4	75	12	250
	TOTAL	9.0		TOTAL	36535	9.0		TOTAL	5616

PEOPLE

Area	No	Hrs/Wk	Infiltration Item	Crack Usage	Figure	Factor	Cfm
Main Store	25	77	4 x 8 Door	150	8 - 15	250	250

DOMESTIC WATER

Use	Temp Hot	Temp Cold	Gal/Meals or Person	People/Meals	Gallons Per Day	Btu/Yr E6	TOTAL Eff. & Conversion	MCF /Yr
Hand Wash	110	60	3	25	75	8.1	68 x 1	11.9

DISTRIBUTION LOADS

Item	Cfm	Gpm	Oa	Hp	Voltage	Amperage Rated	Amperage Opr	Kw	TOTAL Hrs/Wk	Kwh Per Year
A-H Unit No. 1	6078	60	250	7-1/2	460/3	11	10	3.6	99	18533

FIGURE 10-50 OFFICE BUILDING LOAD SURVEY FORM MODIFIED FOR PECO #6

BUILDING LOAD SURVEY FORM

JOB: HB&S Building | AREA: Office | WINTER: 75 - 30% | SUMMER: 75 - 50% | PECO NO: 6 | DATE: 9-9-78

GLASS

Facing	SqFt	Shading Ext	Shading Interior	Type U= 1.13
N	1152	0	Blinds S.F.-.64	Single Metal Casem. Avg. Fit
S	3608	0		
E	4576	0		
W	5456	0		
TOTAL	14792			

WALLS

Facing	SqFt	Shade	Type U= 0.20
N	8368	0	Light 4" Brick 4" Block Plas. Finish
S	4587	0	
E	9464	0	
W	10609	0	
TOTAL	33028		

ROOF OF FLOORS

SqFt	Type U= 0.20
10300	6" Concrete Built-Up Roofing 2x2 Lay-in Ceiling
10300	

LIGHTING

Area	Type	Occupied Kw	Occupied Fc	Occupied Hrs/Wk	Occupied Kwh/Yr	Custodial Kw	Custodial Fc	Custodial Hrs/Wk	Custodial Kwh/Yr
Office - Exterior	2x4-200W Fluorescent	71.2	75	40	148096	71.2	75	10	37024
Office - Interior	2x4-200W Fluorescent	52.3	75	40	114920	52.3	75	10	28730
Halls	150W. Incandescent Spots	10.8	20	40	22464	10.8	20	10	5616
Restrooms	2x4-200W Fluorescent	3.9	40	40	8112	3.9	40	10	2028
	TOTAL	141.2		TOTAL	293592	282.3		TOTAL	73398

PEOPLE

Area	No	Hrs/Wk	Infiltration Item	Crack Usage	Figure	Factor	Cfm
Office - Exterior	241	40	Doors	250 Pas/Hr	8 - 15	400	400
Office - Interior	181	40	Windows	14144 LF	8 - 14	0	0
						TOTAL	400

DOMESTIC WATER

Use	Temp Hot	Temp Cold	Gal/Meals or Person	People/Meals	Gallons Per Day	Btu/Yr E6	TOTAL Eff. & Conversion	MCF /Yr
Handwash	110	60	3	422	1266	136.4	68 x 1	200.6

DISTRIBUTION LOADS

Item	Cfm	Gpm	Oa	Hp	Voltage	Amperage Rated	Amperage Opr	Kw	TOTAL Hrs/Wk	Kwh Per Year
9-Exterior A-H Units	67900	456	1900	87.5	460/3	123	106	39.8	50	103480
8-Interior A-H Units	29300	216	1700	40.0	460/3	61	48	16.6	50	43160
2-Toilet Exhaust Fans	3200			1.0	230/1	8.5	7.5	0.8	40	1602

The new gas energy use is recorded on the Gas Energy Tabulation Form Figure 10-51.

FIGURE 10-51 GAS ENERGY TABULATION FORM MODIFIED FOR PECO #6

ENERGY TABULATION FORM

JOB: HB&S Building FUEL Gas UNITS Mcf/Yr PECO NO (6) DATE: 9-9-78

ITEM / AREA	Drug Store	Office							TOTALS
Domestic Water	11. 9	200.6							212.5
Heating	165.7	2549.3							2715.0
SUB-TOTAL	177.6	2749.9							2927.5
MISC.	10.0	150.0							160.0
REHEAT	-0-	1108.5							1108.5
TOTAL	187.2	4008.4							4196.0

19,696 Utility Bill Energy Use - $\frac{5434 - 4900}{4900}$ Base Year - Avg. Year / Avg. Year Degree Days x 5975 Yearly Building Heating Use = 19044.5 Average Yearly Energy Use — 19044.5

The PECO #6 Energy Tabulation Form is not required as the energy reduction has already been established. All of the other forms remain unchanged except to reflect the change of PECO number to #6.

Copies of PECO #6 forms are made for use in analyzing PECO #7.

(7) The analysis of PECO #7 deals with energy reduction in power systems. Oversize motors are replaced in this case.

PECO #7 (Text ECO #50): Improve Efficiency by Replacing Oversize Motors

THEORY

The reduction in lighting (PECO #2) permits a reduction in air flow (PECO #5). This reduces the load on the fan motors, which makes them oversized for the job. Oversized motors are inefficient. (See Figure #8-55)

PROCEDURE

Determing the proper size motor required to meet the new loads. Motors that are presently in use may need to be transferred to another fan if the resultant loading of the motor is between 60% and 100%.

ANALYSIS

	Loading	Efficiency
Drug Store	$\frac{3.6}{7.0}$ = 51%	77%
Office Exterior	$\frac{39.8}{81.6}$ = 49%	76%
Interior	$\frac{16.8}{37.3}$ = 45%	75%

Increase efficiency to 80% by proper size motors.

Increased motor efficiency results in less kw input to the motor. Reduce the motor kw on the Building Load Survey Forms for the Drug Store and Office by the difference in the calculated efficiency and the new motor efficiency. (Figures 10-52 and 10-53) In the drug store this will be a 3% reduction.

FIGURE 10-52 DRUG STORE BUILDING LOAD SURVEY FORM MODIFIED FOR PECO #7

BUILDING LOAD SURVEY FORM

JOB: HB&S Building AREA: Drug Store WINTER: 75°-30% SUMMER: 75°-50% PECO NO: (7) DATE: 9-9-78

GLASS

Facing	SqFt	Shading Ext	Shading Interior	Type U= 1.13
S	200	0	Blinds S.F. 64	Single Fixed Tight Fit
E	880	0		
TOTAL	1080			

WALLS

Facing	SqFt	Shade	Type U=0.2
S	325	0	Light 4" Brick 4" Block Plas. Finish
E	1145	0	
TOTAL	1470		

ROOF OR FLOORS

SqFt	Type U=

LIGHTING

Area	Type	Occupied Kw	Occupied Fc	Occupied Hrs/Wk	Occupied Kwh/Yr	Custodial Kw	Custodial Fc	Custodial Hrs/Wk	Custodial Kwh/Yr
Main Store	96" Slimline	8.6	175	77	34434	8.6	175	12	5366
Store Room	96" Slimline	0.4	75	77	1601	0.4	75	12	250
	TOTAL	9.0		TOTAL	36535	9.0		TOTAL	5616

PEOPLE

Area	No	Hrs/Wk
Main Store	25	77

Infiltration Item	Crack Usage	Figure	Factor	Cfm
4 x 8 Door	150	8 - 15	250	250

DOMESTIC WATER

Use	Temp Hot	Temp Cold	Gal/Meals or Person	People/ Meals	Gallons Per Day	Btu/Yr E6	TOTAL %Eff. & Conversion	MCF /Yr
Hand Wash	110	60	3	25	75	8.1	68 x 1	11.9

DISTRIBUTION LOADS

Item	Cfm	Gpm	Oa	Hp	Voltage	Amperage Rated	Amperage Opr	Kw	TOTAL Hrs/Wk	Kwh Per Year
A-H Unit No. 1	6078	60	250	7-1/2	460/3	11	10	(3.5)	99	(18018)

FIGURE 10-53 OFFICE BUILDING LOAD SURVEY FORM MODIFIED FOR PECO #7

BUILDING LOAD SURVEY FORM

JOB: HB&S Building AREA: Office WINTER: 75 - 30% SUMMER: 75 - 50% PECO NO: (7) DATE: 9-9-78

GLASS

Facing	SqFt	Shading Ext	Shading Interior	Type U= 1.13
N	1152	0	Blinds S.F.-.64	Single Metal Casem. Avg. Fit
S	3608	0		
E	4576	0		
W	5456	0		
TOTAL	14792			

WALLS

Facing	SqFt	Shade	Type U= 0.20
N	8368	0	Light 4" Brick 4" Block Plas. Finish
S	4587	0	
E	9464	0	
W	10609	0	
TOTAL	33028		

ROOF OR FLOORS

SqFt	Type U= 0.20
10300	6" Concrete Built-Up Roofing 2x2 Lay-in Ceiling
10300	

LIGHTING

Area	Type	Occupied Kw	Occupied Fc	Occupied Hrs/Wk	Occupied Kwh/Yr	Custodial Kw	Custodial Fc	Custodial Hrs/Wk	Custodial Kwh/Yr
Office - Exterior	2x4-200W Fluorescent	71.2	75	40	148096	71.2	75	10	37024
Office - Interior	2x4-200W Fluorescent	52.3	75	40	114920	52.3	75	10	28730
Halls	150W. Incandescent Spots	10.8	20	40	22464	10.8	20	10	5616
Restrooms	2x4-200W Fluorescent	3.9	40	40	8112	3.9	40	10	2028
	TOTAL	141.2		TOTAL	293592	282.3		TOTAL	73398

PEOPLE

Area	No	Hrs/Wk
Office - Exterior	241	40
Office - Interior	181	40

Infiltration Item	Crack Usage	Figure	Factor	Cfm
Doors	250 Pas/Hr	8 - 15	400	400
Windows	14144 LF	8 - 14	0	0
			TOTAL	400

DOMESTIC WATER

Use	Temp Hot	Temp Cold	Gal/Meals or Person	People/ Meals	Gallons Per Day	Btu/Yr E6	%Eff. & Conversion	MCF /Yr
Handwash	110	60	3	422	1266	136.4	68 x 1	200.6

DISTRIBUTION LOADS

Item	Cfm	Gpm	Oa	Hp	Voltage	Amperage Rated	Amperage Opr	Kw	TOTAL Hrs/Wk	Kwh Per Year
9-Exterior A-H Units	67900	456	1900	87.5	460/3	123	106	37.7	50	98020
8-Interior A-H Units	29300	216	1700	40.0	460/3	61	48	15.8	50	41080
2-Toilet Exhaust Fans	3200			1.0	230/1	8.5	7.5	0.8	40	1602

This reduced motor input also reduces cooling as shown in Figures 10-54 and 10-55.

FIGURE 10-54 DRUG STORE ANNUAL HEATING AND COOLING TABULATION FORM MODIFIED FOR PECO #7

ANNUAL HEATING AND COOLING TABULATION FORM

JOB: HB&S Building AREA Drug Store PECO NO: 7 DATE: 9-9-78

ITEM	UNITS	COOLING				HEATING		
	SqFt-Cfm	Fig No 8-	Factor	Shade Factor	Total Btu/Yr-E6	Fig No 8-	Factor	Total Btu/Yr-E6
North Glass	--	16 or 17	--		--	19 or 20	--	--
South Glass	200	16 or 17	135E3	.64	17.3	19 or 20	60E3	12
East Glass	880	16 or 17	135E3	.64	76.0	19 or 20	67E3	59
West Glass	--	16 or 17	--	--	--	19 or 20	--	--
Glass Conduction	1080	18	8E3		8.6	--	--	--
North Wall	--	22 or 23	--		--	25 or 26	--	--
South Wall	325	22 or 23	2.5E3		0.8	25 or 26	12E3	4
East Wall	1145	22 or 23	2.5E3		2.9	25 or 26	15E3	17
West Wall	--	22 or 23	--		--	25 or 26	--	--
Wall Conduction	1470	24	3.0E3		4.4	--	--	--
Roof Solar	--	27	--		--	--	--	--
Roof Conduction	--	24	--		--	28	--	--
Outside Air*	250	30	53.1E6		13.3	31	87E6	21.8
Humidification*	250	--	--		--	32	42E6	10.5
*Factor is Btu/1000Cfm		Sub-Total Cooling			123.3	Total Heating		124.3

Item	Kwh/Yr	Conversion Factor	%Cooling Season/Yr	
Lighting	42152	3413 x 1.0	50	72.0
Fans	18018	3413 x 1.0	50	30.7

People	No	Hrs/Yr	Btu/Person	%Cooling Season/Yr	
	25	4004	450	50	22.5
TOTAL COOLING					248.5

Summary	
Heating	
Conv. Factor	1E6
Efficiency	75%
Mcf/Yr	165.7
Cooling	
Conv. Factor	3413
COP	4.0
Kwh/Yr	18202

FIGURE 10-55 OFFICE HEATING AND COOLING TABULATION FORM MODIFIED FOR PECO #7

ANNUAL HEATING AND COOLING TABULATION FORM

JOB: HB&S Building AREA Office PECO NO: 7 DATE: 9-9-78

ITEM	UNITS	COOLING				HEATING		
	SqFt-Cfm	Fig No 8-	Factor	Shade Factor	Total Btu/Yr-E6	Fig No 8-	Factor	Total Btu/Yr-E6
North Glass	1152	16 or 17	55E3	.64	40.6	19 or 20	85E3	97.[illegible]
South Glass	3608	16 or 17	135E3	.64	311.7	19 or 20	60E3	216.5
East Glass	4576	16 or 17	135E3	.64	395.4	19 or 20	67E3	306.[illegible]
West Glass	5456	16 or 17	135E3	.64	471.4	19 or 20	67E3	365.5
Glass Conduction	14792	18	8E3		118.3	--	--	--
North Wall	8368	22 or 23	1.5E3		12.6	25 or 26	17E3	142.[illegible]
South Wall	4587	22 or 23	2.5E3		11.5	25 or 26	12E3	55.[illegible]
East Wall	9464	22 or 23	2.5E3		23.7	25 or 26	15E3	142.[illegible]
West Wall	10609	22 or 23	2.5E3		26.5	25 or 26	15E3	159.
Wall Conduction	33028	24	3.0E3		99.1	--	--	--
Roof Solar	10300	27	5.5E3		56.7	--	17E3	175.
Roof Conduction	10300	24	1.5E3		15.5	2[illegible]	--	--
Outside Air*	3600	30	32.2E6		115.9	3[illegible]	47E6	169.[illegible]
Humidification*	3600	--	--			3[illegible]	23E6	82.[illegible]
*Factor is Btu/1000Cfm		Sub-Total Cooling			1698.9	Total Heating		1912.[illegible]

Item	Kwh/Yr	Conversion Factor	%Cooling Season/Yr	
Lighting - Exterior	185120	3413 x 1.0	50	315.9
- Interior	181870	3413 x 1.0	100	620.7
Fans - Exterior	98020	3413 x 1.0	50	167.2
- Interior	41080	3413 x 1.0	100	140.2

People	No	Hrs/Yr	Btu/Person	%Cooling Season/Yr	
Office - Exterior	241	2080	450	50	112.8
- Interior	181	2080	450	100	169.4
TOTAL COOLING					3225.1

Summary	
Heating	
Conv. Factor	1 x E6
Efficiency	75%
Mcf/Yr	2549.3
Cooling	
Conv. Factor	3413
COP	4.0
Kwh/Yr	236236

The electrical consumption and demand are affected by the PECO and these reductions are reflected in Figures 10-56 and 10-57. All other forms are unchanged except to indicate "PECO#7".

FIGURE 10-56 ELECTRICAL ENERGY TABULATION FORM MODIFIED FOR PECO #7

ENERGY TABULATION FORM

JOB: HB&S Building FUEL Electrical UNITS Kwh/Yr PECO NO 7 DATE: 9-9-78

ITEM \ AREA	Drug Store	Office	Bldg. & Park.					TOTALS
Lighting	42152	366990	76440					485592
A-H Units	18018	139100						157118
Cooling	18202	236236	19755					274193
CHW Pumps			92962					92962
Condenser Water Pumps			135906					135906
Cooling Tower			72072					72072
Elevators			73000					73000
Exhaust Fans		1602						1602
Boiler Fans			14742					14742
Domestic Water Pump			7207					7207
SUB-TOTAL	78372	743928	492084					1314384
MISC.	8380	108670	32950					150000
REHEAT	0	82082	0					82082
TOTAL	86752	934680	525034					1546466

Utility Bill Energy Use - (Base Year - Avg. Year) / (Avg. Year Degree Days) x Yearly Building Heating Use = Average Yearly Energy Use

FIGURE 10-57 ELECTRICAL ENERGY DEMAND TABULATION FORM MODIFIED FOR PECO #7

ENERGY TABULATION FORM

JOB: HB&S BUILDING FUEL Demand UNITS Kw/Yr PECO NO 7 DATE: 9-9-78

ITEM \ AREA	Drug Store	Office	Bldg. & Park.					TOTALS
Lighting	9.0	141.2	15.5					165.7
A-H Units	3.5	53.5						57.0
Cooling			264.0					264.0
CHW Pumps			17.6					17.6
Condenser Water			26.4					26.4
Cooling Tower			17.6					17.6
Elevators			140.3					140.3
Exhaust Fans		0.8						0.8
Boiler Fans			3.2					3.2
Domestic Water Pump			1.8					1.8
SUB-TOTAL	12.5	195.5	486.4					694.4
MISC.								
REHEAT								
TOTAL								

Utility Bill Energy Use - (Base Year - Avg. Year) / (Avg. Year Degree Days) x Yearly Building Heating Use = Average Yearly Energy Use

The PECO #7 Energy Tabulation form (Figure 10-58) is now completed and provides the following information:

Energy Savings

Electrical - 8663 kwh @ 2.7%	=	$234.00
Electrical Demand 3 kw x 5 x 12	=	180.00
TOTAL SAVINGS		$414.00

Estimated cost for motors (nine existing motors can be reused and nine new ones) = $1350.00

Payback = 3.3 years.

FIGURE 10-58 ENERGY TABULATION FORM FOR PECO #7

PECO ENERGY TABULATION FORM

JOB: HB&S Building

PECO NO 7
DATE: 9-9-78

FUEL	AREA / ITEM	Drug Store	Office	Bldg. & Park					TOTALS
ELECTRICAL	Before ECO Energy Use	87333	942762	525034					1555129
	After ECO Energy Use	86752	934680	525034					1546466
	Kwh Savings	581	8082	0					8663
	Before ECO Demand Use	12.6	198.4	486.4					697.4
	After ECO Demand Use	12.5	195.5	486.4					694.4
	Kw Savings	0.1	2.9	0					3.0
GAS	Before ECO Energy Use	187.6	4008.4						4196.0
	After ECO Energy Use	187.6	4008.4						4196.0
	Mcf Savings	0	0						0

H. Final Report

All of the building data has been gathered, the energy tabulated, books balanced, and the applicable PECO's have been evaluated. It is now necessary to convey all of this information to the proper decision makers.

This is accomplished by a "Final Report", which is in essence a condensation of all of the information that has been gathered in the process of the energy study. Most decision makers are not interested in a lot of detail, however they are interested in knowing clearly what their options are in terms of financial considerations and impacts on their buildings.

Following is a report on the Energy Study made for the HB&S Building, and is typical of the type of report that would be made at the conclusion of the Energy Study. For the purpose of this report, it is assumed that the seven PECO's previously discussed are the only ones found feasible by the Energy Consultant.

ENERGY STUDY FOR HB&S BUILDING--ST. LOUIS, MO.

INTRODUCTION

This report contains the findings of an Energy Study on the HB&S building. High energy costs have created a financial burden on the operation of the building, and it was decided at your board meeting in March of 1990 to have a study made, for the purpose of examining the opportunities that exist to reduce your energy consumption. It was stated at that time that any energy reduction which would have adverse effects on tenant comfort could not be tolerated, and a 10-year payback should be the maximum. This study was made with these facts in mind.

The study was performed in the following manner:

1. The building was surveyed to identify and quantify all of the energy users within the building, as well as all of the items that effect energy use, such as the building envelope, people, etc.
2. The energy use for all of the major categories was tabulated, and the total compared and balanced with your utility bill. This tabulation identified major users such as lighting, cooling, heating, fans, pumps, etc.
3. All opportunities to conserve energy were investigated to determine their feasibility.

INTRODUCTION (cont'd)

4. The feasible energy conservation opportunities were investigated from a cost/benefit, and those falling within the payback period were listed.
5. A future energy goal was established for each of the major energy categories.

SUMMARY

The Energy Study revealed that seven opportunities to conserve energy exist within the building. Some of the opportunities cost little or nothing to implement, while others require capital expenditure. The low cost category would require an expenditure of $8.700.00 to achieve an annual savings of $93,250.00

The capital cost category would require an investment of $36,190.00 to achieve an annual savings of $14,985.00.

All of these annual savings are based on the following utility costs:

- Electrical consumption - 2.7¢ per KWH
- Electrical demand - $5.00 KW per month
- Natural Gas - $1.37 per MCF

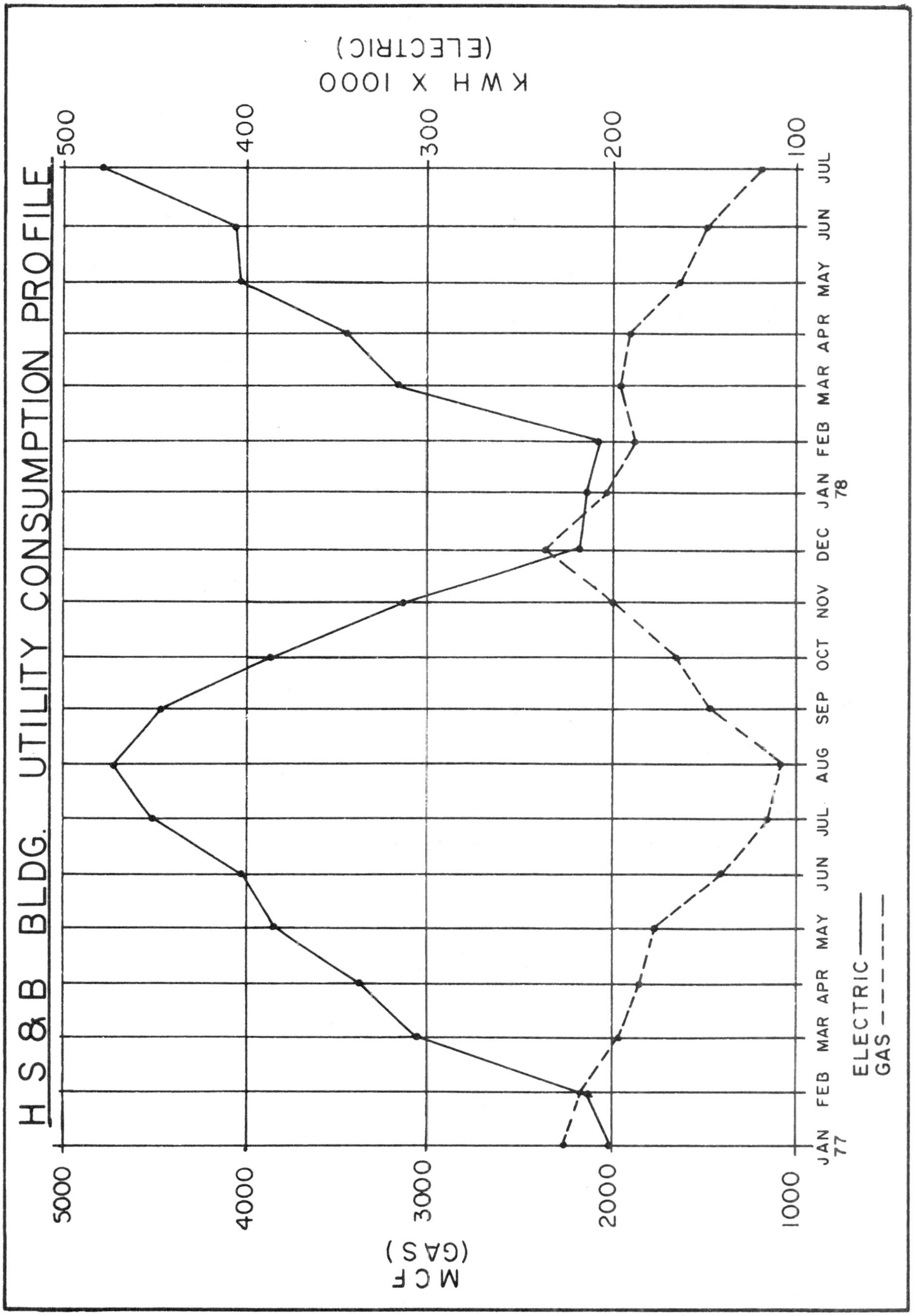
H S & B BLDG. UTILITY CONSUMPTION PROFILE
KWH X 1000
(ELECTRIC)
500
400
300
200
100
MCF
(GAS)
5000
4000
3000
2000
1000
JAN 77
FEB
MAR
APR
MAY
JUN
JUL
AUG
SEP
OCT
NOV
DEC
JAN 78
FEB
MAR
APR
MAY
JUN
JUL
ELECTRIC
GAS

TABULATION OF ENERGY USE

ELECTRICAL	PRESENT CONSUMPTION KWH/YR	FUTURE CONSUMPTION KWH/YR	PERCENTAGE REDUCTION %
Lighting.	1,041,520	485,592	53
A-H Units	870,966	157,118	82
Cooling	560,987	274,193	51
Chill Water Pumps .	157,754	92,962	41
Cond. Water Pumps .	230,630	135,906	41
Cooling Tower . . .	96,096	72,072	25
Elevators	73,000	73,000	0
Exhaust Fans. . . .	6,730	1,602	76
Boiler Fans	19,656	14,742	25
Domestic Water Pump.	15,725	7,207	54
Cooling for Reheat.	938,078	82,082	91
Miscellaneous . . .	150,000	150,000	0
	4,161,142	1,546,466	63

GAS	MCF	MCF	
Heating.	5975.0	2715.0	55
Domestic Water	240.5	164.2	32
Reheat	12,669.0	1108.5	91
	19,044.5	4147.7	78

TABULATION OF ELECTRICAL DEMAND

	PRESENT KW	FUTURE KW	PERCENTAGE REDUCTION
Lighting.	315.8	165.7	48
Air Handling Units.	99.7	60.0	40
Cooling	264.0	264.0	0
Chill Water Pumps	17.6	17.6	0
Cond. Water Pumps	26.4	26.4	0
Cooling Tower	17.6	17.6	0
Elevators	140.3	140.3	0
Exhaust Fans.	0.8	0.8	0
Boiler Fans	3.2	3.2	0
Domestic Water Pump	1.8	1.8	0
	887.2	694.4	22

ECONOMIC ANALYSIS

	PRESENT $	FUTURE $	SAVINGS $
Electrical Consumption. . .	$164,789	$ 94,192	$ 70,597
Electrical Demand	53,232	41,664	11,568
Gas Consumption	33,328	7,258	26,070
	$251,349	$143,115	$108,235

PROJECT ENERGY CONSERVATION

OPPORTUNITIES

MASTER LIST

PECO No.	PECO	Impl. Cost	Savings $/Yr.	Payback Years
1	Reduce Operating Time	$ 3,500	$ 65,385	0.05
2	Reduce Illumination Levels	2,500	22,047	0.11
3	Reduce Outside Air & Infiltration	15,000	1,969	7.6
4	Reduce Air Flow	2,700	5,818	0.47
5	Reduce Reheat	19,600	12,469	1.6
6	Insulate Storage Tank	240	49	4.9
7	Replace Oversize Motors	1,350	414	3.3
	TOTALS	$44,890	$108, 151	0.42

PECO # 1 REDUCE OPERATING TIME

(A) Reduce operating time of the air handling unit fan in the drug store to 22 hours and the office area fans to 10 hours above the occupied hours.

(B) Reduce the "ON" time of office area lights to 10 hours during custodial hours. Reschedule cleaning as required.

(C) Cut off domestic water pump and exhaust fans except for occupied hours.

(D) Cut off chilled and condenser water pumps and cooling tower fans when drug store is secured.

All of these items can be accomplished manually if operating personnel are available. It is recommended, however, that time clocks be provided so that the schedules operate automatically.

Implementation Cost	Savings $/Year	Payback Year
$3,500.00	$65,385.00	2.8 weeks

PECO # 2 REDUCE ILLUMINATION LEVELS

Reduce all lighting by 50 percent by removing lamps and disconnecting ballasts or substituting smaller wattage in the case of incandescent.

Implementation Cost	Savings $/Year	Payback Year
$2,500.00	$22,047.00	5.9 weeks

PECO # 3 REDUCE OUTSIDE AIR AND INFILTRATION

The cost of reducing the door infiltration to 125 cfm is not practical, so reduce drug store ventilation to 250 cfm. The 3200 cfm exhaust in the office area is required, plus 400 cfm of infiltration through the door. Thus, reduce ventilation to 3600 cfm by weatherstripping and sealing the windows.

Implementation Cost	Savings $/Year	Payback Year
$15,000.00	$1,969.00	7.6 years

PECO # 4 REDUCE ENERGY CONSUMPTION FOR FANS BY REDUCING AIR FLOW RATES

Air flow varies directly with the speed, so reduce the rpm of the fan by the same percentage as the desired flow reduction. This speed reduction is accomplished by adjusting the drive pulleys, if possible, or by changing to new pulleys and drives, if necessary.

Implementation Cost	Savings $/Year	Payback Year
$2,700.00	$5,818.00	5.6 months

PECO # 5 REDUCE REHEAT

Connect all of the pneumatic thermostats served by each air handling unit together through a load analyzer relay so that the one calling for the most cooling controls the discharge air temperature. Install volume dampers behind each reheat coil and reduce the air 50 percent before introducing any reheat. This will be controlled by the room thermostat. Turn off heating when outside temperature is 65°.

Implementation Cost	Savings \$/Year	Payback Year
\$19,600.00	\$12,469.00	1.6 years

PECO # 6 INSULATE HOT WATER STORAGE TANKS AND PIPING

Insulate the storage tank with 3 inches of insulation having a "k" of 0.3.

Implementation Cost	Savings $/Year	Payback Year
$240.00	$49.00	4.9 years

PECO # 7 IMPROVE EFFICIENCY BY REPLACING OVERSIZE MOTORS

Determine the proper size motor required to meet the new loads. Motors that are presently in use may be transferred tc another fan if the resultant loading of the motor is between 60 and 100 percent.

Implementation Cost	Savings $/Year	Payback Year
$1,350.00	$414.00	3.3 years

CONCLUSIONS AND RECOMMENDATIONS

All of the energy conservation opportunities listed in the Study fall within the ten year payback period. PECOs 1, 2, and 4 have the best payback at the lowest cost and although the economics are all based on implementing the PECOs in the order listed, the savings would not be effected by implementing these three PECOs as the first phase of the programs.

PECOs 5,6, and 7 represent the next payback and could either be implemented simultaneously or individually, without effecting the economics.

PECO #3 represents the longest payback and could be delayed until the third phase.

PECOs 1, 2, and 4 require an implementation cost of $8,700.00, but would produce savings in excess of $6,000.00 per month by implementing these items immediately. These resultant savings would finance Phase 2 and Phase 3 by the time the necessary design details could be completed, bids taken, and the implementation work finished. Certainly the entire program could be completed within 12 months and would actually represent only an initial outlay of $8,700.00

BIBLIOGRAPHY AND SELECTED REFERENCES

1. The Ad Hoc Committee on Energy Efficiency in Large Buildings. Report to the Interdepartmental Fuel and Committee of the State of New York. Albany: State of New York, March 7, 1973.

2. American Institute of Architects Research Corporation. Energy Conservation in Building Design. Washington, D. C.: American Institute of Architects, May 1, 1974.

3. American Institute of Plant Engineers. Better Ideas for Conserving Energy. Cincinnati, Ohio, August 1, 1974.

4. American Society of Heating, Refrigerating, and Air-Conditioning Engineers. Applications Volume. New York, 1974.

5. American Society of Heating, Refrigerating, and Air-Conditioning Engineers. Design and Evaluation Criteria for Energy Conservation in New Buildings. (Proposed Standard 90-P). New York, 1974.

6. American Society of Heating, Refrigerating, and Air-Conditioning Engineers. Handbook of Fundamentals. New York, 1972.

7. American Society of Heating, Refrigerating, and Air-Conditioning Engineers. Equipment Volume. New York, 1972.

8. American Society of Heating, Refrigerating, and AirConditioning Engineers. Systems Volume. New York, 1973.

9. Blackwell, H. R. "Application procedures for evaluation of veiling reflections in terms of ESI", Journal of IES, April, 1973.

10. Brown, J. L. and C. Mueller. "Brightness discrimination and brightness Contrast", in Vision and Visual Perception. New York: John Wiley and Sons.

11. Boynton, R. M. and D. E. Boss. "The effect of background luminance upon visual search performance." University of Rochester. Report to the Illuminating Engineering Research Institute.

12. Citizens' Advisory Committee on Environmental Quality. Citizen Action Guide to Energy Conservation. Washington, D. C.: Government Printing Office.

13. Commercial Refrigeration Manufacturer's Association. Retail Food Store Energy Conservation. Washington, D. C.

14. Daylighting Committee of the Illuminating Engineering Society. "Recommended Practice of Daylighting". New York: IES, 1962.

15. Departments of the Air Force, Army, and Navy. _Engineering Weather Data_. Washington, D. C.: Government Printing Office, 1967.

16. Dorsey, Robert T., "An Extension of the Energy Crisis: More Light on Lighting." _American Institute of Architects Journal_, June, 1972.

17. Dubin, Fred S., "Checkoff List on Energy Conservation with Regard to Power." _Electrical Consultant_, 89:4.

18. Dubin, Fred S., "Energy Conservation in Building Operations." Address to Regional Directors of the Building's Management Divison of the General Services Administration, April 3, 1973.

19. Dubin, Fred S., "Energy Conservaton Needs New Architecture and Engineering." _Public Power_, March/April, 1972.

20. Dubin, Fred S., "Energy Conservation through Building Design and a Wiser Use of Electricity." Address to the Annual Conference of the American Public Power Association: San Francisco, California, June 26, 1972.

21. Dubin, Fred S., "If You Want To Save Energy." _American Institute of Architects Journal_, December, 1972.

22. Dubin, Fred S. and Margot Villecco. "Energy for Architects." _Architecture Plus_, July, 1973.

23. Dubin-MindellBloome Associates, P.C. _A Study of Air Conditioning, Heating, and Ventilating Design for Veterans Administration Hospital Kitchens and Laundries_. Washington, D.C. Research Staff, Office of Construction, Veterans Administration, February, 1970.

24. Dubin-MindellBloome Associates, P.C. _A Study of Design Criteria and Systems for Air Conditioning Existing Veterans Administration Hospitals_, Washington, D.C.: Research Staff, Office of Construction, Veteran's Administration, March, 1971.

25. Dubin-Mindell-Bloome Associates, P.C. _Energy Conservation and Office Building Design: Research, Analysis, and Recommendations: Office Building, Manchester, New Hampshire_. May 15, 1973.

26. Dubin-MindellBloome Associates, P.C. _Energy Conservation Design Guidelines for Office Buildings_. Washington, D. C.: General Services Administration, Public Buildings Service, January 1974.

27. Dubin-Mindell-Bloome Associates. _Report to Connecticut General Insurance Corporation on Energy Conservation Opportunities_. December, 1973.

28. "Energy System Limits Waste." Design Engineering Journal, February, 1974.

29. Fanger, P.O. Thermal Comfort: Analysis and Applications in Environmental Engineering. New York: McGrawHill, 1972.

30. Federal Energy Administration. Tips for Energy Savers. Washington, D.C.: Government Printing Office.

31. Federal Energy Office. Project Independence Background Paper. Washington, D.C., February, 1974.

32. Gatts, Robert R., Robert G. Massey and John C. Robertson (U.S. Department of Commerce, National Bureau of Standards). Energy Conservation, Program Guide for Industry and Commerce. Washington, D.C.: Government Printing Office, 1974.

33. Givoni, B., Man and Architecture. London: Elsevier Publishing Company, 1969.

34. Griffith, J. W., "Analysis of Reflected Glare and Visual Effect from Windows". Presented to the National Technical Conference of the Illuminating Engineering Society. Detroit, Michigan, September 8-13, 1963.

35. "GSA Sets Energy Saving Steps." Design Engineering Journal. May 1974.

36. Handbook of Air Conditioning, Heating, and Ventilating (Clifford Strock and Richard L. Koral, editors). New York: Industrial Press, 1965.

37. "Heat Recovery Cuts Energy Use". Design Engineering Journal. February, 1974.

38. Heating, Piping and Air Conditioning Seminar on "Energy Conservation for Existing Buildings", Proceedings of New York, December 1974.

39. Henderson, R.L., J.F. McNelis, and H.G. Williams. "A Survey and Analysis fo Important Visual Tasks in Offices". The Annual Conference of the Illuminating Engineering Society, 1974.

40. IES Lighting Handbook. New York: Illuminating Engineering Society, 1972.

41. "Innovative Lighting Fixture Aids." Actual Specifying Engineer, April, 1972.

42. Jennings, Burgess H., Necati Ozisik, and Lester F. Schutrum. "New Solar Research Data on Windows and Draperies." American Society of Heating, Refrigerating, and Air Conditioning Engineers Journal reprint.

43. Kaufman, John E. "Energy Utilization: Optimizing the Uses of Energy for Lighting." Lighting Design & Application. October, 1973.

44. "Life Support Systems for a Dying Planet." Progressive Architecture. October, 1971.

45. Lighting Survey Committee. "How to Make a Lighting Survey." Report to the Illuminating Engineering Society, November, 1963.

46. McClure, Charles J.R., "Optimizing Building Energy Use." American Society of Heating, Refrigerating and Air Conditioning Engineers Journal. September, 1971.

47. Murphy, E.E. and J.G. Dorsett. "Demand Controllers Disconnect Some Loads During Peak Periods." Transmission & Distribution. July, 1971.

48. Northeast Utilities Company. Guidelines for Implementing Northeast Utilities Systems' Energy Management Program. Hartford, Connecticut, 1973.

49. Office of Emergency Preparedness. The Potential for Energy Conservation. Washington, D.C.: Government Printing Office, October, 1972.

50. Olivieri, Joe B. "Heat Recovery Systems Part I." Air Conditioning, Heating, & Refrigeration News, January 1, 1972.

51. Olivieri, Joe B. "Heat Recovery Systems Part II." Air Conditioning, Heating & Refrigeration News, January 17, 1972.

52. Phipps, H. Hary. "Energy systems Analysis Why and How." Building Systems Design. May, 1972.

53. Plant Engineering. "Getting the Most from Your Electrical Power System: Selected Articles." Barrington, Illinois: Plant Engineering Library.

54. Price, Seymour. Air Conditioning for Building Engineers and Managers. New York: Industrial Press, 1970.

55. "The Role of Refrigeration in Energy Conservation and Environmental Protection." Seminar of the Air Conditioning and Refrigeration Institute: New Orleans, Louisiana, January 26, 1972.

56. Ross & Baruzzini, Inc. Energy Conservation Applied to Office Lighting. Washington, D.C.: Federal Energy Administration, April 1, 1975. Contract No. 14-0100011845.

58. Saunders, J. E. "The Role of the Level and Diversity of Horizontal Illuminatin in an Appraisal of a Simple Office Task." Lighting Research and Technology, 1969:1.

59. Shlaer, S. "The Relation Between Visual Acuity and Illumination." Journal of General Physiology, 1937:21.

60. Stewart, John L. "Report on TC 4.5Fenestration." American Society of Heating, Refrigerating, and Air Conditioning Engineers Journal, November 1972.

61. Syska & Hennessy, Engineers and Tishman Research Corporation. A Study of the Effects of Air Changes and Outdoor Air on Interior Environment, Energy Conservation, and Construction, and Operating Costs: Phase II. Washington, D.C.: General Services Administration,December, 1973.

62. "Three Hundred Hints to Save Energy." Congressional Record. October 30, 1973.

63. U. S. Department of Commerce, National Bureau of Standards. Technical Options for Energy Conservation in Buildings. Washington, D.C.: Government Printing Office, July, 1973.

64. U. S. Department of Commerce, National Oceanic and Atmospheric Administration, National Weather Service. Operations of the National Weather Service. Silver Springs, Maryland: October, 1970.

SOLAR ENERGY

65. American Society of Heating, Refrigerating, and Air Conditioning Engineers. "Low Temperature Applications of Solar Energy." New York: ASHRAE, 1967.

66. Barber, E.M. and D. Watson. "Criteria for the Preliminary Design of Solar Heated Buildings." Guilford, Connecticut: Sunworks, Inc., 1974.

67. Colorado State University and University of Wisconsin. Design and Construction of a Residential Solar Heating and Cooling System. Washington, D.C.: National Science Foundation, Report No. NSR/RANN/SE/GE-40451/PR/74/2.

68. Daniels, Farrington. Direct Use of the Sun's Energy. New York: Ballantine, 1974.

69. Dubin-Mindell-Bloome Associates. "Specifications of Solar Collectors for the Federal Office Building, Manchester, New Hampshire." Boston: General Services Administration.

70. Duffie, J.A. and W. A. Beckman. Solar Energy Thermal Processes. New York: John D. Wiley and Sons, 1974.

71. General Electric Company. Final Report of the Solar Heating Experiment on the Grover Cleveland School, Boston, Mass. Washington, D.C.: National Science Foundation, Report No. NSF/RANN/74/064.

72. Report of the Solar Heating and Cooling Committee of BRAB, National Academy of Engineers. Washington, D.C.: National Science Foundation, 1974.

73. Shurcliffe, W.A. "Solar Heating Buildings, A Brief Survey," Solar Energy Digest (P.O. Box 17776, San Diego, California 922117).

74. "Solar Energy Industry Report." (1001 Connecticut Ave., N.W., Washington, D.C. 20035).

75. U. S. Department of Commerce, Naitonal Bureau of Standards. Performance Criterial for the Solar Heating/Cooling of Residences. Washington, D.C.: Government Printing Office.

76. Yellot, John I. "Utilization of Sun and Sky Radiation for Heating and Cooling of Buildings." ASHRAE Journal, December, 1973.

TOTAL ENERGY SYSTEM REFERENCES

77. Ahner, D.J. "Environmental Performance." Based on a paper presented at the General Electric Gas Tubine State of the ARt Engineering Seminar, June, 1971.

78. Barrangon, Maurice. "Conservation of Resources, How Total Energy Saves Fuel." GATE Information Digest. March, 1971.

79. Becker, Herbert P. "The Concept of Total Energy--Lecture No. 1." Fall Lecture Series. ASHRAE at the United Engineering Center, October, 1967.

80. Bjerklie, John W., P.E. "Small Gas Turbines for Total Energy Systems." Actual Specifying Engineer. August, 1971.

81. Dubin, Fred S., P.E. Editorial, "Life Support Systems for a Dying Planet." Available now: "Systems that save Energy" and "Can Building Codes Help Protect the Environment?" Progressive Architecture. October, 1971. "The New Architecture and Engineering," present at the Workshop on Total Energy Conservation in Public Buildings. State University of New York, Albany. January, 1972.

82. "TR-2 Total Energy for Schools and Colleges." A technical report prepared for the Educational Facilities Laboratories of the Ford Foundation. "Total Energy Systems for Mass Housing." Paper delivered at the Urban Technology Conference 2. July, 1972. San Francisco.

83. DubinMindellBloome Associates. "Total Energy." A technical report from Education Facilities Laboratories. May, 1970.

84. Federal Register. "Standards of Performance for New Stationary Sources." Environmental Protection Agency. December, 1971. Washington, D.C. Vol. 36, No. 247, Part 11.

85. Henderson, John O. "Turbines or Engines for Total Energy?" Caterpillar Tractor Company Gas Age. May, 1966.

86. Holdren, John and Herrera, Philip. Energy, Sierra Club, Copyright 1971.

87. Huber, Ernest G., P.E. "Total Energy Concept Applied to Small Diesel Engines and Gas Turbines." Air Force Civil Engineer. February, 1970. File No. 3207.

88. Kennedy, J.J. "Watch the Basic Details When Installing Gas Prime Movers." Plant Engineering. February, 1970. File No. 3202.

89. "Total Energy Evaluated." Energy International. February 1972.

90. Williams, J.R. "Operational Experience with Total Energy Plants in the U.S.A."

GLOSSARY

Absorption Chiller	A refrigeration machine using heat as the power input to generate chilled water.
Absorption Coefficient	The fraction of the total radiant energy incidnet on a surface that is absorbed by the surface.
Absorptivity	The physical characteristic of a substance describing its ability to absorb radiation.
Ambient	Surrounding (i.e. ambient temperature is the temperature in the surrounding space).
Activated Carbon	A form of carbon capable of absorbing odors and vapors.
Air changes	Expression of ventilation rate in terms of room or building volume. Usually air changes/hour.
Ballast	A device used in a starting circuit for fluorescent and other types of lamps.
Blow down	The discharge of water from a boiler or cooling tower sump that contains a high proportion of total dissolved liquids.
British Thermal Unit (Btu)	A heat unit equal to the amount of heat required to raise one pound of water one degree Farenheit.
Building Envelope	All external surfaces which are subject to climatic impact; for example, walls, windows, roof, floor, etc.
Building Load	Heating load is the rate of heat loss from the building at a steady state conditions when the indoor and outdoor temperatures are at their selected design levels (design criteria). The heating load always includes infiltration and may include ventilation loss and heat gain credits for lights and people.

Cooling load is the rate of heat gain to the building at a steady state condition when indoor and outdoor temperatures are at their selected design levels, solar gain is at its maximum for the building configura-ion and orientation, and heat gains due to infiltration, ventilation, lights, and people are present.

Cavity Ratio — Number indicating room cavity proportions, which is calculated using length, width, and height.

Centrifugal Chiller — A refrigeration machine using mechanical energy input to drive a centrifugal compressor to generate chilled water.

Centrifugal Fan — Device for propelling air be centrifugal action. Forward curved fans have blades which are sloped forward relative to direction of rotation. Backward curved fans are generally more efficient at high pressures than for ward curved fans.

Coefficient of Performance — Ratio of the tons of refrigeration produced to energy required to operate epuipment.

Coefficient of Utilization — Ratio of lumens on a work plane to lumens emitted by the lamps.

Cold Deck — A cold air chamber forming part of a ventilating unit.

Condensate — Water obtained by changing the state of water vapor (i.e. steam or moisture in air) from a gas to a liquid usually by cooling.

Condenser — A heat exchanger which removes latent heat from a vapor changing it to its liquid state. (In a refrigeration chiller the components which reject heart.)

Conductance, Thermal — A measure of the thermal conducting properties of a composite structure such as wall, roof, etc., including the insulating boundary layers of air.

Conductivity, Thermal	A measure of the thermal conducting properties of a single material expressed in units of BTU per inch thickness (sq. ft.) (hour) (degree F temperature difference).
Cooling Tower	Device that cools water directly by evaporation.
Damper	A device used to vary the volume of air passing through an air outlet, inlet, or duct.
Degree Day	The difference between the median temperature of any day and 65°F when the median temperature is less than 65°.
Degree hour	The difference between the median temperature for any hour and a selected datum.
Demand Factor	The ratio of the maximum demand of a system, or part of a system, to the total connected load of a system or part of the system under consideration.
Desiccant	A substance possessing the ability to absorb moisture.
Direct Expansion	Generic term used to describe refrigeration systems where the cooling effect is obtained directly from the refrigerant (e.g. refrigerant is evaporated directly in a cooling coil in the air stream).
Disability Glare	Spurious light from any source, which impairs a viewer's ability to discern a given object.
Double Bundle condenser	Condenser (usually in refrigeration machine) that contains two separate tube bundles allowing the option of either rejecting heat to the cooling tower or to another building system requiring heat input.
Dry Bulb Temperature	The measure of the sensible temperature of air.
Economizer cycle	A method of operating a ventilation system to reduce refrigeration load.

Whenever the outdoor air conditions are more favorable (lower heat content) than return air conditions, outdoor air quantity is increased.

Efficacy of Fixtures — Ratio of usable light to energy input for a lighting fixture or system (lumens/watt).

Energy Requirement — The total yearly energy used by a building to maintain the selected inside design conditions under the dynamic impact of a typical year's climate. It includes raw fossil fuel consumed in the building and all electricity used for lighting and power. Efficiencies of utilization are applied, and all energy is expressed in the common unit of Btu's.

Enthalpy — For the purpose of air conditioning enthalpy is the total heat content of air above a datum usually in units of Btu/lb. It is the sum of sensible and latent heat and ignores internal energy change due to pressure change.

Equivalent Sphere Illumination — That illumination which would fall upon a task covered by an imaginary transparent hemisphere which passes light of the same intensity through each unit area.

Evaporator — A heat exchanger which adds latent heat to a liquid changing it to a gaseous state. (In a refrigeration system, it is the component which absorbs heat.)

Foot-Candle — Energy of light at a distance of one foot from a standard (sperm oil) candle.

Heat Gain — As applied to HVAC calculations, it is that amount of heat gained by a space from all sources, including people, lights, machines, sunshine, etc. The total heat gain represents the amount of heat that must be removed from a space to maintain indoor comfort conditions.

Heat Loss — The sum cooling effect of the building structure when the outdoor temperature is lower than the desired indoor temperature. It represents the amount of heat that must be provided to a space to maintain indoor comfort conditions.

Heat, Latent — The quantity of heat required to effect a change in state.

Heat, Sensible — Heat that results in a temperature change but no change in state.

Heat, Specific — Ratio of the amount of heat required to raise a unit mass of material one degree to that required to raise a unit mass of water one degree.

Heat Pump — A refrigeration machine possessing the capability of reversing the flow so that its output can be either heating or cooling. When used for heating, it extracts heat from a low temperature to the point where it can be used.

Heat Transmission Coefficient — Any one of a number of coefficients used in the calculation of heat transmission by conduction, convection, and radiation, through various materials and structures.

Hot Deck — A hot air chamber forming part of a ventilating unit.

Humidity, Relative — A measurement indicating moisture content of air.

Infiltration — The process by which outdoor air leaks into a building by natural forces through cracks around doors and windows, etc. (Usually undesirable)

Insolation — The amount of solar radiation on a given plane. Expressed in Langleys or $Btu/ft.^2$

Langley — Measurement of radiation intensity. One Langley = 3.68 $Btu/ft.^2$

Life Cycle Cost	The cost of the equipment over its entire life, including operating and maintenance costs.
Load Leveling	Deferrment of certain loads to limit electrical power demand to a predetermined level.
Load Profile	Time distribution of building heating, cooling, and electrical load.
Lumen	Unit of luminous flux.
Luminare	Light fixture designed to produce a specific effect.
Makeup	Water supplied to a system to replace that lost by blow down, leakage, evaporation, etc.
Manchester, New Hampshire Project	A demonstration building commissioned by GSA (Isaak and Isaak, Architects) and developed by Dubin-Mindell-Bloome Associates to incorporate energy conserving architectural features and mechanical and electrical systems.
Modular	System arrangement whereby the demand for energy (heating, cooling) is met by a series of units sized to meet a portion of the load.
Orifice Plate	Device inserted in a pipe or duct which causes a pressure drop across it. Depending on orifice size, it can be used to restrict flow or form part of a measuring device.
Orsat Apparatus	A device for measuring the combustion components of boiler of furnace flue gasses.
Piggyback Operation	Arrangement of chilled water generation equipment whereby exhaust steam from a steam turbine driven centrifugal chiller is used as the heat source for an absorption chiller.
Power Factor	Relationship between KVA and KW. When the power factor is unity, KVA equals KW.

R-Value	The resistance to heat flow expressed in units of Sq. Ft. hour Degree F/Btu.
Raw Source Energy	The quanitity of energy input at a generating station required to produce elctrical energy including all thermal power power conversion losses.
Roof Spray	A system that reduces heat gain through a roof by cooling the outside surface with a water spray.
Seasonal Efficiency	Ratio of useful output to energy for a piece of equipment over an entire heating and cooling season. It can be derived by integrating part load efficiencies against time.
Software	Term used in relation to computers normally describes computer programs and other intangibles.
Sol-air Temperature	The theoretical air temperature that would give a heat flow rate through a building surface equal in magnitude to that obtained by the addition of conduction and radiation effects.
Ton of refrigeration	A means of expressing cooling capacity - 1 ton = 12,000 Btu/hour cooling.
'U' Value	A coefficient expressing the thermal conductance of a composite structure in Btu per square foot hour degree F temperature difference.
Veiling Reflection	Reflection of light from a task, or work surface, into the viewer's eyes.
Vapor Barrier	A moisture impervious layer designed to prevent moisture migration.
Wet Bulb Temperature	The lowest temperature attainable by evaporating water in the air without the addition or subtraction of energy.

ABBREVIATIONS

ASHRAE	American Society of Heating, Refrigeration and Air Conditioning Engineers
Btu	British thermal unit(s)
Btu's x 10^6	Millions of Btu's
Btu's x 10^3	Thousands of Btu's
cfm	cubic feet per minute
COP	Coefficient of Performance
CU	Coefficient of Utilization
dbt	dry bulb temperature
dx	direct expansion
E6	10^6
E3	10^3
ECO	Energy Conservation Opportunity
esi	equivalent sphere illumination
hid	high intensity discharge (lamps)
Hz	Hertz
HVAC	Heating, Ventilating and Air Conditioning
IES	Illuminating Engineers Society
kva	Kilovoltampere
kw	kilowatt
kwh	kilowatt hour
Mcf	thousand cubic feet
oa	outside air

ABBREVIATIONS (CON'T.)

PECO	Project Energy Conservation Opportunities
pf	power factor
psi	pounds per square inch
sqft	square foot
td	temperature difference
TE	Total Energy (system)
wbt	wet bulb temperature

APPENDIX A -- TYPES OF HVAC SYSTEMS

TYPES OF HVAC SYSTEMS: INTRODUCTION

Each HVAC system is of course unique and its particular characterisitics can only be identified by inspection and measurement. It will, however, have many characteristics which are common to all systems of its generic group and it is the common characteristics which are discussed in this section.

The minumum information required to understand the operation of an existing system and to decide which modifications are likely to prove beneficial is tabulated below. Each part of the system should be measured at the same time.

Air Flow Rates	total outdoor air total return air total supply air truck ducts terminal units air cooled condenser
Water Flow Rates	through boilers through chillers cooling towers heat exchangers coils and terminal units
Temperatures	outdoor air dbt & wbt return air dbt & wbt mixed air entering coils, dbt & wbt supply air leaving coils, dbt & wbt hot deck cold deck air at terminals conditioned areas dbt & wbt (typical for each functional use) boiler supply & return chiller supply and return condenser supply and return heat exchanger supply and return coil supply and return
Refrigerant Temperatures	hot gas line suction line

Energy conservation must be approached in a systematic manner rather than considering individual items out of context. Systems do not operate in isolation but depend on the react with other systems. It is important to recognize this interaction of systems as modification to one will cause a reaction in another which may be either beneficial or counter-productive.

Buildings frequently use a combination of system types to meet the various requirements of interior and perimeter zones. It is common practice, for instance, to use a VAV system for interior spaces and a mult-zone system and/or radiation for perimeter spaces. There is an interface point between the interior and perimeter zones where mixing and interaction between the two different system takes place. In cool weather, cold supply air from the interior zone can spill into the perimeter zone, resulting in overcooling, a shift to the heating mode of operation, and in hunting with both systems cooling and overheating. To avoid this situation, increase the supply air temperature for interior zones in cold weather.

SINGLE-DUCT, SINGLE-ZONE SYSTEMS

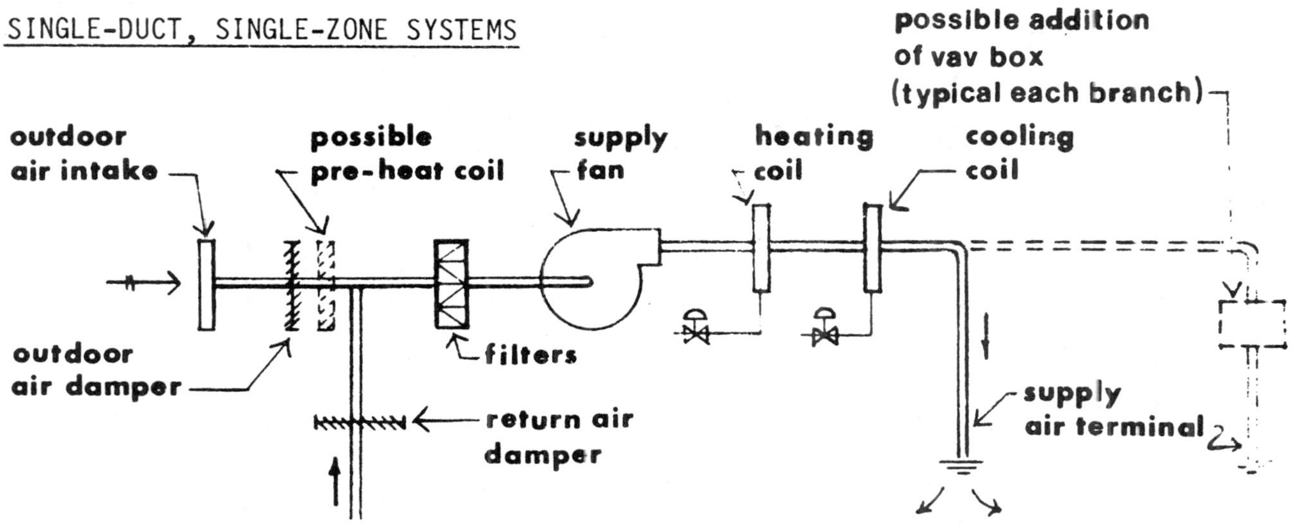

These systems are the simplest and probably the most commonly used. They often comprise just a single supply system with air intake filters, supply fan and heating coils. More complex versions will also have a return air duct, return air fan, cooling coil, and various controls to optimize their performance. Basically, the system supplies air at a pre-determined temperature to one zone or the entire building, the quantity of heating or cooling being controlled either by modulating the supply air temperature or by turning the system on and off.

The energy output of a single duct system to meet a space load is determined by the volume/temperature differential relationship, i.e., to maintain a space temperature of 65°F, the heating load could be met by a system supplying 10,000 cfm at 105°F, or 6,000 cfm at 125°F.

To conserve energy in single zone, single duct systems determine from measurements the system output and the building load. In the heating mode, if the system maximum output exceeds design building load, first reduce the volume of system air, as this will show the greatest energy saving, i.e., 90% of the supply temperature; although this will not conserve as much energy as reduced volume flow.

In the cooling mode, if the system output exceeds the design building load, the saving by reduced air flow must be equated with the saving by increased air temperatures, as increased COP of the refrigeration equipment may yield savings which exceed the fan power losses.

Install controls on the heating and cooling coils to modulate the supply air temperature. Allow an 8 - 10^{0} "Dead Zone" between heating and cooling to prevent simultaneous heating and cooling and rapid cycling. If possible, use an economizer cycle when: (1) the total heat of the outdoor air is favorable during occupied periods; (2) the sensible temperature is favorable in unoccupied periods; (3) the size of the system makes the installation economic. (Refer to ECO# 80 Section 6C.)

The economizer cycle can be a simple arrangement of dampers controlled by outdoor and return air DB temperature or a more complex system based on the total heat or enthalpy of the outdoor air, return air, and the desired enthalpy of the supply air.

Single-zone single-duct systems can be readily converted to VAV by adding control boxes on each branch and substituting outlets with a VAV type. Fan volume should be controlled according to demand, either by installing inlet guide vanes or by installing a variable speed motor.

TERMINAL REHEAT SYSTEMS

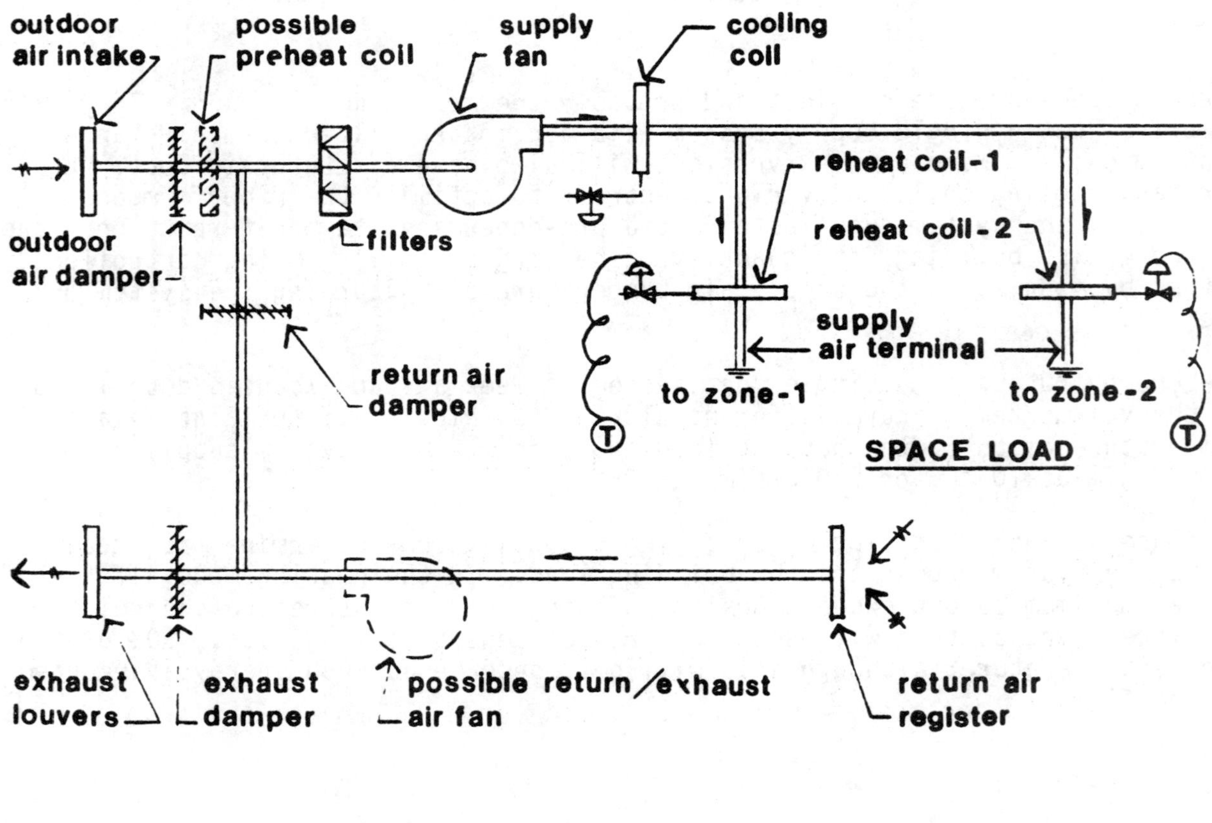

Terminal reheat systems were developed to overcome the zoning deficiencies of single-duct systems and basically comprise a single-duct, single-zone system with individual heating coils in each branch-duct that is connected to zones of similar loads. Reheat systems were also developed to give closer control of relative humidity in selected spaces.

Terminal reheat allows each zone to be individually controlled but wastes energy in the cooling season, as all of the supply air must be cooled to a low enough temperature to meet the most critical load zone and then reheated for zones of lesser loads to avoid overcooling.

As with single-duct systems, energy can be conserved by reducing the supply air volume. Many reheat systems are controlled at a fixed supply temperature of around 55°dbt/55°wbt. To conserve energy, fit controls to reschedule the supply air temperature upward according to demands of the zone with the greatest cooling load. If one zone has cooling loads grossly in excess of all others, the controlling thermostat should be located in that space.

The greatest quantity of energy can be saved by adding variable air volume boxes to each of the major branch ducts. Each VAV box should be controlled by a space thermostat located in its particular zone, and its associated reheat coil should be provided with controls to prevent reheat until the VAV box has reduced the zone supply air volume to 50%.

In buildings where a reheat system supplies zones of different occupancy, add dampers and control valves to enable water to be shut off during unoccupied times.

Reduce the water flow and temperature of hot water to reheat coils.

Use waste heat from condensate, incinerators, diesel or gas engines, or solar energy for reheat; but in all cases, raise the cold duct temperature to reduce refrigeration load.

Modify controls to operate terminal reheat systems on a temperature demand cycle only. Caution: Control of humidity will thus be eliminated and zone control modified.

Add or adjust controls to schedule supply air temperature according to the number of reheat coils in operation. If 90% of the coils are in operation, reset supply air temperatures to a higher level until the number of reheat coils falls to 10%.

Install an interlock between the two valves to prevent simultaneous heating and cooling.

If air conditioning is needed during unoccupied hours or in very lightly occupied areas, install controls to de-energize the reheat coils, raise the cold duct temperature, and operate on a demand cycle.

Under an alteration or expansion program, install variable volume rather than terminal reheat or other systems.

De-energize or shut off terminal reheat coils, raise the chilled water and supply air temperature of the central system, and add re-cooling coils to ducts in areas where lower temperatures are needed.

MULTI-ZONE SYSTEMS

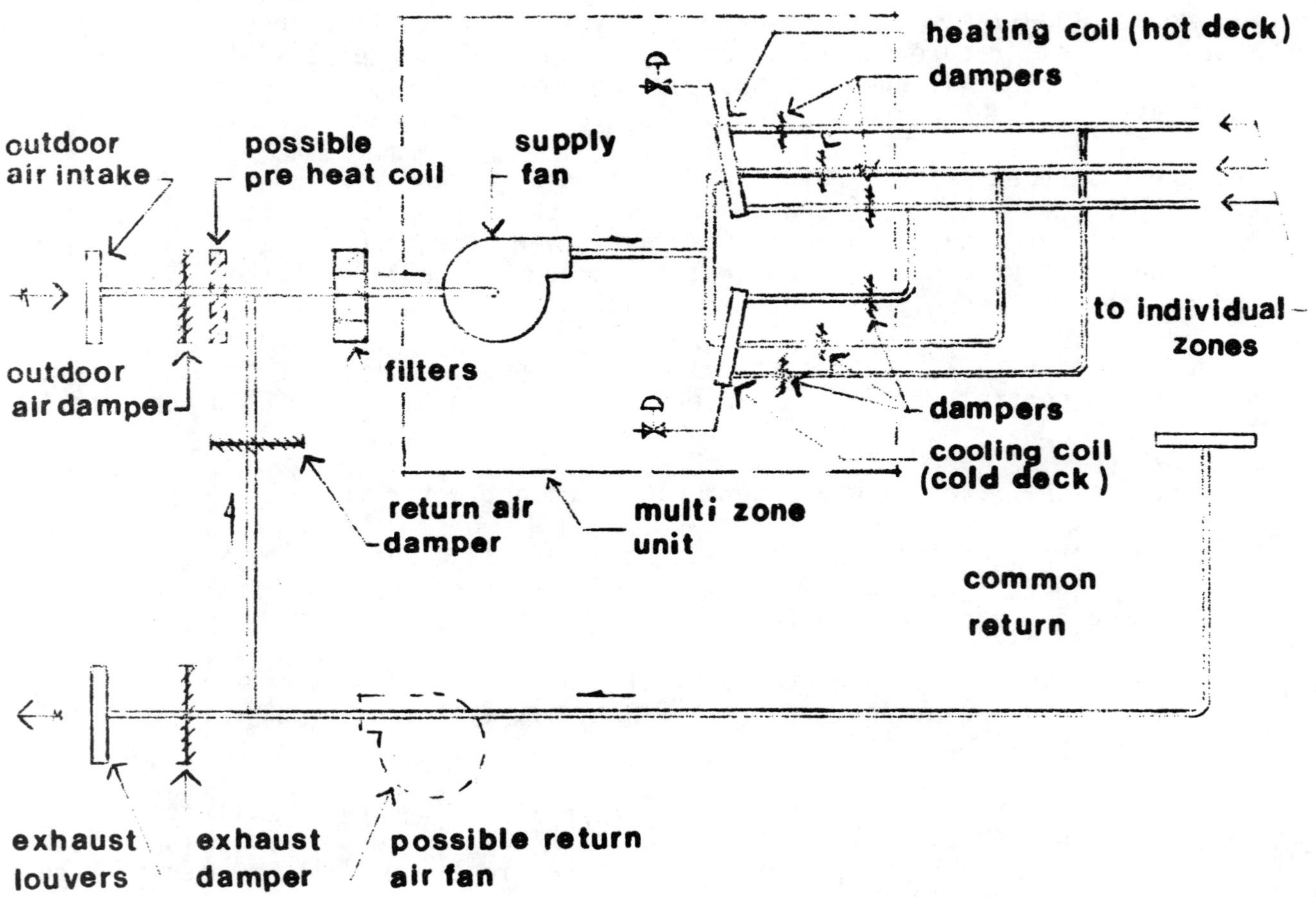

Most existing mult-zone units have a single heating coil serving the hot deck and a single cooling coil serving the cold deck. Each zone supply temperature is adjusted by mixing the required quantities of hot and cold air from these coils. With these types of units, the hot deck temperature must be sufficiently high to meet the heating demands of the coldest zone and cold deck air must be sufficiently low to meet the demancs of the hottest zone. All intermediate zones are supplied with a mixture of hot and cold air wasting energy in a manner similar to reheat systems.

Newer mult-zone units are now available with individual heating and cooling coils for each zone supply duct, sc the supply air is heated or cooled only to meet the zone load demand. These new units use far less energy than units with common coils, and where renovations are contemplated or the existing multi-zone unit is at or near the end of its useful life, replacement should be considered, using a multi-zone unit with individual zone coils.

Analyze muti-zone systems carefully, treating each zone as a single zone system, and adjust air volumes and temperature accordingly.

Hot deck and cold deck dampers are often of poor quality and allow considerable leakage even where fully closed. To conserve energy, check these dampers and modify to avoid leadage.

Install controls or adjust existing controls to give the minimum hot deck temperature and maximum cold deck temperature consistent with the loads of critical zones.

Where a mult-zone unit serves interior zones that require cooling all summer and most of the winter, energy can be conserved by converting to VAV reheat. To achieve this, blank off the hot deck and add low pressure VAV dump boxes with new reheat coils in the branch duct after the VAV box. This system works well in conjuction with an economizer cycle, and reheat energy is minimized. But careful analysis is requred of the existing system and zone requirements to make the correct selection of equipment.

Arrange the controls so that when all hot duct dampers are partially closed, the hot deck temperatures will progressively reduce until one or more zone damper is fully opened and, when all of the cold duct dampers are partially closed, the cold duct temperature will progressively increase until one or more of the zone dampers is fully opened.

Install controls to shut off the fan and all heating control valves during unoccupied periods in the cooling season, and shut off the cooling valve during unoccupied periods in the heating season.

Convert system entirely or in part of variable volume by adding terminal units and pressure bypass and/or add fan coil units in specific areas requiring constant air volume.

DUAL-DUCT LOW-VELOCITY SYSTEM

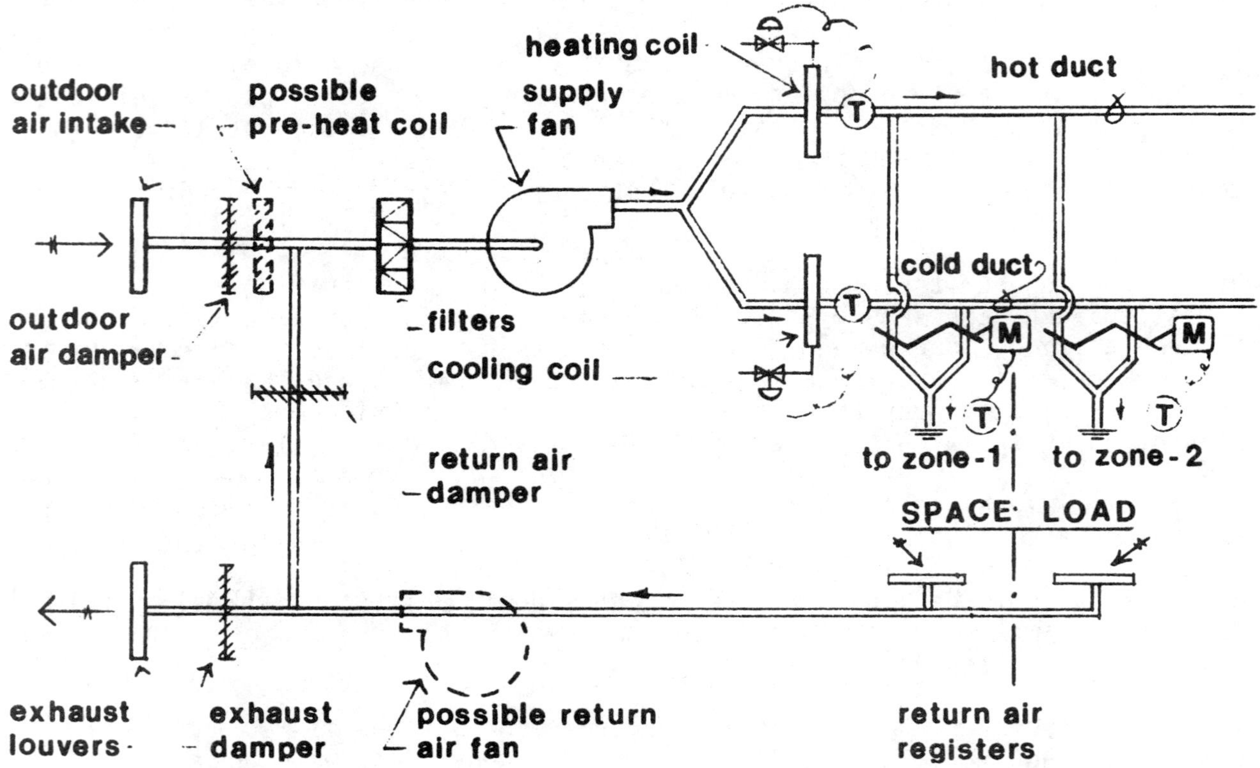

Dual-duct low-velocity systems supply hot and cold air in individual ducts to the various zones of the building which are supplied with a mixture of hot and cold air to maintain the desired zone conditions. Control of mixing is normally achieved by automatic dampers in the branch ducts, each damper being positioned according to the dictates of the space thermostat. Each duct should be analyzed as a single-zone system to determine whether the air volume or its temperature set-point can be advantageously changed to conserve energy. Reduce the temperature of the hot duct and increase the temperature of the cold duct to the point where the heating and cooling loads of the most critical zone can just be met.

Return air is a mixture of all zones and reflects the average building temperature. In some designs of central station equipment, it is possible to stratify the return air and the outdoor air by installing splitters so that the hottest air favors the hot deck and coldest air favors the cold deck. This will reduce both heating and cooling loads.

DUAL-DUCT HIGH-VELOCITY SYSTEMS

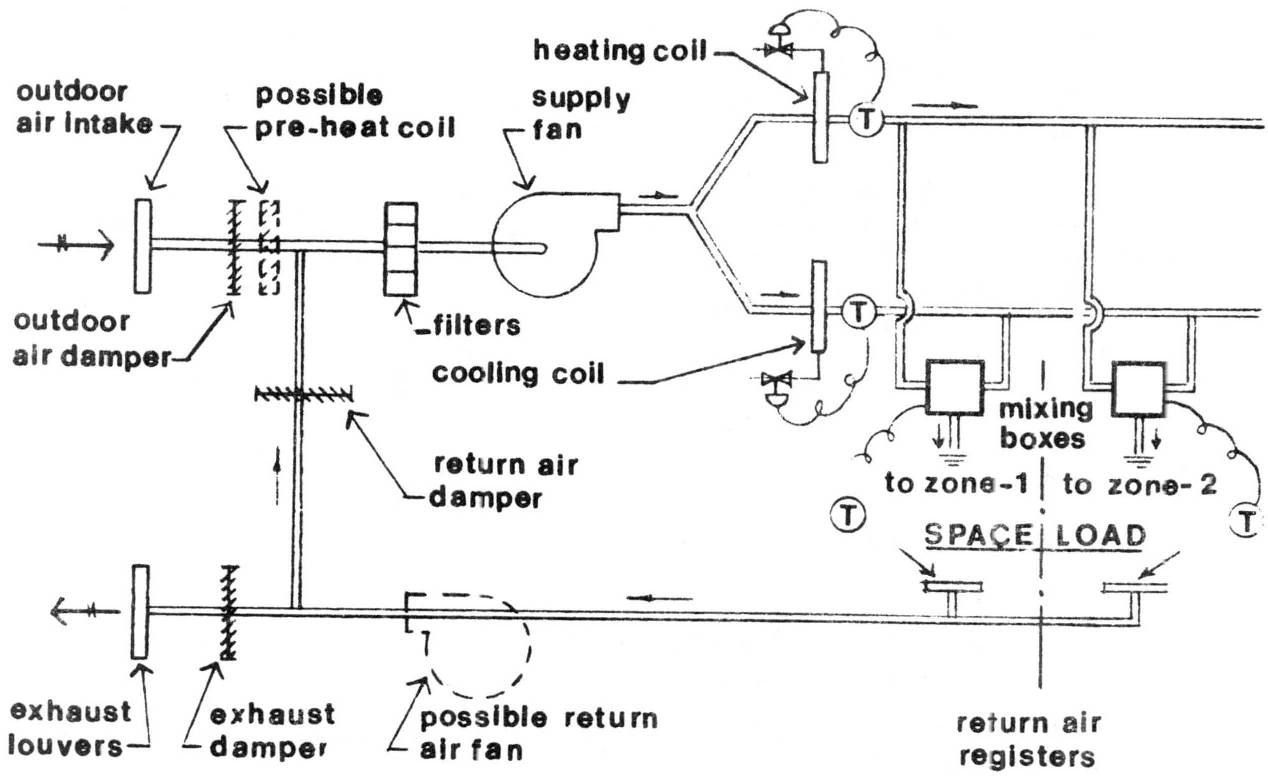

Dual-duct high-velocity systems operate in the same manner as low-velocity systems, except that the supply fan runs at a high pressure and each zone requires a mixing box with sound attenuation.

Considerable quantities of energy are required to operate the fan at high pressure, and a close analysis of pressure drops within the system should be made, along with a reduction in fan pressure to the minimum required to operate the mixing boxes.

To reduce system pressure, replace existing high pressure mixing boxes with lower pressure types.

> When there is no cooling load, install controls to close off cold air duct; de-energize chillers and cold water pumps and operate as a single-duct system, rescheduling the warmer air duct temperature according to heating loads only.
>
> When there are no heating loads, install controls to close off the warm air ducts; shut off hot water, steam or electricity to the warm duct and operate the system with cold duct air only, rescheduling supply air temperature according to cooling loads.

Replace obsolete or defective mixing boxes to eliminate leakage of hot or cold air when the respective damper is closed.

Provide volume control for the supply air fan and reduce capacity, perferably by speed reduction when the the hot deck and cold deck air quantities can be reduced to meet peak loads. Reducing the heat loss and heat gain provides an opportunity to reduce the amount of air circulated.

When there is more than one air handling unit in a dual-air system, modify duct work if possible so that each unit supplies a separate zone to provide an opportunity to reduce hot and cold duct temperatures according to shifting loads.

Change dual-duct systems to variable-volume systems when the savings from such an investment are not overcome by adding VAV boxes and fan control.

VARIABLE-VOLUME AIR SYSTEMS

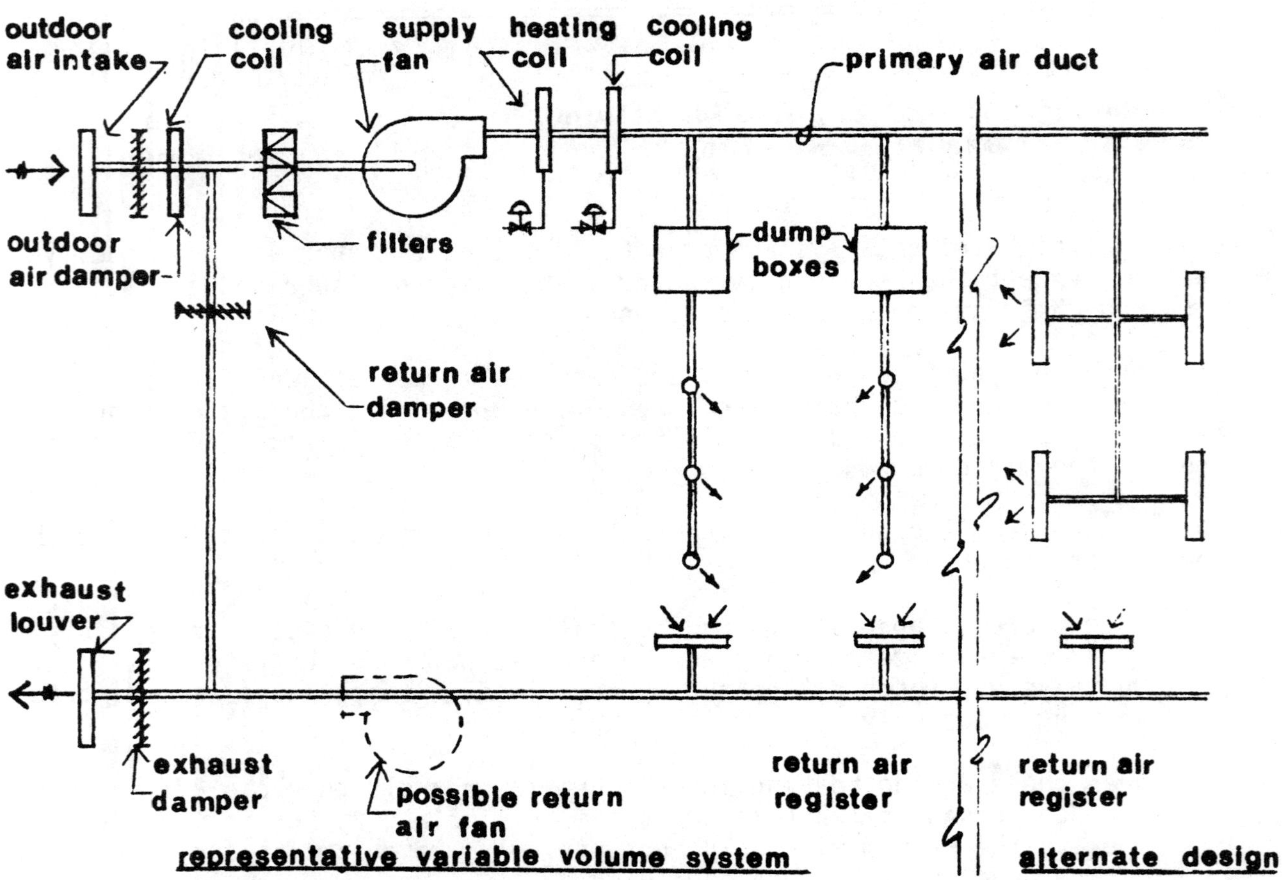

VAV Systems are basically a modification of single-zone, single-duct systems to allow different condition in multiple zones to be met by varying quantities of constant temperature air. The air quantity supplied to the zone can be modulated in two different ways:

(1) Dump boxes installed in the zone supply-duct modulate the amount of air through the outlets by bleeding off unused quantities of air, dumping if into the ceiling void and back into the return air system. This type of VAV system maintains constant fan volume and constant flow in all the trunk ducts and often wastes energy by mixing the treated dump air with the return air. This system does, however, have practical advantages when used with direct expansion cooling coils, since it maintains constant flow through the coils regardless of load, which is necessary to obtain correct operation of refrigerating equipment.

(2) In the second type of VAV system, the air volume is controlled at the outlet by closing down dampers as the zone load decreases. This type of system does, in fact, vary the total supply volume handled by the fan by dampering the air flow under light load conditions.

To conserve energy, regulate the fan volume according to the demands of the system by a variable speed motor with SCR controller or by inlet guide vanes.

Provide a control to modify the fan speed up or down to maintain constant pressure in the supply duct. Reset the supply air temperature in accordance with zone loads.

If not installed, provide either inlet vortex dampers or variable speed drive, such as a multi-speed motor or SCR control, so that fan volume will be reduced when dampers begin to close until one or more VAV dampers is fully opened.

Provide a control to reschedule supply air temperature to a point where the damper on the variable air-volume box with the heaviest zone-load demand is fully opened.

Install a control to reduce the hot-water temperature and raise the chilled-water temperature in accordance with shifting thermal demands.

As with single-duct variable-volume systems with reheat, set the control to delay reheat until cfm is reduced to minimum.

Where possible, eliminate the reheat coil and set the box to full close off.

Variable-volume with reheat systems should be operated and modified urder conditions similar to those described for terminal reheat.

INDUCTION SYSTEMS

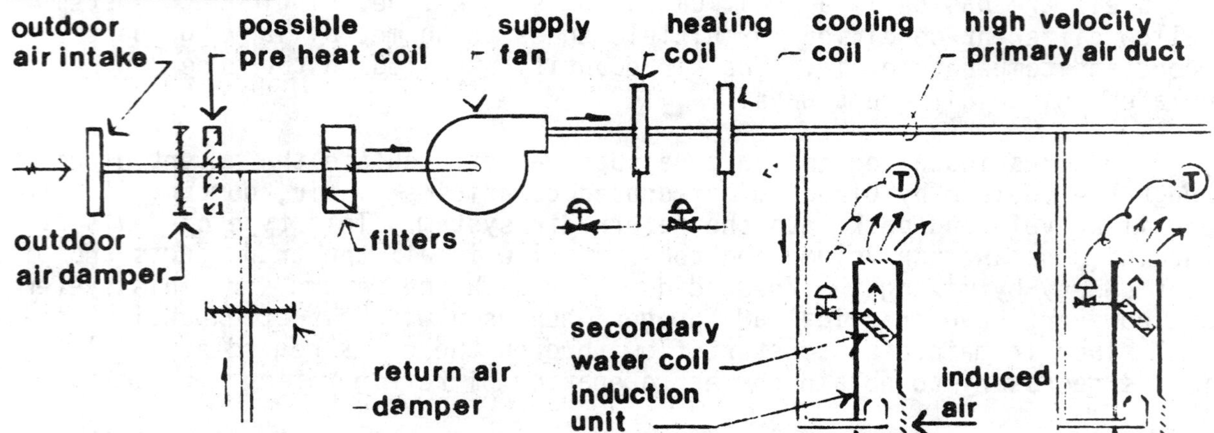

Induction systems are commonly used for heating and cooling perimeter zones where large fluctuations in heating and cooling loads occur.

Primary air is either heated or cooled and supplied at high pressure to the induction units located within the conditioned space. Primary air is discharged from nozzles arranged to induce room air into the induction unit at approximately four times the volume of the primary air. The induced air is cooled or heated by a secondary water coil. The water coil may be supplied by a two-pipe system, producing either chilled water or heated water (but not simultaneously), or by a three-pipe system where separate supplies of hot water or chilled water are continuously available and, after passing through the unit, are mixed into a common return. In a our-pipe system, the supply and return of hot water and chilled water are both continuously available.

The primary supply-air fan operates at high pressures requiring high horsepower input. After careful analysis, reduce the primary air volume and pressure to the mimimum required to operate the induction terminal units.

Induction unit nozzles may be worn due to years of cleaning and operation, causing an increased primary air quantity at lower air velocities with lower induced air volumes. Check each induction unit and either repair the nozzles or replace them.

Provide controls to accomplish the following:

Set the primary air reheat schedule as low as possible without causing complaints in the occupied spaces.

Lower the set-point of secondary heating-water temperatures in winter and schedule it according to outdoor conditions.

During maximum cooling or heating periods, reduce the secondary water flow rate to the minimum necessary to obtain satisfactory room conditions.

Raise the temperature of primary air for cooling to reduce refrigeration load, but weigh this benefit against an increased cooling demand by the secondary coils.

Reduce water temperature to the heating coils during the heating season to reduce piping loss and improve boiler efficiency.

For night operation during the heating season, shut down primary air fans, raise the hot-water temperature and operate the induction units as gravity convectors.

FAN COILS - UNIT VENTILATORS

Fan coil and unit ventilator performance can be improved by careful maintenance which in turn will allow energy savings to be made in the associated heating and cooling water distribution systems. Heating and cooling coils should be cleaned and air and water flow reduced to the minimum required to meet space conditions. Consider changing four-pipe heating and cooling supply systems to a two-pipe system. This will avoid changeover loss and prevent simultaneous heating and cooling within a given zone.

For fan coil systems which have separate coils for heating and air conditioning, install a control to prevent simultaneous heating and cooling.

Install a seven-day timer to shut down fans, close off valves, and shut off chilled water pumps, compressors and cooling towers or air-cooled condensers during unoccupied periods.

Install controls to shut off fans during unoccupied periods during the heating season and operate the fan coil unit as a gravity convector. (The same is applicable to unit ventilators as well.)

Block off outdoor air inlets where no dampers are installed if infiltration meets ventilation requirements.

If fan coil units are not located in conditioned areas, insulate the casings to reduce heat loss or gain.

With three-pipe systems, set controls for a minimum mixing of hot and cold water.

Change four-pipe or three-pipe systems to two-pipe systems where possible to avoid change-over losses.

WATER-TO-AIR HEAT PUMPS

Air coils should be kept clean to maintain the highest possible seasonal COP. Heat pump units are normally controlled by integral thermostats which turn them on and off according to the demands of space load, and these units are frequently left in operation 24 hours a day regardless of whether the building is occupied.

Install a centralized system of on/off controls on a floor-by-floor basis, using a remote control contactor on the power circuit feeding the heat pump. Rewire the power supply to heat pumps in such a manner that 1/3, 2/3, or all of the units in any zone can be turned on and off depending on whether the building is occupied or not and what the expected outdoor conditions are likely to be.

When retrofitting, install one larger unit rather than multiple smaller units for greater efficiency.

All guidelines for condensers, compressors, evaporators, fans, pumps, piping and duct work listed in other sections are applicable to unitary closed loop water-source heat pumps.

AIR-TO-AIR HEAT PUMPS AND INCREMENTAL AIR CONDITIONING UNITS

The guidelines for evaporators and condensers in Section 6C ("Cooling and ventilation") are applicable to these units.

Where possible, direct the warm exhaust air from the building to the inlet of air-to-air heat pumps to raise their coefficient of performance. Waste heat from other processes can also be used as a heat source.

Replace air-to-air heat pumps with water-to-air heat pumps where there is a source of heat such a ground water with temperatures above average ambient winter air temperatures.

For incremental air conditioning units, follow guidelines for the fan coil units during the heating cycle.

Install a seven-day timer to program operation of compressors in accordance with occupied and unoccupied periods.

When replacing incremental air conditioning units which are equipped with electric heating coils, install a heat pump model instead.

Replacement compressors should have an EER rating of 9 or better during the cooling cycle.

EXHAUST SYSTEMS

All the ECO's and guidelines in this section detailing methods of reducing fan power requirements by reducing flow rate and resistance apply equally to exhaust systems. In addition:

Balance exhaust systems so that the exhaust air flow rate does not exceed supply air flow rates of associated systems. Ideally, exhaust should be 10 percent less than supply to maintain a positive building pressure.

Where applicable, modulate exhaust-fan volume in conjunction with an associated VAV supply fan by installing inlet guide vanes or a variable speed control.

Recirculate toilet room exhaust air through charcoal filters to reduce makeup air requirements.

Install controls to operate toilet exhaust fans intermittently for 10 to 20 minutes each hour and to shut them off automatically during unoccupied periods.

Install a motorized damper at inlet grilles and connect wiring from damper to light switches to reduce the air quantity when toilet rooms are unoccupied. Modulate fan volume according to load by monitoring pressure in main exhaust stack.

For new installations or when existing fans and/or motors are replaced, provide variable speed control or inlet van control to operate in accordance with static pressure.

Shut off supply fans which serve only as makeup air for toilet rooms and install new door louvers or cut off the bottom of the door to permit air from conditioned areas to migrate into the toilet rooms as makeup for the exhaust system. Set maximum capacity to provide one cfm per square foot of toilet area.

THERMAL STORAGE

For large buildings with interior zones that require cooling all year round, analyze the potential for conserving energy by capturing the heat from the hot condenser line and storing it in water tanks when it is not required for immediate use in the perimeter zones, but can be used during unoccupied times when the cooling machines are not running.

Analyze the cost benfits of providing chilled water storage to reduce peak demands and permit operation of the refrigeration systems at lower condensing temperatures (which occur at night or early morning). Note: Both of the above options should be considered only when making sizeable additions or alterations to existing buildings. A thorough engineering and cost analysis is required to alter these systems.

APPENDIX B -- BLANK FORMS

CLIMATIC ZONE FORM

Job. ______________________________ Date: __________

Design Temperature: Summer_____ Dbt _____Wbt Winter_____Dbt

Wind Velocity _____Miles/Hr.

MAP	SOURCE	VALUE
Latitude	Fig 8-1	
Heating Degree Days	Fig 8-1	
Solar Radiation in Langleys	Fig 8-2	
Degree Hours Less Than 54F Wbt/68F Dbt	Fig 8-3	
Degree Hours Greater Than 78F Dbt	Fig 8-4	
Degree Hours Greater Than 66F Wbt	Fig 8-5	
Degree Hours Greater Than 85F Dbt	Fig 8-6	

REMARKS

UTILITY CONSUMPTION/COST

JOB:____________________ SURVEY FORM____________________ DATE:__________

Month Year	Elec Kwh	Elec Demand	Elec $	Fuel	Fuel $	Fuel	Fuel $	Degree Days-Htg
Jan								
Feb								
Mar								
Apr								
May								
June								
July								
Aug								
Sept								
Oct								
Nov								
Dec								
Jan								
Feb								
Mar								
Apr								
May								
June								
July								
Aug								
Sept								
Oct								
Nov								
Dec								

BUILDING LOAD SURVEY FORM

JOB: ____ AREA: ____ WINTER: ____ SUMMER: ____ PECO NO: ____ DATE: ____

GLASS

Fac-ing	SqFt	Shading Ext	Shading Interior	Type U=
TOTAL				

WALLS

Fac-ing	SqFt	Shade	Type U=
TOTAL			

ROOF OR FLOORS

SqFt	Type U=

LIGHTING

Area	Type	OCCUPIED Kw	OCCUPIED Fc	OCCUPIED Hrs/Wk	OCCUPIED Kwh/Yr	CUSTODIAL Kw	CUSTODIAL Fc	CUSTODIAL Hrs/Wk	CUSTODIAL Kwh/Yr
	TOTAL			TOTAL				TOTAL	

PEOPLE

Area	No	Hrs/Wk	Infiltration Item	Crack Usage	Figure	Factor	Cfm

DOMESTIC WATER

Use	Temp Hot	Temp Cold	Gal/Meals or Person	People/ Meals	Gallons Per Day	Btu/Yr E6	%Eff. & Conversion	/Yr
							TOTAL	

DISTRIBUTION LOADS

Item	Cfm	Gpm	Oa	Hp	Voltage	Amperage Rated	Amperage Opr	Kw	Hrs/Wk	Kwh Per Year
									TOTAL	

DISTRIBUTION LOAD SURVEY FORM

JOB: ______________ AREA: ______________ PECO NO: ______________

DATE: ______________

Area Served	Item	Cfm	Gpm	O A	Hp	Voltage	Amperage		Kw	Hrs/ Wk	Kwh/ Yr
							Rated	Opr.			

ANNUAL HEATING AND COOLING TABULATION FORM

JOB: ______ AREA ______ PECO NO: ______ DATE: ______

ITEM	UNITS	COOLING				HEATING		
	SqFt-Cfm	Fig No 8-	Factor	Shade Factor	Total Btu/Yr-E6	Fig No 8-	Factor	Total Btu/Yr-E6
North Glass		16 or 17				19 or 20		
South Glass		16 or 17				19 or 20		
East Glass		16 or 17				19 or 20		
West Glass		16 or 17				19 or 20		
Glass Conduction		18				--		
North Wall		22 or 23				25 or 26		
South Wall		22 or 23				25 or 26		
East Wall		22 or 23				25 or 26		
West Wall		22 or 23				25 or 26		
Wall Conduction		24				--		
Roof Solar		27				--		
Roof Conduction		24				28		
Outside Air*		30				31		
Humidification*		--				32		
*Factor is Btu/1000Cfm		Sub-Total Cooling				Total Heating		

Item	Kwh/Yr	Conversion Factor	%Cooling Season/Yr	
Lighting				
Fans				

People	No	Hrs/Yr	Btu/ Person	%Cooling Season/Yr	
TOTAL COOLING					

Summary	
Heating	
Conv. Factor	
Efficiency	
Mcf/Yr	
Cooling	
Conv. Factor	
COP	
Kwh/Yr	

ENERGY TABULATION FORM

JOB:________________ FUEL__________ UNITS__________ PECO NO__________
DATE:__________

ITEM \ AREA								TOTALS
SUB-TOTAL								
MISC.								
REHEAT								
TOTAL								

$$\frac{\text{Utility Bill}}{\text{Energy Use}} - \frac{\text{Base Year - Avg. Year}}{\text{Avg. Year Degree Days}} \times \frac{\text{Yearly Building}}{\text{Heating Use}} = \frac{\text{Average Yearly}}{\text{Energy Use}}$$

PECO ENERGY TABULATION FORM

JOB: ______________________

PECO NO: ______________
DATE: ______________

	ITEM \ AREA								TOTALS
FUEL									
ELECTRICAL	Before ECO Energy Use								
ELECTRICAL	After ECO Energy Use								
ELECTRICAL	Kwh Savings								
ELECTRICAL									
ELECTRICAL									
ELECTRICAL	Before ECO Demand Use								
ELECTRICAL	After ECO Demand Use								
ELECTRICAL	Kw Savings								
ELECTRICAL									
ELECTRICAL									
	Before ECO Energy Use								
	After ECO Energy Use								
	Savings								

INDEX